ENVISIONING

ociety

SUNY Series in Environmental Public Policy
Lester W. Milbrath, Editor

ENVISIONING

A Sustainable Society

Learning Our Way Out

Lester W. Milbrath

State University of New York Press

Published by
State University of New York Press, Albany

© 1989   State University of New York

For information, address State University of New York
Press, State University Plaza, Albany, N.Y., 12246

Library of Congress Cataloging-in-Publication Data
Milbrath, Lester W.
    Envisioning a suitable society: learning our way out/by Lester
W. Milbrath.
        p. cm.–(SUNY series in environmental public policy)
    Bibliography: p.
    Includes index.
    ISBN 0-7914-0162-6. –ISBN 0-7914-0163-4 (pbk.)
    1. Human ecology. 2. Economic policy–Environmental aspects.
3. Environmental policy. 4. Social change.   I. Title. II. Series.
GF41.M53   1989
304.2–dc19   88-37558
                                                                         CIP

10 9 8 7 6 5 4 3 2

This book is dedicated to
Future Generations – Of All Species

Contents

Part III
Transition From Modern Society To A Sustainable Society: Some Possible Scenarios

List Of Abbreviated Terms

DSP — dominant social paradigm

DDT — dichloro-diphenyl-trichloro-ethane, an insecticide

EIS — environmental impact statement

EPA — Environmental Protection Agency

FCC — Federal Communications Commission

GBI — guaranteed basic income

GNP — gross national product

HMO — health maintenance organization

IRG — information routing group

IUCN — International Union for the Conservation of Nature and Natural Resources

LDC — less developed country

MDC — mostly developed country

NEP — new environmental paradigm

NGO — non-governmental organization

NIC — newly industrialized country

OPEC — Organization of Petroleum Exporting Countries

OTA — Office of Technology Assessment

Recombinant DNA— DNA molecules are the basic molecular structure that make up the genetic code of an organism. These molecules can be recombined to alter a genetic code so that the organism will permanently exhibit a new characteristic in succeeding generations (this procedure is sometimes called gene splicing).

SFTC –Systemic and Futures Thinking Capability

UN   –United Nations

UNEP–United Nations Environment Programme

Preface

I have, in the back of my mind, been preparing for the last thirty years to write this book. My thinking took an especially significant turn early in the 1970s when I felt compelled to take up the study of our environmental predicament and allowed my career as a mainstream political scientist to languish. It became clear to me then, and it is even clearer now, that our environmental predicament has its origin in the very nature of modern society and that we can only extricate ourselves by transforming society. Since then I have intensely studied the relationship between environment and society because it is the crucial underpinning of our entire civilization.

Many millions of people now recognize that we must transform society but they have difficulty imagining what a new society, designed to be sustainable, might be like. Envisioning the way we can make the journey from our present society to this new society is even more difficult. Any society, but especially a new society, must first exist in the minds of the people. Every society most fundamentally exists as a set of mental constructs that takes many years to learn. Those images are thoroughly entrenched in our psyches, leading most of us to believe that our present society is so solidly fixed that we have no hope of changing it. We fail to realize that modern society has existed for an exceedingly short time, it is changing swiftly, and it will transform because its present trajectory is not sustainable.

My first intention for this book was, to present a vision of a sustainable society; that is still its major topic. As the book evolved, however, I returned again and again to the recognition that we must improve and hasten social learning as the major avenue to social change. So the central theme of the book became: *learning our way to a sustainable society.* I present my vision for a new society to stimulate you to think deeply about these questions and to enlist your participation in our joint quest for understanding. No one can have all the answers; they will arise from our common learning enterprise. Our goal will be to design a new society that provides a decent quality of life while coexisting in a long-run sustainable relationship with the natural environment that nourishes it.

I sometimes envy painters or photographers who can communicate a complex message to a viewer in one tableau. Writers must communicate one line at a time and portraying a complex thought usually takes many lines. I found this necessity especially vexing this time because the entire book is an extended essay and all parts are interrelated. Ideas developed early in the book need to be known to understand ideas presented later. I ask your patience as they are developed—one line at a time. These ideas are not difficult to grasp; rather, they are multifaceted with many connections to other ideas. I have used this book in manuscript form in six upper division college courses. That experience discloses that when the book is read carefully in the present order of chapters it tends to transform the way a person thinks–the first step to a new life and a new society. In that sense the book really works. The book does not work well if skimmed or if the reader dips into it here or there; doing so misses the interconnections between ideas. Wrestling with the ideas in the book will be rewarding, even to those who eventually conclude that they must disagree, because we all need to think deeply about these fundamental questions.

I have most directly had in mind an audience of students who have persistently asked over the years, "What will this new society be like?" But I also feel I am speaking to all humanity. Learning how to bring a new society into being is a sociopolitical activity in which we all must participate. The book does not require any special background in order to be understood. I hope it will stimulate much discussion and will encourage many others to portray their own vision of a new society. The future will envelope us whether we plan for it or not, but it will be a better future if we thoughtfully prepare to enter it.

This book will not fit neatly into most of the categories that people use to classify books. It is not a textbook that summarizes a field, although I hope it will be read and discussed in many classes. It is not addressed primarily to other academics and does not attempt to comprehensively cite all the relevant literature. It is a book for thoughtful people everywhere who are concerned about the future of the biosphere and society. The ideas in the book have relevance for politics, economics, sociology, philosophy, science, technology, humanity, and all the nonhuman creatures in the ecosphere. I hope that my readers will come to see their civilization in a new light and with new lenses. That is the first step in learning our way out.

Part I contains a critique of contemporary society and an epistomological/ philosophical examination of the thought structure that must underpin a sustainable society. In Chapter 2 I ask, "Can we Sustain our Direction?" and present a many-faceted argument that we cannot. We must recognize physical imperatives that simply cannot be ignored or transgressed if we wish our society to be sustainable.

In Chapter 3 I examine the aggressive thrust for power and domination that is embedded deep in the thoughtways of our culture and conclude that we must transform those beliefs and values in order to find peace and maintain a nourishing ecosystem.

Chapter 4, "An Inquiry into Values for a Sustainable Society," challenges the reader to reason deeply to discover the most basic values that unite us as a species and as a society. I discuss the implications of this analysis for the rest of our values and for the structure of a sustainable society. This will show how value analysis is central to the design of a society and to the policymaking process that guides society. Values also will be shown to be central for providing meaning to knowledge, research, and development.

As I imagined this new society and wondered how we might achieve it, I concluded again and again that social learning is the most viable route to social change. But how does social learning work and how can we assist a society to learn more thoroughly and swiftly? I inquire deeply into those questions in Chapter 5.

Chapter 6, "Stories about the Way the World Works: Belief Paradigms" examines the nature of belief paradigms and shows the way they shape our thoughts and behavior. I also report some findings from a three-nation study of belief paradigms to portray the paradigm makeup as well as the conflict between paradigms in modern industrial societies.

Political ideologies also are important thought structures in modern societies. In Chapter 7 I examine the way ecological thinking relates to left-right ideological thinking. I show that more similarities than differences are found between capitalism and socialism. I relate what we know about human society and behavior to the typical content of ideological thinking.

I conclude Part I by summarizing the main points made in the preceeding seven chapters.

Part II elaborates my vision of a sustainable society by examining the way natural systems work and by designing human institutions and activities to fit as harmoniously as possible with ecological systems.

Chapter 8, "Ecosystem Viability," shows that ecosystems are essential to our survival and their protection must be the primary object of public policy. The essential characteristics of ecosystems are identified, and a program for their preservation is discussed. Various ways that humans relate to nature are critically analyzed.

Modern agriculture uses high inputs of energy and chemicals, but can it sustain our food supply? Can it maintain viable soils? The analysis set forth in Chapter 9 shows that it cannot. Instead, we must move to a regenerative agriculture that works more in harmony with nature.

In Chapter 10 I discuss how we can have "Fulfilling Work" even though our technological society needs fewer employees. We need to distinguish work from employment and restructure work into a more sustainable, yet satisfying, mode.

In Chapter 11 I ask the reader to reflect on the meaning of quality of life and evaluate the limited role of material goods for bringing quality to living. I also suggest numerous ways that we can enjoy life without material indulgence.

What will be the role of science and technology in a sustainable society? In Chapter 12 I discuss the intimate role values play in science and technology. Furthermore, I argue that trying to develop science and technology without guidance from societal values threatens society and the biosphere. I then discuss ways society can guide development and deployment of new technologies.

In Chapter 13 I extend the discussion of technology by seeking ways that society can recognize and avoid the detrimental effects of enticing technologies.

In Chapter 14 I argue that we must design politics and government to help a society *learn* how to become sustainable. I propose that we add a learning structure to governments that would have a Council for Long-Range Societal Guidance as a key new institution. These structures would nurture foresighted assessments of the impact of new policies and technologies on the environment, society, and our values.

How can a sustainable society defend itself against aggressors? In trying to answer this question in Chapter 15, I discuss ways to develop societal thought structures that nurture peace and suggest ways that we can build transnational institutions to preserve the integrity of our global commons and help us live in peace with our neighbors.

We have one biosphere but a world fragmented into 160 nations. In Chapter 16 I discuss the imperative that they work together to protect our global commons. Disparities of development between North and South exacerbate an already very difficult set of problems. Changing our ways of thinking and our institutions will be necessary for solution.

In Part III I examine the way we might make the transition from modern society to a sustainable society. I develop two scenarios– paths–that we might take.

In Chapter 17 I presume that we will strive to maintain our present trajectory and conclude that avoiding change will make us victims of change. Furthermore, I sketch the kind of society that would result if we did achieve what we currently say we are trying to achieve. We probably would not want to go there even if we could get there.

In the last chapter I am as hard-headed as possible in analyzing the probabilities that we will collectively wake up and begin to take corrective action. I come to the conclusion that most of the traditional methods for educating and arousing people will not be sufficient because most people will not be listening. Yet, we also know that occasionally societies make surprising turns in their course, sometimes in a very short period of time. Ideas stemming from some exciting new theories developed in natural science hold out some hope that when things get bad enough to force us to cast about for understanding and solutions, we can make changes that would be beyond the realm of possibility in "normal" times.

I invite my readers to join me in learning our way to a sustainable society.

Acknowledgments

Everyone who writes a book of this scope gets help from many people. The contributions of many writers are noted in the text and bibliography. Special thanks to the following for permission to reprint: the Population Reference Bureau for Figure 2.1; the Worldwatch Institute for Table 2.1; Erik Milbrath for Figure 2.2; the Strawberry Press for "Dogwood Blossoms;" Duane Elgin for figures on pages 92, 111, and 255 of *Voluntary Simplicity*; The SUNY Press for Table 2.1, Figure 2.2, and Table C-3 in my *Environmentalists*; Joji Watanuki for Table 6.4; Gower Publishing Co. for Table 10.1; and George Braziller for Figure 18.1

Of course, many others helped behind the scene, and I wish here to thank them for their contribution. When I first decided to write the book, Rosalyn Taplin and Philip Tighe were visitors in the Environmental Studies Center from Australia. They discussed with me for many hours while we worked out the first outline of the book. They had hoped to also assist in writing the book, but the 12,000 miles between us prevented that. The book developed in a somewhat different way than we had planned, but they would still recognize its general outline.

I am most indebted to Frada Naroll who critiqued and copy-edited the first two drafts of the book—her help went far beyond normal friendly expectations. Louise Prezyna, Hsiao-Shih Cheng, Joachim Amm, and Natalie Waugh also read and critiqued full drafts. Russell Stone, Charles Livermore, Erik Milbrath, and Sue Crafts each critiqued several chapters. Crafts also read and briefed me on other writers and performed a multitude of "housekeeping" chores required in the production of a book. Melissa Forgione and Linda Milbrath assisted by rounding up materials and briefing me on them. Orville Murphy freed some time for my writing by teaching a seminar in the Environmental Studies Center.

When the third draft was completed, other colleagues provided helpful criticism: Betty Zisk, Duane Elgin, David Orr, Sally Lerner, Jeff Fishel, Richard Schwartz, Jogeir Stockland, and Pal Strandbakke.

Colleagues and students discussed with me in seminar for many hours—in myriad subtle ways they contributed to the quality of this book. I gratefully remember the participation of William Conroy, Fred Snell, Paul Reitan, Orville Murphy, Amanda Hiller, Raoul Naroll, Sean Enright, Debbie Abrams, Hsiao-Shih Cheng, Sue Sullivan, Kate Miller, Tai-Sheng Won, and Dave Turkon.

Peggy Gifford, editor at the SUNY Press, fought to get this book published when most other editors refused to take a chance. Finally, I thank my wife Kirsten who has been a "writer's widow" for so long she has begun to think it is a normal way to live. Fortunately, she has her own busy and successful career.

Introduction

Chapter 1

O n a Thanksgiving morning in the latter half of the 1980s I reflected that I have much to be thankful for. My house is warm; it is stocked with food for a feast; dinner guests will soon be arriving. My personal world and my society are functioning reasonably well. Why am I thinking about ways to make society sustainable?

My mind tells me that our hope for a good future is slipping away—even now while things seem to be going well. Our capability for abstract reasoning leads to our sense of unease. We are the only species with a well-developed sense of time; we can imagine a past and a future that is almost as vivid as the present. Our perception of the past and future is enlarged and sharpened by our growing knowledge of the world. Let us use that knowledge to perceive more clearly where we have been, where we are now, where our present path will lead us, and compare that with where we would like to go.

Many people, everywhere in the world, have this same sense of unease. Other animals, like dogs or horses, probably do not feel uneasy. Why do we humans feel it? What aspects of our lives seem to be slipping away? Why is this happening? What do we want to retain? How do we retain it? How can life be good far into the future? Do we mean good for all species? I wrote this book hoping to enlist the reader's help in the quest for answers to these questions for we must share a vision for a new society before we can realize it. Designing a better society and maintaining a good life require deep thought and sustained effort by all of us. Reasoning together is the only way we can bring it about.

Humans In The Perspective Of Geological Time

Our culture leads us to think of time in lifecycles and lifetimes. A lifetime seems like a long time but is it really very long? Geologists speak of geological time and begin their count with the beginning of our planet. Scientists now know much more about what has happened to our planet since it was formed approximately 4.6 billion years ago. Humans have lived on the planet for only about 2.5 million years. Civilizations developed about 10,000 years ago and written histories only cover about 5,000 years. In the last 200 years, a mere .00000044 percent of earth time, our species has wrought more change on the planet than has taken place in the past billion years. How can we gain perspective on these time spans?

Imagine a movie that runs a full year representing all the time since the origin of the earth. Each frame in the motion picture is the equivalent of one year of real time. The normal movie speed of twenty-four frames a second has to be increased about six times to 146 frames (years) per second to fit this movie into a single year. That means that 8,752 years of real time would flash by during each minute of the movie; 525,740 years would flash by in an hour. A day of the movie would represent 12,602,240 years. Imagine that the movie begins on January 1, coinciding with the origin of the earth, and ends with our present time at New Years' Eve the following year. As the movie runs for weeks, no sign of life is seen. The first glimmers of one-celled microbial life do not develop until March. These tiny creatures, visible only by electron microscope, are the only form of life for an additional 2 billion years.

In our year-long movie, more complex life forms (i.e. eukaryotes) do not develop until August and September. Larger and still more complex multicellular organisms do not appear until November. Dinosaurs appear about December 13th and become extinct after about thirteen days. Mammals appear about December fifteen. The genus *Homo* does not develop until five hours before midnight on December 31. *Homo sapiens sapiens* (modern humans) developed only 100,000 years ago; eleven minutes before midnight. Civilization does not appear until one minute before midnight. A lifetime of a modern human would be only one-half of a second.

The industrial era has lasted about two seconds. During that era, humans have used up and scattered a large proportion of the resources in the earth's crust, altered and exploited ecosystems to serve strictly human needs, held all other species at their mercy, and driven many species to extinction. They are now well on the way to poisoning the biosphere and changing the earth's climate. In comparison to the dinosaurs who survived on the planet for thirteen days, can Homo sapiens last even one day? Microbes that long preceded us will still be here long after we are gone.

Let me cite another analogy. Imagine the height of the Empire State Building in New York City, including its television mast, as the equivalent of geological time from the beginning of the earth. On this scale, the time span since Columbus "discovered" America (nearly 500 years ago) would be the equivalent of the thickness of one sheet of paper. The time span of human civilization would be the equivalent of the thickness of ten sheets of paper.

One hundred years ago, my grandfather was a pioneer on a farm in Minnesota. He carved cultivable land out of forests and rocks, using simple machinery and tools, horsepower, and lots of backbreaking physical labor. My father, born in 1886, lived close to the land all his life. He once said to me, "You talk of the beauty of rocks and trees; oh poodle! I fought them all my life." To him, nature was to be struggled against and endured, awesome in its power but not revered. Ever since the beginning of civilization people have struggled to dominate nature. Until recently, however, humans impacted only slightly on nature's well worked out cycles. One hundred years ago, even fifty years ago, nature was resilient and forgiving.

We Learned Too Well How To Dominate

The threatening problems faced by our species stem from our very success as a species. We are a "successful" species in that we have been able to defeat and bend to our will every other species. We are "successful" in that we have been able to appropriate for our own needs a great proportion of the planet's biological productivity—even dipping into the earth's crust to extract accrued productivity (fossil fuels) stored there long before we evolved as a species. We are "successful" in that we continue to reproduce at record rates while we reduce populations of other species, or eliminate them altogether. Human population is now more than 5 billion and will double to 10 billion in only fifty more years. We must, somehow, learn to control our exuberance or our "success" will lead to our extinction. Why is this so?

Our ability to learn enabled us to do better and better those things that we have always tried to do. We wanted our children to survive and live long and good lives. Therefore, we reduced infant mortality, conquered disease, mended broken body parts (or replaced them)—and we live longer and longer. We wanted sufficient food so we appropriated much of the earth's bioproductivity for our own use. Now we are faced with a veritable explosion in human population.

We wanted to live more comfortable lives so we isolated ourselves from extremes of weather. We wanted excitement so we learned to travel swiftly to the far corners of the globe and to bring into our homes a vast array of entertainment. Our science and technology delivered into our hands ever greater power to manipulate the planet's resources to our own enjoyment. Can this kind of success lead to failure?

We have failed to take into account the long-run consequences of just doing what we have always done—but better and better. Doubling the world's population in fifty years will more than double the burden we place on the environment because most of those billions of people will strive to live at increasingly higher standards of living. We are using up the planet's resources at an unprecedented rate, especially fossil fuels, and we surely will encounter severe shortages of many of them. All of those consumed resources will eventually turn to waste and be cast into the environment. The biosphere is disrupted not only by the sheer volume of wastes but also by the fact that many wastes are unnatural compounds that biospheric systems do not know how to absorb and recycle (toxics). In effect, we have built a society, and an economic system, that cannot sustain its trajectory.

How serious is the disruption we are inflicting on planetary systems? The damage will be far greater than most of us currently believe. The drought in the American Midwest in the summer of 1988 signals that it may already have begun. The build up of greenhouse gases, due to burning fossil fuels and other production and consumption activities, has already initiated a global warming that will change climate patterns, perhaps more swiftly and drastically than we thought possible even a few years ago.

Think for a moment about the proportion of investment decisions that depend on the premise of continuity, especially climatic continuity (choosing a place to live, building a house, starting a business, buying stock in a business, accepting collateral for a loan, making contributions to a pension fund, and so on). My quick review of such decisions suggests that about 90 percent of investment decisions are premised on continuity. If climate change forecloses continuity over large proportions of the globe, and scientists are now reasonably confident it will, the socioeconomic disruption will be horrendous. Savings will be wiped out, families will be forced to move, some communities will die out while others will be devastated by uncontrollable hordes of in-migrants. Some localities will be under water and others will turn to desert—former deserts may be able to support plants again but must first build new plant/animal communities. It seems probable that many people will die because we depend for food on plant and animal communities that will lose productivity because they will be ill-adapted to a changing climate regime.

Devastation from climate change will be exacerbated by other global biospheric effects: loss of the ozone layer, acid rain, poisonous red tides of algae, toxic pollution of soils, water, and air, species extinction. Nature may have many additional unpleasant surprises in store for us. When these effects are combined with resource shortages, we may well wonder how we can continue to support even the 5 billion people already living, much less the additional billions that are destined to arrive (even if we strive vigorously to limit population growth).

Today, our power to dominate and injure nature is awesome. We can move mountains, fly to the moon, obliterate cities, slash down jungles, poison large water bodies, create new species. Our ability to control the power we have gained is not well-developed. The industrial disasters at Bhopal and Chernobyl signal that we can expect many more such disasters. A nuclear war followed by a nuclear winter could obliterate nearly all life from planet earth.

The people living 100 years ago could hardly have imagined the changes that this brief span of time would bring. Nearly everyone today expects the pace of change to accelerate. If nanotechnologies, which I discuss in Chapter 12, were developed, they would accelerate our thrust for power and change 1,000 times. We are told that this is progress and that it is good. Labelling an activity as progress implies that we know where we are going and that we can measure our speed in getting there. Actually, we do not know where we are going and we do not know where we wish to go. *Progress* is a meaningless term without this knowledge. We also forget to ask, "Can we sustain what we are building?"

Theory And Practice

The basic nature of *Homo sapiens* has not changed very much in the past 200 years. What has changed is our practice—and the cumulating effects of that practice. We normally believe that theory leads practice, but that is not always the case. Sometimes practice so alters conditions that it forces us to revise our theories. The lag of theory behind practice is especially true of our political and social theories. Most of the theories that guide human affairs today were developed more than a century ago; for example, John Locke and Karl Marx for politics and economics, Adam Smith for economics, Max Weber for society, and Rene Descartes for science. Their theories were reasonably valid for conditions they knew at the time of their development. As with many theorists, however, they did not fully anticipate the cumulative effects of following their precepts. They never imagined a world of 10 billion people; they never envisaged that people would poison ecosystems and change the earth's climate; they never dreamt that humans would gain the power to create new forms of life or to destroy most life.

Peter Wenz, in *Environmental Justice* (1988), examines several theories used to structure the relationships between people in modern society: the theory of property, libertarian theory, efficiency theory, the human rights theory, the animal rights theory, utilitarian theory, cost-benefit analysis, Rawl's theory of justice, biocentric individualism, and ecocentric holism. He asks of each theory if it ensures justice for humans and other sentient creatures. He also asks if it preserves the integrity of the natural environment that these creatures depend on for life—and for realization of their right to justice. He analyzes each theory for its logical consequences in the light of contemporary socioenvironmental conditions. He concludes that all of the theories we have been using in modern society fail to provide environmental justice. In effect, we cannot escape our planetary predicament by relying on past theories. Readers interested in close analysis of these theories should consult Wenz.

I approach the problem of invalidity of theories somewhat differently. I frequently challenge contemporary perspectives on policy questions to show how they are invalid for solving the problems faced by modern society. I especially address the ability of these ideas to maintain our society in a long-term harmonious relationship with nature. In the next chapter, I challenge the conventional wisdom that growth is good, especially the idea that it can be sustained. Next I dispute the popular notion that competition, winning, and acquiring power are likely to lead to a good life. After that, I challenge the conventional wisdom that the process of knowing that we call science is different from the process of knowing that we call valuing. Learning how to reason together about values is crucial to saving our species. As a society, we have to learn better how to learn—I call it *social learning*; it is the dynamism for change that could lead us to a new kind of society that will not destroy itself from its own excesses.

Part I

The Predicament of
Modern Society:

An Inquiry Into Why
It Is Not Sustainable

Many people today, in many lands, have a gnawing feeling that something is profoundly wrong with modern society. This is more than just a desire to make some modest alterations in the way we do things. It is normal for most citizens to want to change something about the way things are done in their community, state or nation. Furthermore, communities or nations are constantly tinkering with new legislation, programs, and structures, trying to make things work better. This process is like maintaining and repairing a car to keep it in good running order.

Is not this approach adequate for the problems of modern society? After all, things seem to be going along reasonably well. Could we not solve most of our problems by thoughtful reform? This question cannot be answered by merely taking a current assessment of how things are going in one's community or nation. It is not sufficient to ask, as former President Reagan did in the 1984 election campaign, "Aren't you better off than you were four years ago?" We must take a longer-range view of the consequences of continuing our course.

Most natural and social scientists who have examined the long-run prospects for modern society are not sanguine about our future. While they are not unanimous in their judgment about the viability of our society, the overwhelming weight of the evidence and analysis indicates that society, as presently constituted, cannot continue on its present trajectory. The kinds of changes required are so drastic that, when implemented, they will constitute a new society. The old car cannot be fixed up any more; we must design not simply a new model but a new kind of vehicle, one that is not on display on any showroom floor.

Let us make the fair assumption that people will choose survival over demise; that they desire to live in a society that will sustain not only the species *Homo sapiens* but also their own individual lives. Let us assume further that they would like their lives to be of reasonably high quality—so long as their lifestyles are compatible with sustainability. With these two assumptions our discussion can proceed.

Can We Sustain Our Direction?

Chapter 2

"We cannot command nature except by obeying her. — Francis Bacon

The Dilemma Of Growth

G rowth is an honorific word in modern society. We are told constantly that we *should* be growing in economic output, in population, in prestige, in strength, in stature, in complexity. Growth is associated with development, health, and progress. Nongrowth is associated with decline, illness, and lack of progress. Progress, defined as growth, is believed to be inevitable and good. Some people even believe that if we do not grow we will die.

Wachtel (1983, p. 92) traces some of our fascination with growth to our preoccupation with reducing all economic activity to a common metric (money) and trying to express how well a country is doing with a single number (the gross national product or GNP).

Life holds infinite variety, but if everything is reduced to one number, if how we are doing can always somehow be added up, then the only real value can be "more," and growth in some quantity is the only acceptable sign of progress or doing well. And so we experience an imperative to grow. The word *"growth"* becomes for us synonymous with the good and is trotted out in an enormous range of contexts. Economists seek to bring about growth; psychologists help their patients to grow; our presidents are expected to "grow" in office. With "growth" such an omnipresent symbol of the good, it is very difficult for us to accept any idea of a limit to growth as implying anything other than stagnation. Our emphasis on growth leads us to equate contentment with complacency. Our more general values parallel the assumptions that underlie the workings of our economy. Even those of us opposed to unchecked growth of industry tend to endorse the broader societal trend of seeking, questing, restlessness.

This perspective on growth applies to Third World countries as well as to developed countries. Third World countries have adopted a "development ideology" that equates development with growth. (Woodhouse, 1972; Orr, 1979) This developmental growth is expected to lift people from poverty and provide them with the good life that they see in the developed part of the globe.

In the early 1970s, a great debate got underway about whether growth has limits. A study of growth was initiated by The Club of Rome (a prestigious group of approximately 100 scientists, businessmen, and national leaders). The club commissioned a work group at The Massachusetts Institute of Technology (MIT) to estimate with a new computer model how long it would be before modern society would run out of critical resources. The MIT group published *The Limits to Growth* (Meadows, et al., 1972) that was eventually translated into about 30 languages. It sold millions of copies. This analysis so strongly challenged the conventional wisdom about growth that it ignited a storm of controversy. Some research teams challenged the validity of the computer model. Some analysts using new models confirmed the original analysis that growth has limits (Mesarovic and Pestel, 1974). Other analysts claimed that growth has no serious limits (Simon, 1981). The debate has raged ever since and is far from concluded. National survey studies in the early 1980s showed that within the U. S. public, about as many people believe there are limits to growth as deny there are limits, whereas in West Germany, nearly everyone believes there are limits to growth (Milbrath, 1984).

We can usefully distinguish organismal growth, which naturally limits itself, from growth in numbers that must be limited externally. Organisms experience phased growth, much as a child growing into adulthood and eventually dying. Growth in early stages is good but is no longer desirable when adulthood is reached. Growth also cannot forestall death; in fact, death from cancer results from growth. This kind of growth was not discussed in the debate over limits to growth although it might have been usefully considered.

Growth in numbers of a species is different. The offspring of adults soon become adults who, in turn, have more offspring. When growth is exponential, it can accelerate very rapidly. Two doubles to 4 doubles to 8, 16, 32, 64, 128, 256, 512, 1024, 2048; ten doublings increase 2 to 2048. Analysts of exponential growth speak of doubling times; if the doubling time is long, growth is slow; if it is short, growth can be very swift. This principle is illustrated by a classic riddle: "Imagine a pond with a water lily growing in it. The lily doubles in size every day. At thirty days it covers the entire pond. On which day did it cover half the pond?" The first response of most people is the fifteenth day; only after closer reflection do they recognize that the lily does not become large enough to cover one-half of the pond until the twenty-ninth day. It takes only one additional day to cover the remainder of the pond.

The speed of doublings can be roughly calculated by the "law of sevens": The percentage rate of growth per year is divided into seventy to give the doubling time. For example, a species increasing in numbers at the rate of 1 percent per year will double in seventy years. If it is increasing at the rate of 2 percent per year, it will double in thirty-five years. At the rate of 3 percent per year it will double in less than twenty-four years. At the rate of 7 percent per year, it will double in ten years. At 10 percent per year, it will double in seven years. At the rate of 10 percent per year, ten doublings would take only seventy years; that would mean more than a thousandfold increase in only seventy years. Money invested at 10 percent interest, where the interest is also invested, would show such an increase. During the 1960s, energy consumption in the United States increased at the rate of 7 percent per year. This increase led a utility executive to exuberantly claim in the early 1970s that the United States would consume as much energy in that decade as in all its previous history. This did not happen because of energy shortages and changes in consumption patterns. The executive's projection illustrates the foolhardiness of predicting the future from simple trend lines.

All species have the potential for exponential growth if the resources and other conditions needed for speedy reproduction are present. Biologists speak of fast exponential growth of a species as an *irruption*. Algae, bacteria, and other small organisms may "bloom" to many millions in a few hours if conditions for reproduction are favorable. For example, an algae bloom devastated near-shore marine life in the North Sea in late May 1988; some people called it a *red tide*; in Norway, they called it the *death algae*. I was in Norway at the time; most observers were ascribing the bloom to nutrients washed down from overfertilized farmlands and carried into the North Sea by rivers. The algae so densely concentrated in near-shore waters (many billions in a cubic meter) that they choked off all other forms of marine life. Ocean fisheries, a veritable lifeline in the Norwegian economy, were wiped out for a time.

Nature has several mechanisms for keeping the number of members of a species within acceptable bounds. Once the available food supply is exhausted, the species must die back. It is not uncommon in the natural world for a given species to reproduce so rapidly that it goes into "overshoot"; its consumption exceeds the resources it needs to sustain life. The faster the rate of reproduction, the greater the overshoot and the more swift the dieback. (The red tide in the North Sea in 1988 disappeared about as swiftly as it bloomed.)

Disease also reduces numbers rather quickly. Natural predators are another check on species growth. For example, rabbits reproduce rapidly; as they do so, they provide more food to their predators (foxes) who also reproduce more rapidly. As the number of foxes increases, the number of rabbits declines, and eventually the number of foxes also declines.

With these natural checks working, nature's systemic processes eventually work out a balance in which the reproduction and survival rates of a diversity of species are maintained at a sustainable level. An ecosystem with many species living in balance will have a certain "carrying capacity." If a certain species reproduces so quickly as to exceed its carrying capacity, the ecosystem probably will be sufficiently damaged so that its overall carrying capacity will be reduced; this is what happened in the North Sea in 1988. When the offending species is as powerful as *Homo sapiens*, the reduction in carrying capacity can be devastating to all creatures.

Garrett Hardin wrote a classic essay in 1968 titled, "The Tragedy of the Commons." In it he illustrates how humans who act freely in their own interest can exceed the carrying capacity of an ecosystem. He invites us to imagine a pasture (such as the Commons in an English village) that is open to all villagers. It is worthwhile for each villager to graze as many cattle on the commons as possible. Even though adding an additional cow might lead to overgrazing, it is in the interest of each villager to add it because he receives all the benefits from its feeding while the losses in grazing capacity are shared by all. Therein is the tragedy. The system encourages each villager to increase his herd without limit—but the limited commons is destroyed. Under conditions of overpopulation, freedom in an unmanaged commons brings ruin to all.

With these concepts in mind, we can evaluate the meaning of what has happened to the population of humans in the last few centuries. Figure 2.1 shows the growth curve of human population since 1750 with projections for an additional century (this span is a mere two seconds in our year-long movie). Clearly, *Homo sapiens* have experienced an irruption in population in the last two centuries. Population, worldwide, grew at a rate exceeding 2 percent per year in the 1960s. Recently, it has declined to about 1.7 percent. Remember, at the 2 percent rate, world population doubles in about thirty-five years. Note also in Figure 2.1 that most of the recent and projected growth is occurring in the less developed countries where they have the least capability to provide a decent life for people. If we allow two more doublings to occur, our current population, which exceeds 5 billion, will become 20 billion before the next century has run its course. In my judgment, the human population growth rate must be brought to zero within the next fifty years, as is also shown in the figure. Growth beyond one more doubling would seriously diminish quality of life and also would be likely to destroy some of the carrying capacity of ecosystems. If we do not learn how to plan for and voluntarily limit human population in the next fifty years, nature's own limits will force decline through human deaths.

Figure 2.1
Population Growth, 1750-2100: World, Less Developed Regions, and
"European"/More Developed Regions

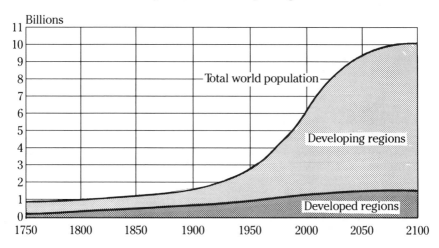

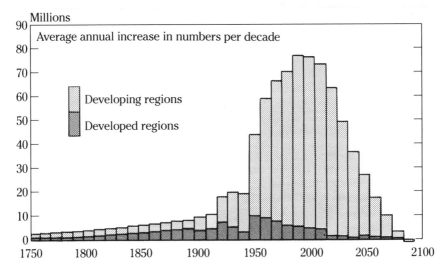

Sources: 1750-1900: Based on estimates by Alexander M. Carr-Saunders, John D. Durand, and Walter F. Wilcox cited in United Nations. *The Determinants* and *Consequences of Population Trends* (New York: 1973). After 1900: Based on medium variant estimates and projections by the United Nations. Population Division, as assessed in 1963, 1980, and 1984, and projections of the World Bank, published in My T. Vu, *World Population Projections 1985* (Baltimore: Johns Hopkins University Press, 1985). Although the use of various sources introduces small inconsistencies in the time series, the resulting general pattern of past and projected population growth should give a satisfactory picture of world population trends.

Note: Three less developed regions = Africa, Asia (minus Japan), and Latin America: Two "European"/More developed regions = Europe, U.S.S.R., Japan, and Oceania (including Australia and New Zealand) combined and North America (Canada and the U.S.).

When this century began, scarcely one lifespan ago, world population numbered 1.6 billion. Assuming an average per capita income of $400 per year (1986 dollars), the gross world product was $640 billion, just slightly more than France's 1986 national product of $550 billion. Over the next half-century, world population grew by nearly a billion, bringing the total to 2.5 billion. Modest progress in raising per capita income brought the gross world product to roughly $3 trillion in 1950.

Though impressive by historical standards, this growth was dwarfed by what followed. Between 1950 and 1986 human numbers doubled to five billion, expanding as much during these 36 years as during the preceding few million. Per capita income also roughly doubled, pushing the gross world product over $13 trillion. Within a generation, the global output of goods and services quadrupled. . . . Between 1950 and 1986 world fossil fuel consumption also increased fourfold. . . . Stanford University biologist Peter M. Vitousek, and his colleagues, estimate that nearly 40 percent of the potential net primary productivity on land is now used directly or indirectly by human populations. . . . The portion remaining to sustain all other species, and to maintain the integrity of natural systems, gets smaller and smaller as the size and demands of the human population mount. (Brown, et al., 1987, pp. 5, 9, see Table 2.1)

Table 2.1
World Population, Economic Output, and Fossil Fuel Consumption, 1900-1986.

	Population (billions)	Gross World Product (trillion 1980 dollars)	Fossil Fuel Consumption (billion tons of coal equivalent)
1900	1.6	0.6	1
1950	2.5	2.9	3
1986	5.0	13.1	12

Source: Population statistics from United Nations; gross world product in 1900 authors' estimate, and in 1950 from Herbert R. Block, *The Planetary Product in 1980: A Creatuve Pause?* (Washington, D.C.: U.S. Department of State, 1981), with updates from International Monetary Fund; fossil fuels consumption in 1900 from M. King Hubbert, "Energy Resources," in *Resources and Man* (Washington, D.C.: National Academy of Sciences, 1969); for remaining years, Worldwatch estimates based on data from American Petroleum Institute and U.S. Department of Energy.

How did it come about that humans were able to reproduce so swiftly? Two developments in recent history enabled humans to increase their numbers dramatically.* About 400 years ago, Europeans discovered and began colonizing North and South America; later they colonized other large land masses like Australia and Africa. These continents, sparsely populated by

---

*I am indebted to Catton (1980) for many of the ideas and concepts in this section.

humans who made their living mainly by hunting and gathering, provided a bountiful habitat that Europeans could expropriate as their own niche. The new methods of warfare, agriculture, and industry they brought to the new lands enabled them not only to displace the indigenous humans but also to so modify the habitats as to encourage growth in human numbers. Humans spurted in population growth.

The second major development was the flowering of the Industrial Revolution which was accelerated by the swift development of science and technology. New tools and techniques enabled humans to do many things that formerly were impossible when they possessed only simple hand tools and beasts of burden. Now swamps can be filled, mountains levelled, and roads constructed into remote places. Humans can now live in very diverse habitats, tolerating a wider range of climate and terrains than any other species. As our habitat range has expanded, we have been reducing the carrying capacity of these habitats for other species, and may soon reduce it for ourselves.

The development of medical science and technology was even more important for population growth. Modern medicine was carried to the far corners of the globe by colonizers, missionaries, and international agencies. Infant mortality has dropped dramatically throughout the world and human life expectancy has doubled. Our ability to keep people alive has temporarily upset one of nature's methods for keeping human population in check. Either we will learn to voluntarily adopt our own checks or nature will eventually provide the checks, perhaps in a fashion we will regret.

The tools and techniques of modern industrial man may be thought of as prosthetic devices that have so extended human capabilities that our species has been transformed from being *Homo sapiens* to being *Homo colossus* (Catton, 1980). We have shown that we can conquer any other species; we have not yet learned that, in so doing, we may destroy ourselves. Herein lie the similarities and dissimilarities between humans and other animals; both are bound by ecological carrying capacity, but mankind alone is able to *temporarily* transcend the bondage.

Now, we should think about a third kind of growth—the speed with which we consume resources. We normally recognize that science and technology have provided us with new prosthetic devices such as tools, tractors, trucks, computers, trains, airplanes, automobiles, and boats. We sometimes forget that their development and use has been subsidized by fossil fuels. We use fossil fuels to extract materials from the earth's crust, manufacture them into goods, and process them into wastes. Many of these devices also are powered by fossil fuels. These same fossil fuels are used as a raw material to make many consumer goods (plastics) and also to make fertilizer.

Fossil energy has made such a contribution to the capacity of the planet to support humans that it may be thought of as "ghost acreage." Catton

(1980, p. 276) defines it as, "the additional farmland a given nation would need in order to supply that net portion of the food or fuel it uses but does not obtain from contemporary growth of organisms within its borders." American agriculture, much heralded for its productivity, requires approximately ten calories of fossil energy input for each calorie of food delivered to a dinner table. How long can this continue? This ghost acreage, on which we are so dependent, was built up through billions of years of plant and animal growth and cannot be thought of as renewable within the few generations that are the primary concern of people living now. The rate of extraction (misleadingly called *production*) is 10,000 times the rate of renewal (Catton, 1980, p. 49).

We humans have become so adept at extracting resources from the earth to serve not only our vital needs but our pleasures that we have become exuberant and wasteful, multiplying prodigiously in numbers and even more expansively in resource consumption. Note in Table 2.1 that gross world product and fossil fuel consumption have increased much faster than population. Metallic resource reserves are being depleted almost as swiftly as fossil energy reserves. The average modern American consumes approximately sixty times as many resources as the average citizen of India (Catton, 1980, p. 205). The resource consumption ratio is even more expansive when we compare ourselves to prehistoric humans. I invite readers to pay attention for just one day to the number of prosthetic devices we command and the energy we consume. If we reflect on our growth in numbers and our growth in resource consumption, we can readily see why we humans are facing overshoot. "We are living on four parts of phantom carrying capacity for every one part of permanent (real) carrying capacity" (Catton, 1980, p. 46).

How long can this last? No definite answer can be given because it depends on what new resource reserves are discovered, on our cleverness in conserving, on developing new technologies that are less resource consumptive or that turn new materials into resources, on our ability to limit births, and on our lifestyles. If we continue in our present ways, most analysts doubt that this bounty can last for another 100 years. Many think we already are feeling shortages and that they will become severe in about twenty years. Will we be foresighted, compassionate, and conserving as we adjust to these shortages? Probably not. Historical experience suggests that *Homo colossus* will continue to draw down the ghost acreage in the earth's crust. The earth's carrying capacity will be reduced considerably below the level that we might have made sustainable had we used more foresight.

Summary

All growth in population and in resource consumption is exponential. The swifter the growth of a population in a finite environment, the greater the likelihood of overshoot and dieback. Overshoot is likely to reduce the carrying capacity of the ecosphere. The physical limits to growth in human uses of a finite planet indicate that we cannot sustain our present trajectory. In order to change our trajectory, it is imperative that we change our society. If we do not plan ahead and change thoughtfully, nature will force change upon us through pain and death.*

Human Reaction To The Dilemma Of Growth

Humans react to the dilemma posed by growth in several typical ways, the most common of which is to ignore it. Naturally, we cannot react to something we know nothing about. If people do not know about the problem, it is not because the information is not available. Many studies have been conducted, the results have been publicized, and the information is literally thrust at people who will listen. People may be ignoring the problem because their dominant beliefs tell them growth is a good thing. Prestigious public figures assure them that their belief is correct; it is official dogma in most societies. Many people are too preoccupied with the problems of everyday living to be able to think about the distant future. Whatever the reason, the reality is that most people are, consciously or unconsciously, choosing not to listen. They do not wish to hear about problems; they have had enough of gloom and doom. They are not interested because the problem is not pressing heavily upon them right now. They would rather live for the day and let providence take care of tomorrow.

Despite the strong tendency to deny the problem, studies show that a large proportion of adults in developed countries are more or less aware that growth in population and resource usage presents a problem (Milbrath, 1984). Among those who are aware of the problem, many desire to deny that the problem is serious. The arguments in support of denial take several forms. We should seriously evaluate these arguments for their validity. As I examine all sides of these issues, I summarize well-known scientific knowledge; I urge readers who are already familiar with the information to be patient. The arguments may be summarized as follows:

1. Humans have shown that they are clever and resourceful; that will be sufficient to solve the problems posed by growth.

2. Modern science and technology will discover and deploy a form of energy that is virtually unlimited.

---

*This discussion of growth has touched only the high points. See Ophuls (1977) and Catton (1980) for thorough and well-documented discussions. These authors also discuss the implications of growth for societal change.

3. Market mechanisms will find a way to limit consumption and develop new resources not known to us at present.

4. We will emphasize renewable resources that will be so developed and managed as to allow us to continue growing.

The Humans-Are-Clever Argument

It is obvious that humans have been very clever in learning how to extract materials from and dominate nature. If we have been this clever so far, surely we can be equally clever, or more so, in the future. We should be able to find new resource reserves in places that we are not aware of presently. We should be able to make resources out of materials that we do not know how to use at present. We will be able to develop new kinds of energy (fusion power is most frequently suggested) to replace fossil energy when it runs out. There is no need to worry, our cleverness will pull us through (Simon, 1981). The belief in human cleverness is central to the belief that we can continue to grow; it is an element in all the arguments I discuss.

The belief that cleverness will pull us through is based on extrapolation from the past to the future, not on analysis of all the relevant factors. How sure can we be that our cleverness will be up to the task? Does cleverness have limits? Of course it has limits. The disasters that befell the spaceship Challenger, the nuclear plant at Chernobyl, the chemical plant at Bhopal, and the fire at the pesticide plant in Switzerland that polluted the Rhine have all been traced to failures of mechanisms that human cleverness did not foresee. Larger and more complex machines have many parts that are integral to their continued good functioning. Only one part of a complex structure need fail to stop the entire mechanism. Experts now commonly declare that mechanisms cannot be made foolproof and that we can expect similar disasters in the future. Another consideration is that a simple mechanism that fails, like a push lawnmower, can be repaired by most users whereas a more complex mechanism that fails, like a riding power mower, requires expert help.

A growing population with high expectations places increasing demands on its sociotechnical environment—which becomes more vulnerable to breakdown. It is impossible to make products that will not break down or be improperly used. The more our lives are dependent on human cleverness, the more they will be hostage to the threat of breakdown and injury. The more we become clever, specialized, and interdependent, the less free we will be to take charge of our own destiny. The Amish, who have deliberately turned their back on modern technology, have a special freedom that most modern people do not know. I am not advocating that we choose an

Amish lifestyle, but their example teaches us much about self-reliance and freedom.

As you might imagine, we first discovered minerals and fossil fuels in the most accessible places. Our search for new supplies will be in places that are much less accessible; the extraction will be much more difficult and costly requiring additional substantial costs for transportation. The probability of serious or catastrophic accidents rises steeply as we try to exploit these areas. Almost daily reports of such accidents testifies to the validity of this hazard; as these pages were being written, an explosion on an oil platform in the North Sea killed 160 men. While this book was in production, the oil tanker Exxon Valdez struck a reef and spilled 11 million gallons of crude oil into Prince William Sound, Alaska. Accidents also occur frequently in manufacturing, in normal transport of people and goods, as well as in the use of products.

Searching for minerals or energy in difficult places costs much more – not simply money but also energy. The concept of "net energy" is useful here. We should calculate how much energy it requires to extract energy from a source and deliver it to the consumer. If it costs more energy to extract and deliver it than the amount that is available for use when it arrives, the energy is not net. It makes no sense to spend more energy than one gets in return.

As another example, considerable fossil energy lies in oil shale rock found abundantly in the mountains of the western United States. Many leases to explore the feasibility of developing this resource into oil that could be marketed were auctioned off by the U. S. government to energy companies in the 1970s. After a flurry of initial activity, almost all of this developmental effort has ceased, at least for now. The oil is there and can be extracted; but it costs so much money and energy to dig out the rock, crush it, boil off the oil, and dispose of the mountains of waste that the derived energy may not become net. Such activity also has a devastating impact on the environment.

The last point suggests another aphorism: *the further our cleverness departs from nature's well worked out patterns, the greater the likelihood that the clever action will have unintended consequences* – ones likely to injure humans and the environment. We have not had time to find out what those consequences might be, whereas nature has had hundreds of millions of years to experiment and to find out what works. Given that we simply cannot foresee every possibility, we should temper our cleverness by studying and emulating nature. The current scientific thrust to develop new life forms using biotechnology is inherently dangerous because it violates that aphorism. I discuss this problem more thoroughly in Chapter 12.

As we cleverly try to support more and more people using a finite resource base, we are almost certain to increase our injury to the environment. According to the first law of thermodynamics, matter can neither be created nor destroyed; it can only be transformed. Environmentalists extract a cardinal principle from this law: *everything must go somewhere.* Every material that is extracted, processed, consumed, and discarded injects some waste (pollution) into the environment. Habitats are destroyed, ecosystems are disrupted. Many wastes cannot be assimilated and may be poisonous to plant and animal life.

Why cannot we be clever enough to avoid injury to the environment? The problem is not so much with our lack of cleverness as it is with lack of foresight about the cumulative consequences of just doing what we have always done. Another cardinal principle of environmentalism is that, *we can never do merely one thing.* As I indicated in Chapter 1, by simply reproducing swiftly and using technology to bring us more comfort and thrills, we have initiated climate change that may have a greater devastating effect on our society and economy than resource shortages. If bioproductivity declines severely because of climate change, we have no hope of growing. Such massive changes in biospheric systems cannot be reversed by our technological cleverness–although using better technology (to harness solar energy for example) could help mitigate the severity of biospheric disruption.

All the elements of nature are systemically related via cycles so complex that we can never know enough to intervene without doing more than we expect to–and possibly making things worse. Garrett Hardin (1978) makes a distinction among literacy, numeracy, and ecolacy. We used to think a person was educated if he was literate. We later came to see that mere literacy was insufficient, people also needed numeracy. We should expect people to develop the ability to handle numbers and the habit of demanding them. Hardin believes we now should demand ecolacy: a third level of education in which a person develops a working understanding of the complexity of the world and is prepared to ask, *And then what?* Almost everything we do when we intrude in nature has unintended consequences. We had better develop a habit of mind (ecolacy) that features more respect for nature's intricacies and a humble awareness of our own limitations in anticipating all the consequences of our actions.

Summary

As we contemplate using our cleverness to expand the availability of resources and increase the number of humans on the planet, we need to keep the following cautions in mind:

1. a belief that cleverness will save us is not based on thorough analysis;

2. the more our lives are dependent on "clever" technology, the more vulnerable we are to breakdown and injury;

3. extracting resources from hard to reach places is costly and risky;

4. it may consume so much energy to find more energy that the energy product is not net;

5. the more our cleverness departs from nature's well worked out patterns, the more likely it is to have unintended consequences;

6. everything must go somewhere, hence, interventions in nature to support more humans almost always injure the environment;

7. we cannot be clever enough to avoid environmental injury because we can never do merely one thing; and

8. we should develop an ecolacy habit of mind of always asking, "And then what?".

The We-Will-Develop-Unlimited-Energy Argument

Most persons who believe growth has no limits argue that we will be successful in finding new sources of energy. Some believe this energy will be so vast as to be virtually unlimited. Success in this realm is so crucial to the belief in unlimited growth, I devote special attention to this component of the humans-are-clever argument. If we had massive amounts of energy, we could purify water and move it to wherever we need it. We could even distill the fresh water we need from sea water. With heavy energy subsidies, we could grow much more food. We could build beautiful heated and air-conditioned cities. We could clean up all kinds of pollution. We could make machines do most of our work. The list goes on. With such a resource, modern society could continue on its present path and even support in luxury many more billions of people.

An astrophysicist advanced a theory that primordial gas may lie in pools much deeper beneath the earth's crust than we have ever drilled before. Unlike natural gas which is generated from decaying plant and animal matter and is found fairly close to the surface, primordial gas is believed to be made from carbon trapped in rocks. In the mid-1980s, a well more than five miles deep was drilled in Sweden to try to find a pool of primordial gas; the prospectors had found none by the time they had to give up due to lack of funds. We can expect many innovative searches for energy in the coming years; most will fail, but a few might succeed.

Our most likely prospect for massive amounts of new energy is fusion power. Very heavily funded efforts are underway to find ways to control this awesome power generated by the same nuclear reaction found in the hydrogen bomb. Deuterium, a molecule of water with an extra hydrogen atom attached, which is found naturally in sea water, is the source of this energy. M. King Hubbert, a noted geologist, has calculated that there is as much energy in the deuterium in a cubic mile of sea water as in all of the petroleum in the earth's crust. The problem is that the energy in deuterium can only be released in a thermonuclear reaction that is the same temperature as the sun. No vessel can contain energy at such high temperatures and channel it to a place and at a voltage where it can do useful work. Scientists, so far, have been able to sustain a fusion reaction for no more than a fraction of a second. However, the prevalent belief in human ingenuity knows no bounds. Many scientists are confident that the breakthrough will come some day. Some claim that the energy would be clean, but the tritium needed in the reaction is radioactive for about 100 years; it also is scarce. No one can know at this time if it will be possible to harness this potential energy. While this book was in production, in March 1989, two scientists at the University of Utah claimed to have developed fusion power at room temperature. This soon came to be known as "cold fusion." The world was agog at the promised bounty. Other scientists failed to replicate their study, however, and the public euphoria soon dissipated.

Let us assume that fusion power can be developed and that humans have at their disposal massive amounts of new energy. At this point we should ask Hardin's ecolacy question, "And then what?" One part of the answer is that we could do all kinds of useful things to make it possible for more and more people to live increasingly richer lives. But, how about the heat buildup resulting from use of this new energy? Major metro centers such as New York City or Los Angeles already are heat islands with temperatures between 4°C and 6°C higher than the surrounding countryside. The biosphere has a limited capacity to absorb heat without disrupting ecosystems, changing climates, and diminishing rather than enhancing the earth's carrying capacity. Ophuls (1977, p. 110) claims that "there is no possible technological appeal from this heat limit. . . . The maximum period of continued energy growth at current rates before unacceptable climatic consequences are unleashed is on the order of half a century." Once more we learn that, "We can never do merely one thing."

Drexler (1986) proposes nanotechnologies that would operate differently from today's "bulk technologies" and would, if realizable, constitute a reprieve from the heat limit that Ophuls claims. Bulk technologies use heat energy to fabricate products and transport them. Nanotechnologies would fabricate products molecule by molecule and would draw most of the energy they need from the sun; therefore, much less heat energy would be required for fabrication and transport. (I discuss the fascinating and frightening prospects of nanotechnologies in Chapter 12.) If developed, they will have the potential to change everything in the biosphere and our civilization—they also could bring about the extinction of *Homo sapiens*.

Will we find the unlimited energy we think we want? No one can answer that question with certainty. Meanwhile, we would be wise not to count on it and continue wasting energy and polluting the biosphere. Preparing for an energy shortage by conserving our usage and developing nonpolluting sources, such as solar, is prudent. Additionally, we should ask if we would not be better off by not growing in population and in our consumption of goods.

The-Market-Will-Take-Care-of-It Argument

Many people have great faith in market processes. They say there is no need to begin a plan of action now and start saving resources. They believe that, as resources become scarce, prices will rise, which will lead people to conserve. The rise in prices will also make it worthwhile for entrepreneurs to risk searching for more resources. The more widespread the search, the greater the likelihood of finding usable resources. Scarcity will also be an incentive for discovering and developing substitutes. Market theory cannot predict what form the substitutes will take, but it asserts that people will manage with whatever is made available through the market.

Clearly, market forces stir people to action. Both socialist superpowers (Russia and China) have turned to market mechanisms to boost productivity. Here are some thoughts to keep in mind while evaluating this argument. Market pressures *do* stimulate learning. They can help us learn to conserve and to find substitutes. They do not always incline us in those directions, however; they also can lead us to be shortsighted and wasteful, because markets have no capability for anticipating the future. The needs and demands of future generations cannot be registered in today's market.

The market for petroleum illustrates this problem. It is remarkably cheap to extract oil from the earth's crust in places such as Saudi Arabia. During most of the 1950s and 1960s, gasoline was exceedingly inexpensive. Americans built and used large gas-guzzling cars that were, nevertheless, inexpensive to operate because gasoline was so inexpensive. Americans also became accustomed to an energy intensive lifestyle that brought many comforts and conveniences. During the 1970s, the oil-extracting nations came to realize that the oil in their ground would be much more valuable in the future when oil is scarce than it would be if they extracted at full capacity and sold it in the present. They formed the Organization of Petroleum Exporting Countries (OPEC) cartel that restricted exports, raised prices above the level they would have fallen to in a free market, and saved some of their reserves for the future. The oil-consuming nations soon learned how to conserve energy better – the OPEC cartel actually did them a favor by stimulating some quick learning.

After ten years passed, however, the demand for oil dropped sufficiently that some OPEC members concluded they would lower prices in order to capture a greater share of the market. The agreement among OPEC members to limit extraction fell apart; output has risen, prices have fallen, and Americans are once more buying gas-guzzling cars. The market succeeded in lowering prices but it failed to reflect the limited reserves and ultimate scarcity of this vital resource. It cannot plan for the future.

In times of scarcity, markets are especially weak vehicles for providing social justice. (I define *justice* as a process or condition by which people receive what they are entitled to—are treated fairly; it is not simply the process by which courts decide cases.) As prices rise, some people have to drop out of the market and experience a drop in their standard of living; surely they will feel that they are not being treated fairly. Waiting until prices rise before we begin conserving will result in future generations being unjustly deprived of the availability of certain resources because we wasted them. Because markets do nothing to limit human population growth, we will encounter scarcity and high prices at the very time that a denser population must be served. When this happens, it will probably increase the proportion of poor people in society.

Markets also fail at handling what economists call *externalities*, which are aspects of economic processes that are not incorporated into market considerations. Pollution is a good example of an aspect of economic activity that typically is excluded from the market. Firms that manufacture and sell a product such as steel typically try to cut operating costs so as to sell as cheaply as possible and capture as much of the market as they can. It is much cheaper to spew out pollution onto the neighborhoods around a plant than it is to install expensive equipment to capture the pollution. It is in the competitive interest of firms to discharge their pollutants into the environment.

If a local or state government forbids the pollution discharge, another market comes into play, the competitive market among states to attract and retain industry. A local manufacturing plant can threaten to move or shut down if local and state governments propose to establish tight controls over pollution emissions. The only way that local and state governments can be freed from being held hostage in this fashion is for the national government to establish uniform pollution emission standards across the country. Firms then know that all of their competitors will have to pay similar pollution control costs; they are deprived of using the costs of pollution control as a bargaining chip when localities try to entice them to locate in their community. Because of externalities, government has had to intervene in the market to protect human health and the integrity of the environment.

This solution only works effectively if the market is national. It breaks down if the market is international (which is increasingly the case these days.) Nations also compete to attract manufacturing firms. Certain nations promise freedom from pollution controls as they try to attract firms to locate in their country. Protecting the environment in a tightly competitive market that cannot be regulated by an appropriate level of government is exceedingly difficult.

While markets are remarkably effective in pricing and distributing goods that can be privately owned (for example, cars, clothes, and land), they cannot provide for public goods. A public good is owned by the whole community (parks, clean air and water, public schools, and so on); everyone typically has access to a good if it is public. Only public institutions, such as governments, can provide public goods. If you want clean air, you cannot go to a store and buy $50 worth. You can only have clean air by working in concert with others to get government to protect the cleanliness of the air. Public goods are just as important for quality of life as private goods. Think about how much enjoyment we receive from parks, museums, sporting arenas, schools, churches, libraries, symphony orchestras, operas, and parades—not to mention clean air and water. While reliance on market mechanisms may lead to a plentiful supply of private goods, it is likely to lead to a shortage of public goods. Americans are often said to live in private affluence and public squalor because of their exceptional dependence on markets.

The Roman Catholic bishops of the United States established an "Ad Hoc Committee on Catholic Social Teaching and the U. S. Economy," which prepared a Pastoral Letter on the Economy. This famous and controversial letter was highly critical of the great difference between the rich and poor in the United States. In 1985, they invited public comment on their first draft. Four economists with a concern for the way a society might become sustainable responded to the bishops, expressing a perspective the bishops may not have sufficiently considered: the need to design a sustainable economic system. The following discussion is paraphrased from their letter (which is marked, "not for reproduction").

If each recommendation in the first draft of the Pastoral Letter were carried out, the sustainability of the economic system would still be a fundamental problem. *Modern economies follow a dynamic by which they inexorably expand.* Eventually, that expansion will encounter limits imposed by depleted natural resources. Adam Smith showed how a pin factory could gain in productivity by division of labor. Bertrand Russell discovered a drawback in Smith's pin factory. If someone invented a machine that made it possible for the same number of workers to produce all the pins the world needed in four hours each day instead of eight, would we not find all the pin workers working only four hours per day? In actuality, they would continue to work eight hours per day; some of the employers would go bankrupt; and one-half of the workers would become unemployed. This would ensure hard work for some, unavoidable leisure for others, and misery all around.

But Russell's model reveals only part of the total dynamic. This kind of productivity gain has been operating for many decades, yet we do not find one-half of the workers unemployed. We get our next clue from Ford's Model T assembly-line methods, which were so efficient that he halved the price of his car—even though he profited only $100 per car. His share of the market rose so dramatically (from 10 percent in 1909 to 48 percent in 1914) that he was able to double the wages of his workers and still make enormous profits. These lower prices, plus greater profits and wages, eventually led to new jobs for displaced workers.

Consider an economy at momentary equilibrium. The most likely events to upset the equilibrium are the efforts of owners or managers to increase profits by introducing an innovation such as a new production technique or a new product that does a job faster or more inexpensively. A specific innovation usually results in simultaneous changes in five variables: 1) costs will decrease, 2) profits will increase, 3) output can be increased, 4) employment can be reduced, and 5) the remaining employees can receive higher wages. But what about demand? If all demand was being met at previous production levels and prices, management will reduce prices to make its product more attractive to consumers. Profits and market share will still rise. Within this first firm, a new equilibrium is attained with output, profits, wages, and market share being higher, but employment lower.

Outside the firm, however, unemployment has increased. Competitors will try to match or exceed its innovation to remain competitive. Because the product is now cheaper, consumers are saving money. Most savings eventually become demand for new products. An entrepreneur can now hire the workers laid off by the first firm and try to meet this new demand. A new momentary equilibrium is reached at which: 1) the price of at least one product is lower, 2) a new business with a new product has arisen, 3) total employment remains at the same level as before, 4) more natural resources are being extracted, made into consumer goods, then "consumed," and converted into refuse (not always right away, but eventually). This dynamic structure characterizes not only competitive economies but also Marxist economies. Workers and consumers the world over measure economic progress in terms of increased wages and buying power.

The most significant aspect of the structure is its *inherently expansive nature*. There is no way within this structure for new technology or other labor saving innovation to permit workers to reduce their work time correspondingly. *The system must grow in order to create new jobs for workers displaced by productivity increases. But this growth is more akin to cancer than to organic growth that achieves maturation.* Ironically, standard economic theory functions as a flywheel on the engine of growth as it tries to make it run more steadily.

As the pressure for growth continues, no distinction is made between jobs that truly benefit society and those that do not. Many new jobs function merely as tickets giving holders honorable access to the necessities of life. A proliferation of jobs that treat the damage our economy inflicts on personal, social, and environmental well-being also occurs. It is a perversity that our economic system requires damage-control jobs.

If this economic system is not amended, it will end its long expansion when natural resources become scarce enough (from pollution as well as depletion) that the expense of extracting them increases faster than new efficiency measures can reduce costs. When these events occur, real prices will rise, but profits will shrink, owing to reduced consumer buying power. The economy will begin an irreversible period of contraction that will probably be seen at first as merely another recession. Massive deficit spending, largely funded by savings, may briefly forestall this contraction. Some argue that substitutes will be found for scarce resources. Even if we accept this dubious position, what shall be done with the geometrically increasing quantities of refuse? *Maintaining the expansive economy means eventually starving from a dearth of resources or choking on a superabundance of garbage!*

Summary

Although markets can be an effective societal mechanism for allocating goods and services, and for stimulating social learning, they fail to do certain important things:

1. they cannot anticipate and plan for the future;

2. while not inherently unjust, they cannot correct injustice;

3. they fail to protect us from dangerous externalities;

4. they undervalue nature;

5. they cannot provide for public goods; and

6. they cannot restrain our growth or provide quality of life in a society that is overcrowded and experiencing shortages.

We can only turn to government to do all of these things.

The Maximizing-Productivity-From-Renewable-Resources Argument

Those making this argument usually admit to limits of nonrenewable resources (such as metals and oil), but they also believe we can maximize productivity from renewable resources to support many additional persons. Better technology and crop-growing methods have produced many doublings of human food supply; surely such development can continue. We can find more productive plants. We can develop mariculture (seafarming). We can grow protein-producing algae in special ponds and turn the algae into human food. Recombinant DNA technology (popularly called gene splicing) promises to raise the exploitable potential of biological resources. Many such ideas are being turned into everyday practice. What cautionary arguments should we consider as we evaluate this denial of the need to change?

Concentrating our tillage on especially productive plants means simplifying an ecosystem. Recall the aphorism, "We can never do merely one thing." Monocultures and especially genetically homogeneous crops are very vulnerable to environmental change. An infectious disease or an unfortunate change in climate can destroy an entire crop. Ophuls's axiom, "Nature Abhors a Maximum" is apt for this situation. "The survival of any system depends on a subtle and incompletely understood balance of many variables. Maximizing one is almost sure to alter the balance in an unfavorable way. So complex is every natural system that the cascade of consequences started by an ill-advised maximization of a single variable may take years, or even generations, to work itself out." (Hardin, 1978, p. 55)

Another difficulty in maximizing yields from special crops is that they require large inputs of fertilizer and water. Both inputs are achieved by using fossil fuels—which, as we have already seen, are being depleted rapidly. In the United States, we now input about ten fossil calories for every human food calorie we derive. Biological resources may be renewable, but the natural systems that sustain them are not. Depleted soil regenerates so slowly that it should be considered a nonrenewable resource.

Some people have tried to capture additional energy from the sun by developing plants capable of more efficient photosynthesis. Only about .025 percent of the solar flux that strikes the earth's surface is converted by plants into food energy. If the efficiency of this conversion process could be doubled to just .05 percent, our food supply would receive an enormous boost. (F. C. Andrews, 1984) Will scientists succeed in doing this? Some suggest that if it were possible, nature would already have done so. (H. T. Odum, 1971) We should move cautiously in this direction and keep in mind that we can never do merely one thing. Doubling output from the solar flux on plants still draws down water, nutrients, and soil.

If we concentrate on maximizing food output, will we not deplete the soil? That is exactly what is happening. Soil scientists in the United States are so concerned about soil depletion that they are urging Congress to take swift and decisive action. We try to compensate for soil depletion by adding artificial fertilizer, but that is only a stopgap for it is made from scarce fossil energy and also fails to rebuild the soil. Soil formation requires many decades of natural action that includes decaying vegetable matter as a key factor. If most of the vegetative product from a field is harvested and taken away for food or energy, little is left to rebuild the soil. Fields or slopes without much vegetative cover also are very prone to erosion. Soil loss withdraws a fundamental underpinning from ecosystems and therefore from society; it is a treasure that no society, no ecosystem, can afford to lose.

A growing world population needing food and desiring industrial goods is encountering yet another limit, *water*. A few centuries ago, water was thought of as a free good; now it has become a scarce resource. Lack of water is limiting further growth in population and economic activity in many parts of the world. The American southwest may be approaching, or perhaps has even surpassed, the number of people it can comfortably support with its current water supply. Humans typically try to extend this limit by clever technology, but once more they are encountering the principle, "We can never do merely one thing." The following three stories graphically illustrate these follies:

The Ogallala is a fossil underground aquifer that underlies much of the Great Plains in the United States (see Figure 2.2). It is estimated to have contained as much water as in Lake Huron at the time the white settlers moved into the Great Plains. It used to be recharged by runoff flowing eastward from the Rocky Mountains. Land uplift and climate change have redirected most of the mountain runoff to the west, leaving the aquifer with a very slow recharge rate from natural rainfall (Little, 1987), which is why it is called a *fossil aquifer*.

The rainfall in this high plains region is sparse and unable to support crops, such as corn, requiring lots of water. A few decades ago, farmers in this region discovered that they could place a central pivot over a well and distribute water along pipes to the perimeter of a field where a large wheel slowly rotates the entire distribution rig about the pivot. Looking down from an airplane, one sees multiple circles of green against a grey/brown background. Some circles are a mile in diameter. With a dependable and sufficient supply of water, the crops were bountiful. As other landowners witnessed these initial successes, they too sank wells and began irrigating. The rate of withdrawal has been calculated to exceed the annual flow of the Colorado River (Jackson, 1980, p. 72) and far exceeds the rate of renewal. The farmers of this region are literally mining the aquifer and are now experiencing the tragedy of the commons.

The aquifer is completely dry in some regions and swiftly being depleted throughout the remainder. Because farmers in the region face the loss of a large financial investment, they are turning to the federal government to help them out. They want the government to build a gigantic canal to

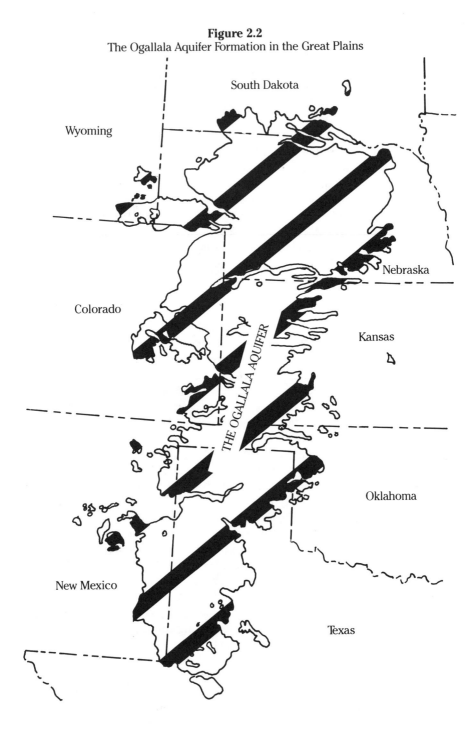

Figure 2.2
The Ogallala Aquifer Formation in the Great Plains

South Dakota

Wyoming

Nebraska

Colorado

Kansas

THE OGALLALA AQUIFER

Oklahoma

New Mexico

Texas

draw water from the Great Lakes to replenish the aquifer. This requires enormous amounts of scarce fossil energy. Political and economic leaders in the Great Lakes basin are fully aware that fresh water is their most precious resource which could attract industry to their rundown "rustbelt" away from the "parchbelt." In February 1985, the governors from the eight Great Lakes states and the premiers from Ontario and Quebec, Canada signed a Great Lakes Charter pledging to resist any further withdrawals of water from the basin; they simultaneously set standards governing inflows and outflows.

If climate change brings hot dry summers to the interior of North America, parched lands will require even more water and the Ogallala would dry up quickly; meanwhile, the Great Lakes also would diminish in size. In that context, any major engineering scheme to relieve the Ogallala would be folly.

The second story illustrates such a folly; it began thirty years ago in Uzbekistan (central Asian part of the U.S.S.R.) when authorities decided to expand irrigation in this semiarid region by drawing water from the Amu-Darya and Sir-Darya Rivers. These rivers emptied into the Aral Sea, but when irrigation expanded from 2 million to 7 million hectares, the inflow to the Aral Sea dropped drastically. The irrigation planners committed three serious errors: 1) they planted cotton, which is an extraordinarily thirsty crop for irrigated arid lands requiring ten waterings a year, 2) instead of directing the water to the Aral Sea after it had irrigated the cotton, they allowed the sea to shrink, sharply increasing its salinity, and 3) they scooped the canals out of the earth without lining them with concrete thus losing much water.

Now the irrigated land is so salty it can hardly produce crops unless regularly flushed with even more water; however, the flushing carries salt into artesian water deep underground. The Aral Sea has lost 60 percent of its water over the past thirty years; the level has dropped forty feet; the shore has receded sixty miles exposing a new salt desert. The fishing industry that used to employ 10,000 men has been destroyed. Wind storms drop crop-killing loads of salt and sand onto cultivated land. The catastrophe is changing the climate of the region and will have detrimental effects on weather as far away as India. Summers are rainless, dry and hot. Winters are snowless, severe, and cold.

What to do now? A leading ecologist calls for discontinuing the cropping of cotton, shifting factories to production that uses less water, and resettling populations that no longer can be supported in the region. Some scientists and engineers propose a huge project to divert rivers that now empty into the Arctic Ocean so that they can replenish the Aral Sea. In all likelihood, so huge an engineering scheme would have many unintended consequences. The local people are resentful that national needs have destroyed their environment. More than 1,000 articles in the Central Asian press have portrayed their resentment. Much of the political unrest in this region in 1988 and 1989 stems from their sense of being victims.

The third story takes place on Long Island which extends east from New York City into the Atlantic Ocean. It has few natural streams and lakes; therefore, most communities on the island depend on wells that tap the underground aquifer for their water supply. As the population on the island has increased, the rate of withdrawal from the aquifer has exceeded the rate of replenishment from natural rainfall. Salt water from the Atlantic Ocean is now intruding into the freshwater aquifer. The residents on the island now must either: reduce their use of water, find ways to remove the salt from the water, find other sources of fresh water, or reduce their population.

At the same time that demands for fresh water are rising steeply, the other activities of *Homo sapiens* are polluting much of it, thereby sharply limiting its use. The seepage of toxic chemicals into water is one of the most devastating of recent environmental calamities. Much of this contamination results from improper disposal of chemical wastes into landfills. Recent Environmental Protection Agency (EPA) studies suggest that about one-half of the underground water supplies in the United States are contaminated. The water in Lake Ontario is so polluted from toxic chemicals that fishermen are warned not to eat their catch. Once more, we discover that, "We can never do merely one thing."

Maximizing the output of food and other goods for an increasing number of humans also threatens the forests on the planet. The editors of *The Ecologist* estimated in 1987 that, "The world's tropical forests are being destroyed at the rate of 100 acres every minute." In many Third World countries, forests are cleared for cultivation and to provide fuel. The consequences of loss of forests have been so great in China (erosion, silting up of hydroelectric reservoirs, climate change, and so on) that the government has launched a massive reforestation effort despite the necessity to meet simultaneously the basic food needs of a billion people.

Entrepreneurs, encouraged by their governments, have slashed down large tracts of tropical forest in the Amazon basin to make cultivatable farms. Much of the timber was not even used, but simply burned. Later they discovered that the soil in tropical forests is exceedingly thin. Most of the nourishment in a tropical ecosystem is carried in the leaves, stems, and mosses that were eliminated when the land was cleared. Extended exposure to the sun and oxygen induces a complex chemical reaction in the soil that produces an end product called *laterite*, which is rock-like and unable to support vegetation. After only four or five years of cultivation, many of the farms were abandoned. It is unclear how vegetation ever can be restored on these lands. This intervention once more proved the aphorism, "We can never do merely one thing." Despite this bitter lesson, forest clearing continues.

Loss of tropical forests is a worldwide concern. The editors of *The Ecologist* circulated a petition calling on the United Nations (UN) to convene an extraordinary session and seek agreement on a world plan to save the forests. A conference was held at Penang Maylasia in 1986 to devise a joint plan of action to save tropical rainforests. A study by the U.N. Economic and Social Commission for Asia and the Pacific disclosed that one-half of the forest area in developing countries was cleared between 1900 and 1965. *If present rates of destruction continue, the tropical rainforest will be completely consumed in another fifty years.* Conference participants denied that fulfilling the needs of multitudes of hungry peasants was the main cause of forest loss. Everywhere the hand of multinational corporations could be seen. Logging projects are an important source of export revenue. Most of the tropical hardwoods exported from the Asian-Pacific area are imported by Japan. The cleared land may not be used to feed people; 40 million hectares of land in developing countries are used to grow three export crops: coffee, tea, and cocoa.

Some of the forest land cleared in the Amazon was used to raise beef to supply the lucrative fast food market in the United States – the so-called hamburger connection. The World Bank and bilateral lending agencies become parties to the destruction by funding large developmental projects. In 1987, the World Bank bowed to pressure from environmentalists and the World Commission on Environment and Development (WCED) by adopting a policy of giving as much weight to environmental consequences as economic considerations when weighing projects. Worldwide coordinated action is necessary to stem the tide of deforestation; but achieving that is exceedingly difficult.

What are the lost benefits (costs in other words) when forests are cleared? Obviously the habitats of many plant and animal species are destroyed. The U.N. Environment Program (UNEP) estimates that more than one million unique species will become extinct in the Amazon Basin alone over the next twenty years. E. O. Wilson, a distinguished professor of science at Harvard University, was recently asked what our great-grandchildren, looking back on our society, would most regret that we did or failed to do. He replied that he is certain they would most regret our failure to protect the diverse gene pool embodied in the unique species of the planet – especially in the Amazon Basin.

As we all learned in high school science, forests also are important for the carbon cycle. Animals consume oxygen as they metabolize their food; they exhale carbon dioxide (automobiles do the same thing). The carbon dioxide exhaled by the animals is used by plants to make food; they transpire oxygen needed by animals. This cycling of carbon is essential to both plants and animals; in the past, a stable balance existed between the two elements. Forests are especially important for using up excess carbon dioxide and for replenishing oxygen. Recent studies also show that trees can help clean up pollution.

In modern industrial society, we burn prodigious amounts of fossil energy. Postel and Heise (1988) estimate that burning contributed between 150 and 190 billion tons of excess carbon to the atmosphere since 1860. They estimate that forest clearing contributed an additional 90 to 180 billion tons. Both actions build up carbon dioxide; Newell and Marcus (1987) calculated a correlation of .9985 between rise in carbon dioxide levels and rise in human population over the last one-quarter of a century, which strongly suggests that carbon dioxide build up mainly results from humans doing better and better those things they have always done.

Many scientists believe that we are beginning to experience a "greenhouse effect." According to this theory, certain gases in the atmosphere, principally carbon dioxide, trap some of the long-wave heat from the sun that normally would be radiated back into space, thereby causing a gradual warming of the surface temperature of the planet; this would be likely to induce climate change. In recent years, an intensive international research effort has been launched to study the greenhouse effect. Frequent conferences are held to share findings. Cautious scientists used to believe the climate change would not be significant before the last half of the twenty-first century; now they believe it will happen sooner; some believe it is already underway.

In 1985, fifty experts from thirty nations met to assess "greenhouse gases" and their possible effects on climate change (Bolin, et al., 1986). The scientists estimated that atmospheric carbon dioxide concentrations have increased by about 25 percent over the last century; other greenhouse gases such as methane, dinitrogen, and chloroflourocarbons also have been building up due to human activities. They expect warming to be greater in high latitudes during late autumn and winter than in the tropics; the longer summers and warmer winters will increase water run off to the rivers, lakes, and oceans in these latitudes. Simultaneously, they expect greater summer dryness over the middle-latitude continents in the Northern Hemispheres, which would devastate agricultural productivity in the American farm belt (we had a foretaste of that with the drought of 1988). In addition, they expect sea levels to rise (by 20 cm. to 140 cm.), primarily due to warmer sea water. Eventually, major ice sheets are likely to melt, raising sea levels even further. Human settlements near the ocean will gradually become flooded.

As I suggest in Chapter 1, these changes will profoundly affect everything about the way we live. We count on climatic continuity for nearly all of our activities. Regretfully, we can do very little now to stop climate change; all we can do is planfully mitigate some of its worst effects. Once more we discover the truth of two aphorisms: "We can never do merely one thing," and "Nature abhors a maximum."

Summary

Humans will undoubtedly continue striving to increase the output of food and other goods from renewable resources. These limit extenders are likely to be partially successful and can relieve some human suffering; but the limit extenders have limits. In summary, several reasons are found for being cautious about pushing hard to expand the output from renewable resources:

1. such efforts tend to simplify ecosystems and make them more vulnerable;

2. they require heavy subsidy from fossil energy that is scarce and swiftly depleting;

3. they deplete the soil that is slow to regenerate;

4. they cut down forests and destroy habitat of other species;

5. they will seriously reduce the planet's gene pool;

6. they contribute to the buildup of greenhouse gases, inducing climate change; and

7. they are likely to deplete or contaminate precious fresh water supplies.

Do We Really Want To Grow?

Our long-range analysis of schemes to make it possible for us to grow indefinitely shows that every route eventually is blocked by some unexpected limit. No matter how hard we try to stay on our present path, we shall have no choice but to change. A profound set of questions still remains. Why should we wish to grow? Why fill up the world with people? Why produce more and more material goods? What values are served by growth?

I anticipate the discussion of values in Chapter 4 and declare here my belief that maintaining the integrity and good functioning of its ecosystem should be the most fundamental value in the value structure of a sustainable society. Without a viable ecosystem, life cannot be sustained, society cannot function, and it will be impossible to realize quality in living.

Let us examine a belief in the sanctity of life in the context of recognizing that ecosystem viability is our fundamental value. Declaring something to be sacred places the concept beyond discussion; everyone must believe it to be true–absolutely. Sanctity of life applies only to human lives, however; the lives of other creatures are not sacred. Sanctity asserts that every human life ought to be preserved at all costs. Is the possible loss of a life not even available for discussion? Given this belief, if technology permits it, comatose patients must be kept alive for months or years, even though they have lost all normal functioning. If technology permits, we are obliged to spend millions of dollars to implant artificial hearts in patients. Those threatened by famine must be saved by relief efforts even though the famine results from their own reproductive rate that has carried their population into overshoot. Our sense of empathy and compassion incline us to help out in all these cases.

Every religion holds a reverence for life, but does that mean that every life is sacred? Can there be no birth control or abortions? Must we continue to grow in population as fast as we can? Is human life so sacred that our numbers must grow even if it means exterminating other animal and plant species?

Death is just as much a part of nature as life, and it comes to every creature; no life can be preserved indefinitely, no matter how sacred. We cannot imagine nature with no deaths; all deaths nourish new life. Mother nature places no special value on either lives or deaths of any species. One can hold a reverence for life and still recognize that human population must be limited. Reverence for life does not demand that persons who cannot live a decent life must be kept alive at all costs. Suppose that emphasizing the sanctity of human life would have the consequence of destroying the ecosystem's carrying capacity, should we still give top priority to the sanctity of life? Hardin (1978, pp. 57-58) has a powerful answer:

> Cherishing individual lives in the short run diminishes the number of lives in the long run. It also diminishes the quality of life and increases the pain of living it. In terms of its implicit goal–maximizing the number of lives and decreasing pain–*the concept of the sanctity of life is counterproductive.* To achieve its goal, the concept of sanctity of life must give precedence to the concept of the sanctity of carrying capacity. . . . Paradoxical though it may seem to you (dear Heart!) sanctifying carrying capacity will, in fact, better serve the end you seek when you speak of the sanctity of life. To achieve the end you want you must give up the intuitive ideal with which you began. (Emphasis in the original.)

I can state my point from another angle. Ecosystems function integratively and well without humans but humans cannot live at all without a well-functioning ecosystem. Societies cannot function well without well-functioning ecosystems, but ecosystems can function splendidly without human society. Humans require a well-functioning society in order to live a high quality of life, but the good functioning of society does not depend on any given individual. By following the logic of these statements we have now developed a value hierarchy. In order to live a good life, humans must place their highest value on maintaining a well-functioning ecosystem (this is equivalent to Hardin's concept of the sanctity of carrying capacity).

> We must begin by accepting the fact that the life community, the community of all living species, is the greater reality and the greater value, and that the primary concern of the human must be the preservation and enhancement of this larger community.... The earth as a bio-spiritual planet must become, for the human, the basic reference in identifying what is real and what is worthwhile. (T. Berry, 1987a, pp. 76-77)

We should place the second highest value in a sustainable society on achieving and maintaining a well-functioning society (sited in a well-functioning ecosystem). Societal carrying capacity is based on a society's natural ecological carrying capacity. Societies existing within stressed ecosystems are likely to experience sharp group conflict; control of this conflict may require stronger and more centralized governmental institutions. (Gurr, 1985) Only in the context of these two well-functioning systems is it possible for individual humans to seek and find high quality of life. The remainder of this book is mainly devoted to specifying what I mean by *well-functioning*. In fact, most attention will be given to designing a well-functioning society, since ecosystems function quite well if there are no humans in them.

This analysis has answered our question, "Why should we grow?" If we think the matter through, I believe we will conclude that we do not really wish to grow. The ecosystem will function better and humans will live better lives if there are fewer humans, not more. The ecosystem will function better and we will live higher quality lives if we produce fewer material goods (with care so as not to injure the biosphere), not more. Society will function better if it nestles into a well-functioning ecosystem and carries out its activities without stressing the ecosystem. Sanctity of life is not an operative value; in fact, it would diminish life. Growth is not a value, it is destructive if pursued vigorously. Pursuit of many of the honorific words in modern society (productivity, progress, power, biggest, winning, superiority, and so on) actually turns out to be counterproductive for achieving our deepest values. We need to reexamine those premised values. We also should take a closer look at the way we come to know things and the way we decide on our values (these topics are discussed in a larger context in Chapter 4).

As we think about how deeply embedded in our society the values of growth, power, competitiveness, and domination have become, we feel overwhelmed in contemplating the awesome task of trying to turn such a society around. Our puny words, which would not be sufficient in any case, will be amplified by the massive inexorable changes our species has already set in motion. Nature will be our most powerful teacher because it leaves us no alternative but to change. We have tried to build a society disconnected from its biocommunity and it cannot sustain itself. If we wish our species to survive, we must change. I will suggest directions our change should be taking. We will have to learn quickly, deeply, and wisely if we desire to save our society and our species.

Transforming The Dominator Society

Chapter 3

Dogwood Blossoms

It's a question of bright stars
and of four petals cupped
to catch sun and reflect
a hovering circle of white

Here where the big trees
were so recently logged off
and the jagged teeth of stumps
and broken arms of branches
question the meaning of sanity

A slide of mud and stones
advances down the ravine
and dogwood and maple are bowed
under weight of future burial

It's a question of the last act
before man-made dying
that hundreds of blossoms
shout a final triumph
for earth and sky to behold

Were we to be an armless
legless race of creatures
belly-crawling through life
perhaps we could learn of beauty

but instead we cut down
the very answers we seek
in torn earth, and the secrets
remain unseen by us, as we
plunge forward, blindly,
    brushing aside blossoms

Peter Blue Cloud, in *White Corn Sister*
Strawberry Press, 1979

Humans, among all species, can invent their pattern of life. But if that is true, why do we seem unable to shape our destiny? Why are wars so frequent? Why do men dominate women? Why is there so much suffering and privation? Why do we wreack havoc on our ecosystem? Why do we fear that the next war will annihilate most life on the planet?

Nearly all societies are dominator societies in which some people rule other people. All of us grew up in a dominator society. Many take domination to be characteristic of all societies. Is that true? Is it inherent in human nature for some people to dominate others? Men are taught to be competitive and aggressive; if they do not do this, others perceive them as not being *real* men. A window sticker on a van parked in my university's parking lot proclaims: "God, Guns, Guts Made America Great; Let's Keep All Three." Is that true? Most women have experienced male domination; they are even socialized to admire and reward it. Capitalist doctrine proclaims the competitive society as the best. Is that true? People in the United States are so competitively oriented that a football coach has been eulogized for proclaiming, "Winning isn't everything, it's the *only* thing."

Many social problems can be traced to the competitive struggle for money, power, prestige, and control. Why do we join this struggle and why is it so difficult to avoid it? How does this struggle interfere with our efforts to develop a society that can live in a long-run harmonious relationship with nature and in which people can live in peace with each other? How did we get into this predicament? Perhaps we can gain insight by taking another look at history.

A New Perspective On Human History

Most of us do not think much about history and have a poor understanding of it. Perhaps for that reason we are insufficiently aware of the extent to which our beliefs about the way the world works are embedded in presumed truths from our past. Most of us also know that history is rewritten from time to time, sometimes because of new knowledge but more frequently to serve the purposes of those who control a society.

The historical perspective sketched here is quite different from the one I was taught as a schoolboy. My new perspective has been stimulated by three recent books: Schmookler's *The Parable of the Tribes* (1984), Eisler's *The Chalice and the Blade* (1987), and Margulis and Sagan's *Microcosmos* (1986). These reinterpretations of history reflect very recent scholarship; none were written by historians. Their new data come mainly from archaeological, anthropological, fossil, and geological records. They show us that traditional history, which is derived primarily from written records, can lead us to misinterpret the essence of human nature and human society.

The reader will recall from the story of the year-long movie, sketched in Chapter 1, that the first glimmerings of microbial life on planet earth did not appear until March. It was the only kind of life until August (2 billion years of earth time). More complex animals did not evolve until November. *Homo sapiens* evolved into an identifiable species only in the last five hours. Civilization did not dawn until about one minute before midnight. The industrial era of civilization took place only in the last two seconds. Nearly all of the devastation humans have wrought against each other, against other species, and against the ecosystem has taken place in the last minute, most of it in the last two seconds. Maybe the trouble is with our civilization.

Popular belief holds that history began with written records about 5,000 years ago. Some people think that history should cover civilization; that would include about 10,000 years. Schmookler (1984) shows us that in order to understand the world as we find it, we must begin with the origin of the planet, 4.6 billion years ago, and study the way life evolved. The magnificent creation of biological evolution never was able to look forward. The selective process did not "know" where an evolutionary experiment would ultimately lead; it selected for what worked. "Biological evolution, one may say, employs no author but only an extremely patient and effective editor." (Schmookler, 1984, p. 64.)

Although many species have an ability to learn, the human species developed a special ability to incorporate a time perspective, giving it the ability to plan. In contrast to evolutionary change that is not controlled by planning, human changes were guided by a vision of where they might lead, and with hindsight we can see that many of those estimates were faulty. Even though humans probably used planning as soon as their capability to do so evolved, the human experiment with learning advanced for more than 2 million years without disrupting the stability of the ecosphere because people lived in harmony with nature. People who lived by hunting and gathering were deeply aware of their intimate dependence on nature. They nurtured it, even worshipped it, for nature nourished them in return.

But then humans began to develop culture; that is, they began to learn from each other and could accumulate this learning for their group. Knowledge could be stored and widely shared, and the learning capability of the species accelerated. With domestication of plants and animals about 10,000 years ago, humans truly departed from the natural order. At first, the economy of domestication was merely an appendage to the ongoing hunting and gathering economy, but the new way of life gradually supplanted the old. The creature with the freedom to choose became unpredictable; it could hunt to extinction as well as nurture reproduction of "desirable" species. Cultural "inventions" not only enabled human population to grow more swiftly but also provided a surplus that allowed some people to pursue other trades and occupations; knowledge accrued at an accelerating rate and humans began striving for material wealth. More and more, the human animal exploited an unprecedented opportunity. It could create its own way of life.

Eisler (1987) summarizes considerable archaeological evidence showing that relations between people in some of the cultures created by this newly enhanced capability for learning may have been very different from those we know today. She characterizes them as "partnership" societies in contrast to the "dominator" societies of today. Because they were not patriarchal, we should not assume they were matriarchal; rather, the archaeological evidence suggests that neither sex dominated the other—they lived together in partnership. Artifacts indicate that they worshipped a goddess representing nature—woman symbolized love, fertility, caring, and nurturing. "This theme of the unity of all things in nature, as personified by the Goddess, seems to permeate neolithic art." (Eisler, 1987 p. 19) Nature was for nurturance; it was not an object to be dominated—as so many think of it today. Their view of power was the feminine power to nurture and give; they seem not to have been attracted to power to take away or dominate. Eisler (1987, p. 28) calls it *actualization* power as distinguished from *domination* power.

Goddess-worshipping partnership societies apparently existed in many places in the Near East that we know today by such names as Palestine, Lebanon, Syria, Turkey, Greece, Romania, Bulgaria, Cyprus, and Crete, as well as across the southern border of the U.S.S.R. that is shared with Iraq and Iran. These were separate societies, not one empire, but they were similar in their worship of a nature goddess and in partnership relationships between the sexes.

Probably the highest expression of what has been termed a "golden age" culture, among these partnership societies, was developed on the island of Crete. Archaeological records show that Crete was colonized about 6,000 B.C., and its culture persisted for 4,000 years. Eisler (1987 p. 32) reports these quotations from archaeologists and art historians: "the most inspired in the ancient world"; "the most complete acceptance of the grace of life the world has ever known." Wealth was shared, even the standard of living of peasants was high.

If these societies brought such happiness, why did they not survive? Around 3,000 B.C., humans learned to control fire and to use it to make metal, especially swords and other weapons. One by one, these peaceful, equalitarian, and creative societies fell to conquerors. The dominance of the blade was forcefully demonstrated by fierce warriors that swept down on horseback out of the steppes of Russia. As an island, Crete was protected from the conquerors for some time, but their society was apparently weakened by catastrophic volcanic eruptions that created huge tidal waves. So devastated by these cataclysmic events, they were easy prey to conquerors. Their society disappeared and was even forgotten–except in folk legends. Many scholars today believe that the legendary lost civilization of Atlantis may actually have been the Cretan civilization, the remains of which archaeologists are now uncovering.

Thomas Hobbes in his classic *Leviathan* portrayed the state of nature as a "war of all against all," but that characterization does not stand up against biological and anthropological evidence. It is true that the living order has no ruler, but it is not anarchic. There is struggle in nature but the struggle is part of an order; the separate interests of individuals and species have been formed by selection into a tightly ordered harmonious system. Hunting and gathering societies were governed by this natural order. Minoan civilization on Crete apparently also sought harmony with nature. But, the conquering patriarchal civilization sought domination over all creatures. "Out of the living order there emerged a living entity with no defined place." (Schmookler, 1984 p. 20)

Both Schmookler and Eisler dwell on the way the civilization that emphasizes domination has pervaded all aspects of life and spread to the remotest parts of the globe. Both agree that the struggle for power and domination is not inevitable; other successful societies existed where people felt equal and did not have to struggle for position. Why did partnership societies give way so completely to the dominating conquerors? Schmookler's "parable of the tribes" (1984) provides an explanation.

Schmookler's parable deals with tribes, or societies, that do not have a governing arrangement that regulates their relationships; it is analoguous to anarchy among individuals. As human numbers grew and tribes enlarged, societies confronted each other. Each society in such a confrontation faced an unpleasant choice. If it willingly stopped its growth so as not to infringe on its neighbors, it could foresee that death would catch up and overtake it. If it continued to expand, it committed aggression.

As civilization developed, humans confronted a situation in which the play of power was uncontrollable. "In an anarchic situation like that, no one can choose that the struggle for power shall cease. But there is one more element in the picture: *no one is free to choose peace, but anyone can impose upon all the necessity for power.* This is the lesson of the parable of the tribes." (Schmookler, 1984 p. 21, emphasis in the original) Evolution under civilization developed a new selection principle: POWER. The evolutionary principle discovered by Darwin asserted that species survived because they found their niche. (Darwin's principle is often misinterpreted as "survival of the fittest" and is used to justify aggressive domination of others.) However, in the evolution of civilization, those cultures survived that best deployed power. The selection for power applies mainly to the struggle among humans, although that struggle often resulted in great harm to other species as well.

Schmookler spends most of his book showing the numerous ways that the struggle for power has enslaved us all and permeates every part of our life. He sees power as a contaminant, a disease, which once introduced will gradually yet inexorably become universal in the system of competing societies. Power dominates because it can prevail; but what prevails may not be best for people and other creatures. The continuous selection for power closed off many humane cultural options that people might otherwise have preferred. Power came to rule human destiny. (Schmookler, 1984 pp. 22-23)

I have room here only to summarize the many consequences for society, the ecosystem, and the quality of individual lives that flow from the selection for power, most of which is drawn from Schmookler:

1. The warlike eliminate the pacifistic and content. A tribe (society) that is confronted by an aggressive power-maximizing neighbor has only four options:

   a. It may suffer destruction; in a struggle for power the surviving society will be the one that employs power most effectively.

   b. It may be absorbed by the aggressive power and become transformed into a power-maximizing society.

   c. It may escape the compelling pressures of the intersocietal system by withdrawing beyond the reach of other societies. Only in the least accessible regions of the planet have the equalitarian and peaceful societies been able to survive into our time. All other societies were drawn into the power contest, or were eliminated by it.

   d. If it chooses to defend itself against the aggressor, the peace-seeking society becomes the imitator of the power-maximizing society. "The tyranny of power is such that even self-defense becomes a kind of surrender. Not to resist is to be transformed at the hands of the mighty. To resist requires that one transform oneself into their likeness. Either way, free human choice is prevented. *All ways but the ways of power are blocked.*" (Schmookler, 1984 p. 54 emphasis in the original) A society may also choose to escalate its power, thus start arms races; this might be thought of as a fifth option. It constitutes total capitulation to the power struggle.

2. Selection for power favors those who exploit nature and discards those who revere it. A society that exploits its resources quickly accrues more power than a similarly based society that husbands its resources and protects its biosphere. The resource-exploitative society may then overpower the more nature-protective society and seize its resources

for additional quick exploitation. The drive for power produces an ever-widening gap away from the natural. "Man's power over Nature means the power of some men over other men with Nature as its instrument." (C. S. Lewis, 1946, p. 178)

3. Larger populations and greater land areas contribute to power; therefore, there will be a tendency toward larger societies. "Size confers power and power facilitates expansion." (Schmookler, 1984 p. 82)

4. There will be a tendency toward complexity, specialization, and efficiency because they enhance power. Centralization is needed in order to manage this complexity effectively. Central control enforces unity of purpose, directs coordination of the parts, and induces the parts to sacrifice. The division of labor in a complex system enhances power but it detracts from the wholeness of humans. The demand for power creates a need for drudges.

5. There is a need to be constantly on the alert. We have become an adrenalin society, always organized and oriented toward the requirements of maintaining power and being ever alert to ward off threats from other societies. Even in the times between wars there is no peace.

6. Intersocietal competition has produced in modern times a single global competitive system that is bringing all cultures toward convergence. Increasingly there is but one way into the future, the technological way. We compete in technological development because that is the key to power. Similarly, we are urged to work hard, to save, to produce capital, and strive to become wealthy because this mode of living is superior for accumulating power. This thrust for power builds an ethic that overvalues productivity and undervalues conservation and protection of nature.

7. Reason especially is brought into service to enhance power. Reason, via science, creates economic and military power. The intellectual sphere, in service to power, is permeated with the paradoxical teaching of the value of "value neutrality." Knowledge and technology created by a "value-free science" become the captive of the power maximizing system. "People who cannot experience their own ultimate purposes provide a vacuum to be filled by the purposes inherent in their systems. ... The way is cleared for the purposes of the systems to be adopted by the people as their own purposes, rather than vice versa. Technology emerges as the trend of the modern world: rule of the tool." (Schmookler, 1984 pp. 202-203)

8. In a synergistic system the interaction of the parts contributes to the good functioning of the whole; but the ceaseless struggle for power in modern civilization creates a societal environment that is unsynergistic:

   a. Conflict gains an ever increasing sway even though humans do not wish it.

   b. The competitive intersocietal system produces a minus-sum game in which all the parties lose.

   c. Power is corrupting; it does not serve an essential life sustaining function for the collectivity as a whole.

Schmookler closes his book by offering only a hint of the steps that might be taken to extricate humanity from its horrible predicament. He does look to government as the only social device available for possibly extricating ourselves. "Anarchy enthrones force and only government can place anything else on the throne." But how do we rise above the race for power in most of our nations? How can we find peace in the anarchy of our nation-state system? No one, currently, can answer those questions.

Schmookler's analysis uncovers the deep and fundamental flaws of modern civilization. We can never live in peace, love, and justice in a clean and nourishing environment, we can never have high quality of life, as long as we retain our power-maximizing, competitive society. We cannot solve the problems of war and peace, or protect the environment, by developing more and better technology. Societal tinkering with new laws or new policies may ameliorate some problems, but will not deflect us from the abyss *as long as the present power-maximizing system is retained.* The only choice that has any hope of saving our species and providing humans with a reasonable quality of life is to transform the dominator society—we must redesign the most fundamental relationships in our civilization.

Eisler dwells much more than Schmookler on the tendency of societies since the dawn of civilization to develop dominator/submissive relationships, especially a patriarchal relationship within families and between the sexes. The societies that conquered the partnership societies emphasized the power that takes away more than the power that gives life; their patriarchal system forcibly asserted male dominion over women, even to deciding their life and death. Patriarchy also used the power of the blade to enthrone a male diety in place of the female goddess.

The new dominators were not content until they completely transformed the way people perceived reality. As this thrust proceeded, the power of the blade became idealized. Both men and women were taught to equate true masculinity with violence and domination and to regard men who did not conform to this ideal as weak or effeminate. Even today in the United States the most desired characteristic of a President is that he should be a *strong leader*. For those brought up in this system (which includes most of us), believing there is any other way to structure human society is difficult.

Technologies of destruction were given highest priority in the dominator society. Material resources were increasingly channelled into more sophisticated and lethal weaponry. The strongest and most brutal men were honored and rewarded for pillaging and conquering. Force or the threat of force controlled the allocation of material goods. Force was used to determine social ranking and soon resulted in a highly stratified social structure. Brute force clearly established male dominance over women. By skillfully creating a new myth, even educating women was made sinful. Natural aspects of womanhood such as birthing were made dirty.

Now, perhaps nowhere as poignantly as in the omnipresent theme of Christ dying on the cross, the image of art is no longer the celebration of nature and of life but the exaltation of pain, suffering, and death. For in this new reality that is now said to be the sole creation of a male God, the life-giving and nurturing Chalice as the supreme power of the universe has been displaced by the power to dominate and destroy: the lethal power of the blade. And it is this reality that to our day afflicts all humanity—both women and men. (Eisler, 1987, p. 103)

Given the pervasive and growing power of the dominator culture, it is surprising that Christ emerged and gained such a following that Christianity became the dominant Western religion. Christ preached love, peace, forgiveness, mercy, meekness, and treated women as equal to men; nearly all of these are partnership rather than dominator values. Of course, the dominator culture could not tolerate such unsettling ideas; as Socrates before him, and as Ghandi and Martin Luther King in our own century, he was put to death for these ideas. Early Christianity tried to live by peaceful partnership ideals; women played as large and central a role in the early church as men.

Eventually, however, the dominator culture had its way and transformed the religion that was trying to transform it. As the Christian church gathered strength, women were systematically excluded from church leadership. Gospels that portrayed women as equal to men were excluded from the official Bible. Societal rankings displaced the equality that Christ preached; ironically, the church leadership demanded the very highest rank. Militaristic rulers working in close collusion with the patriarchal Catholic Church dominated every aspect of life in medieval Europe. The Holy Roman Empire reached the epitome of a totally patriarchal society.

Thomas Berry (1987b) (himself a Jesuit) identifies four patriarchal establishments: the classical empires, the ecclesiastical establishment, the nation-state, and the modern corporation.

The difficulty with patriarchy is precisely that it is something more than a social or political arrangement. It reaches far back into the cosmological structure of existence and into the ritual, moral and belief commitments of religion, even in the sacred writings of the Bible. . . . The sense of the sacred in any civilization is precisely that which cannot be questioned. For the sense of the sacred is the unquestionable answer to all questions. . . . To realize that so much of the accomplishments of these four establishments has been misdirected, alienating and destructive beyond anything previously known in human history is a bitter moment. . . . Only now in the twentieth century have the awesome dimensions of our cultural and institutional pathologies become clarified. . . . Nothing has prepared us for what we must confront. . . . Everything is at stake. . . . This is something more than feminine resentment. . . . It is possibly the most complete reversal of values that has taken place since the neolithic age.

Berry is right, the dominator society must be transformed to a new set of beliefs and values—to a new worldview.

Increasing Our Awareness Of The Subtle Tyrannies Of Dominator Power

We can only free ourselves from dominator beliefs by becoming aware of the myriad ways they control us. The discussion of Schmookler's and Eisler's perspectives on history has already made us more aware how our civilization has been structured by the struggle for power and domination. Now, let us turn our attention to the effects of the struggle for power and domination in contemporary society.

We often hear the phrases, "We live in a competitive society," or "We live in a competitive world." These phrases carry the connotation that this condition is normal and that it is even good; furthermore, we can do nothing about it even if it is not good. Additionally, it is often implied that one had better attain and maintain competitiveness or one will not survive. Males are taught that they must compete if they wish to be considered *real* men and that they must beat out other competitors if they wish to be honored and rewarded. Beginning in early childhood, boys are told that they are not allowed to cry, to appear "weak," or to shun competition. Ironically, girls are taught to admire and love strong and aggressive men. Competition may lead to certain high achievement, but how aware are we of the insidious consequences of fierce competition?

Competition is extolled as the very lifeblood of our economic system. Competitive capitalism is widely believed to be far superior to any other system. Yet, it also is a strong expression of patriarchal domination. Gilder (1981) extols "the male's superior aggression" as a great social and economic value for capitalism. Nearly all industrial corporations perceive that they are struggling in a dog-eat-dog competitive world; they must grow and become strong by defeating others or they, in turn, will be devoured. Corporations are hierarchiacally governed by a small patriarchal elite pursuing male values: money, power, domination, and control. Their control over serfs, workers, women, and children has been mitigated somewhat by growing community awareness of the inhumanity of such domination, but their encroachment, enslavement, and mutilation of other creatures—all of nature—is still virtually unchecked. "The rights of the natural world of living beings other than humans is still at the mercy of the modern industrial corporation as the ultimate expression of patriarchal dominance over the entire planetary process. The four basic patriarchal oppressions are rulers over people, men over women, possessors over non possessors, and humans over nature." (T. Berry, 1987a, p. 79)

Everyone would expect military institutions to be suffused with patriarchal dominance, as they certainly are. It is not so obvious that those inhabiting that realm use a whole array of psychological and linguistic ploys to isolate their psyches from the massive forces and potential destructiveness of their work. Carol Cohn (1987) spent a year immersed in the world of Washington defense intellectuals. She found that feminist critiques such as Helen Caldicott's concept of "missile envy" were not overly simplistic but had never reached the men who speak "technostrategic language." Their psychological defenses were so subtle and pervasive that they were hidden to the participants. It required an outsider, coming from a very different perspective, to discern their ploys. These psychological structures also defend against outsiders who are not given credibility or a hearing until they learn to use the special language of the in-group. These people are far removed from reality. She concludes that language is a significant part of the arms race. Cohn's analysis is so subtle and many-faceted that it is nearly impossible to summarize, but her conclusion is notable:

I believe that those who seek a more just and peaceful world have a dual task before them. . . . Deconstruction requires close attention to, and the dismantling of, technostrategic discourse. The dominant voice of militarized masculinity and decontextualized rationality speaks so loudly in our culture that it will remain difficult for any other voices to be heard until that voice loses some of its power to define what we hear and how we name the world. The reconstructive task is to create compelling alternative visions of possible futures, to recognize and develop alternative conceptions of rationality, to create rich and imaginative alternative voices—diverse voices whose conversations with each other will invent those futures. (Cohn, 1987, p. 24)

How far can patriarchy dominate our lives? Nature itself is beginning to rebuke the aggressor humans. Toxic poisons return to us in our air, water, and food, raising the incidence of pain and disease. Abused and exhausted soils refuse to provide further bounty. Blighted landscapes repel us by their sheer ugliness. Exhaustion of resources will restrict economic output, raise costs, and spread poverty. Even the climate will change and ruin our investments and productivity. Nature's rebellion is not simply against technology, as is sometimes argued, but a rebellion against the exploitative and destructive uses that dominating men put it to. Somehow, men believe they must keep conquering—be it nature, women, or other men.

The patriarchal dominator society is so obsessed with accruing power that it cannot recognize the threat of population growth. Societal leaders perceive a large population as providing many soldiers and productive workers to maximize national power. In a recent book, *The Birth Dearth*, Wattenberg (1986) laments that population growth is so low in the West it detracts from the ability of the West to remain supreme. The Reagan Administration has pursued policies to encourage births and has denied that population growth is a worldwide problem. Other patriarchal organizations, like the Catholic Church or Ayatollah Khomeini's resurgent Islam in Iran, have taken the same path.

Mao Tse Tung pursued population growth policies in China and their population doubled from 500 million to one billion in thirty-five years. Instead of building national power, this massive growth created incredible problems. It was only when Chinese leaders perceived that one more doubling to 2 billion was as imminent as thirty-five years away, and would bring disaster, that they instituted a drastic one child per family control policy.

Policies that would weaken male dominance—and most policies that offer any hope for the human future will—cannot be implemented. Even if they are formulated, such policies must be shelved, given inadequate funding, or otherwise diverted from being effective." (Eisler, 1987, p. 179, emphasis in the original) Curiously, this pattern can be observed in most universities. Women's studies, peace studies, and environmental studies have been forced onto the teaching and research agendas of some universities by activists struggling against the dominator system. Reluctant administrators and faculty have generally refused to give these areas of inquiry mainstream status; instead they are kept on the margin, inadequately funded, and closed down at the first opportunity. Only when their inquiries have turned to technological solutions that do not threaten the dominator system are they looked upon with favor.

As everyone knows, the promotion of science is the supreme value—a virtual godlike entity—at most universities. It is not so obvious that science, especially big science, is embedded in and a promoter of the dominator system. Science proclaims that it only seeks knowledge and disclaims any other values. Achieving a value-free science is a major objective of the enterprise. As will be developed further in the next chapter, a science that tries to be value-free serves the values of the leaders of major societal institutions. A large proportion of scientific effort in modern societies, especially in the larger countries, supports the accrual of power in the military. Large corporations also are able to command the services of science to enhance their growth and wealth. Universities use science to accrue prestige, money, and more students—for more growth.

I recognize that science does many good things for people and I also recognize that many scientists have the singular motive to try to understand puzzling phenomena. My critique is of the uses to which science as an organized enterprise is put. Almost all organized scientific efforts have a dominator motive: control of nature, military power, economic growth, economic power, beating the competition, maximizing prestige and honor, making money. Science would do well to think much more seriously about values. *"To disconnect virtue from power is to ensure that virtue will be powerless and licenses power to be without virtue."* (French, 1985, p. 534, emphasis in the original) Science will always serve some value. We would become wiser if we debated what those values should be, rather than give over the funding, direction, and control of science to the holders of power—the male dominator system.

Few practitioners in science recognize the male dominator mores that govern its practice. Prestige and preference in science is not achieved simply by discovering, although search and discovery certainly are an essential part of the effort. Most scientific findings are open to interpretation and their very interpretability means that more than one perspective is possible. Human fallibility and values creep into the process when meaning is read into the findings. Therefore, scientific findings must be defended against challenges and the challengers are many, for a scientist can make a reputation by knocking others down. Many scientific papers are studied not for what one can learn from them but are given close scrutiny to find areas of weakness that can be exploited in a challenge. Successful scientists, those who achieve prestige, honor, and support, must vigorously defend their interpretations of meaning against challengers. I recall listening to a lecture many years ago by a world famous psychologist. He was discussing a dispute over meaning that he had with another psychologist and declared, "The very fact that he opposes me on this matter makes me all the more certain that I am right."

Competition in science, then, is not unlike competition in sports, or in military prowess, or in the economic marketplace. In order to win respect, each scientist must stand tall and strong as he confronts his challengers. That respect is essential if he wishes to have his research papers or books published in prestigious places. Respect is essential if he wishes to obtain grants or contracts to support his research. Respect is essential to obtaining a good working position with good laboratories, space, and research assistants. Respect leads to invitations to prestigious conferences, to election to professional office, to appointment to prestigious committees, or to an invitation to join the prestigious National Academy of Sciences. In short, standing up to challenge and gaining respect is the way to get admitted to the priesthood of modern society.

The male dominator aspects of science become clearer when I recite the things one must not do for fear of losing respect. One must not appear to value things strongly; leave valuing to the philosophers or priests. One must never appear to be doubting or emotional. One must not take strong political stances. One must not challenge the dominant socioeconomic-political establishment. One must stick to one's narrow discipline. Avoid interdisciplinary areas of study such as peace, feminist, or environmental studies—typically these are perceived as soft areas of inquiry that are difficult to defend against challenge. It is little wonder that relatively few women choose careers in science. A woman that does choose science must be superstrong in standing up to challenge (that is, be manlike) in order to win respect.

Recognizing That The Dominator System Is Maladaptive

The male dominator system seems supremely effective; it has driven out partnership societies; its central objective, power, has guided cultural evolution; it has penetrated to the far reaches of the globe; it has shown extreme resiliency in defending against the occasional resurgence of feminism. However, it has not recognized that it has created a civilization that cannot be sustained, and it has not been able to prevent learning that does not please it. In the previous chapter I set forth the arguments why our present dominator society cannot be sustained. In short, the ways of power and dominance have now accrued so much destructive capability that we can destroy most life on the planet. Even if we avoid that catastrophe, but continue on our present course, we will devastate the ecosphere on which we all depend for life.

Rather than proceed blithely on to breakdown, we must strive in numerous ways to break through to a new, deeper, and more sensitive understanding of our predicament and of what it requires of us. A rising awareness of our predicament, and of patriarchy's role in creating the predicament, is a necessary first step in overturning the dominator society. The strong feminist movement of the last two decades has accomplished a great deal in raising awareness of the subtle and pervasive effects of patriarchy throughout society. That awareness is still confined primarily to women, however, and now needs to be spread to men. Many men will be receptive, especially if they perceive that a change in beliefs and values is essential for saving our ecosystem and society. We can be sure, however, that many other men will feel threatened by partnership ideas and values and will vigorously oppose the change.

The struggle is inevitable, but our aim is to transform conflict rather than to suppress it or explode it into violence. Unlike the struggle of 4,000 years ago when partnership was twisted into patriarchy, this one can probably be won. What has changed? *First*, the world is now very different; the threats of nuclear war, overpopulation, resource depletion, ecosystem devastation, climate change, famine, and disease are so real and imminent that billions of people are aware that drastic changes are needed. System collapse has a remarkable way of freeing one's mind from old conceptions. Nature will not leave us content with our old ways. *Second*, our society has learned to abandon feudalism, slavery, and colonialism. We no longer permit men to brutalize women and children. Brute force cannot be used to ensure continued male dominance over women.

Third, in more and more societies decisions are made by votes instead of guns. In the last two decades, dictatorships have more or less peacefully given way to popularly elected governments in Spain, Portugal, Argentina, The Philippines, Brazil, and South Korea. Attempts by superpowers to impose their ideologies by military might in Vietnam and Afghanistan have failed. Furthermore, huge communist dictatorships in China and Russia are loosening up and allowing more dissent and public discussion. Similar loosening is apparent in smaller communist countries such as Hungary, Poland, the German Democratic Republic, Nicaragua, and Cuba.

Fourth, in nearly all countries that permit elections, women have an equal chance with men to vote and they are learning how to use their vote. The most recent turnout figures in the United States show that women are more likely than men to vote. *The Polling Report* of April 6, 1987 reported a study by Heidepriem and Lake showing that women provided the decisive margin of victory in nine Senatorial races in the United States in 1986. Overall, a pro-democratic gender gap of 9 percent appeared in the Senate races. They concluded, "Women voters returned control of the United States Senate to the Democrats." My own studies show that the pro-environmental protection stance on most policy issues is more likely to be supported by women than men; on average, 15 percent more women than men would support environmental protection.

"In contrast to men, who are generally socialized to pursue their own ends, even at the expense of others, women are socialized to see themselves primarily as responsible for the welfare of others, even at the expense of their own well-being." (Eisler, 1987, p. 189) In my judgment, women have a much better chance than men of saving the planet. We men who perceive the crucial importance of female participation should do all we can to support their efforts. All of us, both sexes, have both male and female feelings and needs but each sex feels unfree to express both sides of its personality. Liberation from the dominator belief system and social structure will be freeing to both men and women.

Speaking Truth For Learning

Occasionally, we all hear aphorisms that purportedly express truth about human nature and human society. For example, I have often heard these phrases: "We have always had wars and we always will have"; "People are naturally selfish and competitive." Probably the personal philosophy and worldview of many people is a mere collection of such aphorisms. Aphorisms have a surface truth quality that may not survive a close analysis, but they persist as part of the myth structure of society because they are not clearly falsifiable. They also serve as excuses for not acting to make a better society.

The analysis presented above provides a basis for falsifying some of these aphorisms, and it presents an opportunity for each of us to make a contribution to social learning by challenging their presumed validity when they are brought up in conversation. Let us examine the validity of a few of the more common aphorisms:

1. *"We have always had wars and we always will have."* A variation on this is, *"Human nature being what it is, we will always have war and conflict."* These statements erroneously assume that wars have been present throughout human history and that being warlike is rooted in human nature. Eisler's evidence discloses that civilizations existed for thousands of years with few wars. Schmookler helps us to see that war is more rooted in the structure of the dominator civilization than in the essence of being human. Humans have a strong desire and need for peace that far outweighs their urgings toward war. If we desire to reduce the probability of war, we should try to change our beliefs and our social structure so that humans are not pressed into going to war.

2. *People are naturally aggressive and competitive.* This statement also implies that aggression is rooted in our biological makeup. Is this so? As much evidence is found that people are naturally loving and cooperative as that they are aggressive and competitive. Social structures

and sex roles put people into positions where the only posture that makes sense is to be aggressive and competitive. The behavior demanded of people in business competition is a good example of structurally fostered aggression; failing to be aggressive results in being eliminated from the competition. But, we should keep in mind that much of life does not require aggression and that we can be successful in achieving what we want by acting cooperatively and lovingly. As a matter of fact, recent research shows that cooperation is more likely to be successful over a long time than aggressive competition (for example, Axelrod, 1984). Some societies emphasize competition on the presumption that it will lead to a better society, while other societies emphasize cooperation for the same reason. The one emphasis is just as compatible with human nature as the other. It is difficult to imagine any society being totally competitive or totally cooperative, but we can play some role in tilting the emphasis toward cooperation.

3. *Men will always dominate women.* Again Eisler's and Schmookler's reviews of history show that partnership societies existed for thousands of years before male domination became the norm. Continued domination of men over women mainly results from a belief on the part of both men and women that domination is justified. Those beliefs and their supporting social structure can be changed.

4. *People first look out for themselves.* The implication is that people are basically selfish and will always put their own interests first; that asking people to subordinate their personal interests to those of the group or community is contrary to human nature. The economic theory of modern competitive capitalism is based on this premise. While it holds for much economic behavior, this premise certainly does not explain the following kinds of human behavior: parents sacrificing for their children; citizens giving time and money to group or community endeavors; taking the trouble to vote in elections when you know your vote cannot affect the outcome; working to restore and maintain a clean environment when everyone benefits from your efforts, even those who do nothing; soldiers sacrificing their lives in war; terrorists gladly sacrificing their lives for their cause. Social scientists have conducted extensive experiments based on the "prisoner's dilemma" game, which assumes tht people will act in their personal self-interest. These experiments have repeatedly shown tht people will sacrifice their personal interest for the good of the group, especially if the opportunity is provided for knowing and understanding what the group interest is (for an example see Van de Kragt, Orbell, and Dawes, 1983).

5. *The System is so big, powerful, and unyielding that there is no use trying to change it.* This belief is widely held and based on considerable evidence; it also is supported by Schmookler's analysis of power. Yet, as indicated above, history provides many examples of social systems that changed: feudalism, slavery, and colonialism have virtually disappeared; in recent decades, several societies peacefully moved from dictatorships to democracy. Much of this book is devoted to examining the possibility for system change and to suggesting ways for ordinary people to help further that change. Finally, our analysis shows that we have no choice but to change. Facing that prospect, it is wiser to believe that ordinary people can help bring about change than to deny cynically that change is possible.

Summary

Major points to keep in mind:

1. Even though the dominator model of society has shown great strength and staying power, it is not the only possible model. History shows us that partnership societies also are viable and have persisted for thousands of years.

2. The parable of the tribes shows us that power has become the principal basis for selection in cultural evolution when there is anarchy among societies.

3. The pursuit of power is a cultural contaminant that deflects humans from their most sublime aspirations.

4. The pursuit of power in the dominator society will devastate the environment.

5. If we persist in the pursuit of power in the dominator society, we are likely to annihilate billions of people in a nuclear war.

6. The pursuit of power has enthroned and enforced male domination over women.

7. Patriarchal dominator structures are especially strong in governments, in the military, in business corporations, in organized religion, in sports, even in universities.

8. Becoming aware of the way the dominator society enslaves us is the first step toward winning our freedom.

9. The dominator culture is maladaptive and cannot be sustained.

10. Feminism's insights and guidance should now be transmitted to men.

11. In some places women already possess the potential to guide the future direction of their society.

12. Challenging the false elements of aphorisms would accelerate social learning.

An Inquiry Into Values For A Sustainable Society: A Personal Statement

Chapter 4

In this chapter, I discuss how we come to know what we believe we know, including values. Our belief in a value-free science interferes with our understanding of science, values, and knowing. There is insufficient space for an exhaustive justification of the many ideas I put forward; while that might better satisfy philosophers, it would greatly lengthen the book and discourage many other readers. My aim is to enlist your participation in this inquiry and encourage you to rethink your own value structure. I seek a structure of values that will make achieving and maintaining a long-run sustainable and harmonious relationship with nature easier for our society.

What Are Values?

Values are fundamental to everything we do. The values we hold govern the way we behave and what we expect from our society. Values differ from preferences in that they are held strongly and generalize readily to many situations, whereas preferences are held less strongly and do not generalize very readily. For example, you might tell me that you prefer ice cream to melon for dessert but, the next time I ask you, I would not be surprised if you were to tell me that you prefer melon to ice cream. Furthermore, that I hold a similar preference would not be important to you. On the other hand, we do strongly desire that other people hold our most important values. If I say I value honesty, it means that it is not only important to me that I am honest with you but that you are honest with me.

Valuing is a uniquely human activity; other animals have preferences but they do not conceive and share values. Only individuals conceive and hold values. We often speak of societal values, but societies do not have a mind and cannot value. Societal values are those that most people in a society believe are important, but they are still held only by individuals. People are linked into belief and value structures that take many forms, growing and changing with time, enabling humans to adapt to an enormously wide range of living conditions. Our superior reasoning ability gives us a great advantage over other species; we can control and/or exterminate them; we can even destroy the biocommunity that nourishes all life (but would destroy our own in the process). This same reasoning ability enables us to forge values; it obligates us morally (as no other species is obligated) to protect the integrity of the biocommunity that supports all life.

Values, unlike preferences, make the most sense if they are arranged into hierarchies. Developing hierarchical structures for preferences is useless; one preference is about as good as another, but that statement should not be made about values. Most values serve other more central values. This process of reasoning through all the steps of a value hierarchy so that instrumental values are linked to a small core of fundamental values is referred to by philosophers as a process of "reasoning to ground." A value may be thought of as defined and understood once we have established its linkages to other values below and above it in a value hierarchy; we have established its "ground."

Preferences cannot provide the social glue to make a society work. Values, in contrast, constitute the fundamental cohesiveness that does make a society work. Whenever we assert that one course of action has higher priority than another, we are indirectly asserting a value hierarchy. Values are the very essence of politics; all political/governmental decisions are a societal statement that certain values will be pursued. Value hierarchies that shape a society are built up by trial and error over many decades, even centuries, of human experience that is often expressed as folk wisdom, custom, and tradition.

It also is useful to distinguish *values* from *beliefs*. For both values and beliefs, we begin with a cognition and attach a feeling to it. When we value something, we attach a valence to a cognition of that thing—a feeling of like or dislike. We only value things that we like and we disvalue things we dislike. I can cognize many things that I do not value but I do not value things I cannot cognize. Liking and disliking also applies to preferences, but keep in mind the differences between values and preferences discussed above.

We also attach a feeling of credulity or incredulity to most cognitions. When a cognition *feels* credulous, we believe it; when it *feels* incredulous, we disbelieve it. I can cognize many things that I do not believe, but I do not believe things I cannot cognize. For example, I can imagine myself flying as a bird, but I do not believe I will ever do so. Our feelings of credulity vary from cognition to cognition; we believe some things very firmly, even fervently, while we hold many other beliefs tentatively and stand ready to change them on the basis of new evidence.

Beliefs and values are related, but distinguishable. We tend to believe things that we value and disbelieve things that we do not value. Yet, we value things that we do not believe. I have often felt it would be wonderful to live in a peaceful world, but I do not believe I will ever do so. Similarly, we believe things that we do not value. For example, I believe there was a

Nazi holocaust against the Jews, but I totally reject the way the Nazis treated the Jews.

How Do We Know Facts And Values?

The process by which a cognition becomes credible is so subtle and commonplace that we are seldom aware of it. A few years ago, a dinner companion asserted his belief that our economy could continue to grow indefinitely. When I probed deeply how he knew what he believed he knew, he finally blurted, "I just have a feeling." He had put his finger on how we come to know things. Our society nourishes a host of commonly accepted beliefs. They shape everything about the way we think and solve problems. Most people do not give much thought to the way they come to know things; they just "know" what they know. Typically, those beliefs are such an integral part of our culture we seldom bring them into question.

In contemporary society where science plays a large role, most of us are taught to believe that the process by which we come to know facts is fundamentally different from the way we come to know values. I argue that the process of inquiry by which we come to know both facts and values is fundamentally the same, which means rational thinking is just as valid for discovering values as it is for discovering facts. In both cases, we seek consistency among our views before we accept a new cognition as credible. In the same vein, I argue that all inquiry utilizes values; that it is misleading to contend that science is value-free. Science, conducted by scientists who believe they need not be concerned about values, usually ends up serving the values of those who control money and power. We need public debate and public decisions about the values science should serve.*

The scientific method has become the honored way to observe and come to know "facts." When we observe scientifically, we measure phenomena as carefully as possible. We rely on other observers to replicate and confirm those observations. This process produces data that are meaningless until our belief structure invests them with meaning—only then do they become facts. Careful scientific processes facilitate agreement among observers concerning the characteristics of a phenomenon; that gives scientists an increasing sense of credulity concerning their observed "facts." For example, in June 1988, a leading meteorological scientist testified before a committee of the U. S. Congress that he was 99 percent sure that the drought in the American midwest was a result of the greenhouse effect. Was this belief a fact or not? Ascribing the greenhouse effect as the cause of the drought is

---

*I am indebted to Wenz (1988, Ch. 12), for helping me to think about these matters.

embedded in a much larger theory of climate change. If nearly all other meteorologists also believed it, we would be more inclined to believe the relationship to be factual. If succeeding summers were also hot and dry in the midwest, we would more easily believe the greenhouse effect to be the cause of climate change. *Facts are not absolutes; they are beliefs that we hold more or less strongly.*

When observers agree about a phenomenon, we usually say they are dealing with it objectively; it would be more accurate to say they are achieving intersubjectivity—they have reasonably similar beliefs about it. Careful measurement of observable phenomena facilitates intersubjectivity about data and about their meaning—observers intersubjectively agree on the facts.

Most of us believe that we come to know our values by a different process. The doctrine of science rejects the idea that values have a place in science. (Science is often taught as doctrinally as most religions are taught.) Many people presume that values become known to us by an essentially irrational process. Values seem to flow to us from tradition, religion, or societal authorities. Embracing a religion seems to be so emotion-laden, so beyond rational inquiry or persuasion, that many societies have excluded religion from politics and government in order to have civil peace; this is embodied in many constitutions as "freedom of religion" and as "separation of church and state." It seems that most people believe we adopt values in the same way we adopt a religion. We may say such things as: "Values are subjective and vague," "There is no use in arguing about values," "One value is as good as another." Perceiving values that way is a fundamental error. Some values *are* more important than others, and it is possible to "reason to ground" about values. That is, we can use rational argument to persuade each other of the validity of a value position or of the meaningfulness of a value hierarchy. I am not arguing that we always use this process to arrive at values, only that rational argument can be persuasive concerning values.

The scientific method facilitates agreement about physically based facts; therefore, it is easier to agree about facts than to agree about values. Values are more contestable because they do not have a physical referent. Each of the values I discuss in this chapter is contestable. Even when people agree that a value should be upheld, they may disagree about its expression in actions or structures. This important distinction between facts and values does not change the reality that all forms of knowing, facts as well as values, have a feeling component. Whenever we observe something and decide that it is believeable, or factual, there is a cognition to which we attach a feeling of credulity or incredulity. The feeling of credulity or incredulity is not automatic with the sensation but rather derives from its fit, or lack of fit, into an existing structure of beliefs. The physical stimuli available to our five senses are very elementary things such as wavelengths of light and color, sound frequencies and intensities, temperature, tactile sensations, and so forth. These constantly changing sensations that are being monitored by our senses could be interpreted in numerous ways.

The meaning of observations is not inherent in the sensations but derives, instead, from the *interaction* of the individual's nerve messages and thought processes with a laboriously built up structure of beliefs and values about how the world works. People do not just see lights, colors, and shapes. They see *things*. People automatically assimilate what they "see" into familiar categories. This "reality" is *common* in the sense that it is perceived similarly by most people in a community; it serves as a fundamental underpinning for their communication. All of us participate in, and more or less accept, the belief and value context of the society/culture in which we live. All observations, then, are belief-laden and theory-laden—they all have a feeling component. Heisenberg (1977) reports that Einstein had remarked to him that, "it is the theory which decides what can be observed."

Personal experiences of success or failure also influence the learning of beliefs and values. These personal encounters with reality, however, constitute the basis for "knowing" only a small part of the totality of things we believe we know. Most of what we "know" has been accepted on faith from authorities such as scientists, teachers, parents, and public figures. I believe, for example, that the earth is one of several planets that orbit around a particular star that we call the sun. I have never observed the other planets, and I have never bothered to confirm with my own senses that the earth orbits the sun; yet I feel confident that, if I took the trouble to do so, my beliefs would be confirmed by my personal observations. I "know" this "fact" only because I have accepted the belief structure of my culture. My belief has no grounding in personal observation.

All interpretations of sensation, all meanings, all "facts," are based on feelings of credulity. Science does not tell us facts; rather, it provides us with data that give us a sounder basis for agreeing about the feeling that something exists or the feeling that something works in the way that we have theorized it works. *Science provides us with a basis for intersubjective agreement about meaning; it facilitates the clarity and ease of communication about meaning. Science is not free of feeling, or emotion, or values; science derives meaning only in a cultural context.* No observation contains information that is a pure, exact reflection of reality because all observations are conditioned by the observer and the meaning context within which he operates. Objectivity is really intersubjectivity; there is no ultimate authority to validate a fact or a belief. (Bar-Tal and Bar-Tal, 1988)

The dependence of knowing (and science) on belief and value structures is best demonstrated when we receive information for which we have not yet worked out a meaning. We gravitate toward an explanation that best coheres with our stock of background beliefs; we wish to modify as few of those beliefs as possible. People are least willing to modify those beliefs that are most important to them. Replacing a belief is usually resisted more than modifying it. A central belief/value is not likely to be replaced unless and until a great deal of evidence against it has come to a person's attention. The important point is that some sort of motivation, perhaps a strong shock from reality, is usually required for people to alter or abandon strongly held convictions. The mere existence of evidence that is inconsistent with the belief is usually insufficient. Some of the most intense of modern day quarrels are between scientists who disagree about the meaning of evidence.

When scientists make choices between possible explanations about phenomena, they employ values; four of these will be discussed here. *Conservatism* was just alluded to. A new theory is more acceptable if it conflicts with as few beliefs as possible. Why repair a belief if it is not broken? *Modesty* is a second value. One theory is more modest than another if it assumes that the observed events are of a more usual sort, more ordinary, hence more to be expected.

The third value is *simplicity*. The simplest theory usually wins out over more complex theories because it is easier to use, and accept. *Generality* is the fourth value. Theories that explain more phenomena within the same framework are likely to be preferred over theories that explain a smaller segment of reality. When theories serve the values of simplicity and generality, we say that they are *"parsimonious."* Serving the value of parsimony can conflict, however, with the values of conservatism and modesty. We may be reluctant to surrender old and comfortable beliefs despite the attraction of parsimony. Science cannot escape these valuational components of the way it functions. Scientists also are motivated to serve their values. The personal satisfaction and excitement that the inquirer derives from learning something new is the most important reason for most scientists to pursue their inquiries.

For many purposes, we find it useful to distinguish descriptive beliefs, that portray facts about how things exist and work, from prescriptive beliefs that guide people's conduct. I do not deny the usefulness of the fact/value distinction. However, such a distinction does not mean that "facts" and "knowledge" are value-free. The point for this discussion is that the same structure of inquiry is used for ascertaining both phenomena.

Rawls (1971), Taylor (1986), and Wenz (1988) discuss a method of inquiry they call *reflective equilibrium*. Imagine yourself in the place of some other creature and empathize with what that creature must feel. For example, Wenz 1988 (p. 298) invites us to participate in this thought experiment:

imagine that our planet has been taken over and is now ruled by extraterrestrial beings that are as far advanced over our human intelligence and capabilities as our human capabilities are advanced over squirrels, frogs, or chickens. These extraterrestrials dominate our lives, sacrifice us in their scientific experiments, and feel free to kill us for food (or at any whim), just as we dominate and allow ourselves to injure or destroy other animals in our biocommunity. Obviously, we would consider their domination to be unjust. Similarly, our domination of nature is unjust. Imagining ourselves in this circumstance will give us a feeling for our moral obligation to animals. This exercise removes any lingering fiction that we currently are acting in a morally justifiable way toward other creatures in our biocommunity.

I pointed out earlier that we always use consistency as a criterion when we build a structure of knowledge. We use consistency in the same way when we build a value structure. "The process of dealing with inconsistencies in our thinking about what is right and wrong is *structurally identical* to the process of arriving at consistent views about what things exist and about how those things interact with one another." (Wenz, 1988, p. 270, emphasis in the original) We do often hold values that are inconsistent, but usually this is because our belief and knowledge structure is held too far in the background of our awareness to be able to highlight the inconsistency. It may also be because we are inadequately self-conscious about our knowing. Then too, our belief/knowledge structure may be poorly developed.

Lack of awareness of value inconsistency can be illustrated from American history. Our Declaration of Independence and many other public documents declare that "all men are created equal." Equality became a cornerstone value of the new American society. Despite the high-sounding rhetoric that most people professed to believe in, slavery continued to be a well-established institution and women were not given the right to vote when the new government got underway in 1787. It took more than a century of struggle—and a civil war—before slaves and women were given the right to vote. They are still struggling for complete equality. The recognized inconsistency between a valuation on equality and the reality that blacks and women were not considered to be equal was, and is, a major psychological force within most people, leading them to bring their values into congruence by treating blacks and women as equals. Many subtleties of conception and language artfully hid the inconsistency from white men; a growing self-conscious awareness uncovered the inconsistencies within many of them and lead to social change.

In conclusion, judgments about the justice of particular actions and policies, and judgments about general principles and theories of justice, are made through the same structure of inquiry that is used in science. In all cases, judgments are affected by the inquirer's values, and by her other beliefs. In all cases good sense is needed. In no sphere of inquiry can judgments often be arrived at or defended solely by recourse to rules of mathematical or logical reasoning. So ethics is not endemically more subjective than science, and beliefs about environmental justice need be neither more nor less certain and objective than beliefs in any other area of environmental studies. (Wenz, 1988, p. 270)

These points summarize my thesis so far:

1. Society plays a crucial role in knowledge development, in development of beliefs about how the world works, and in value clarification (or obfuscation).

2. The structure of inquiry is identical in all three knowing processes.

3. Beliefs are crucial for shaping values and values are crucial for shaping beliefs.

4. The processes by which we come to know facts and values will more likely be under our logical control (rational) if we elevate them to full self-consciousness. In fact, becoming more self-consciously aware of our knowing processes is crucial to the survival of our species.

The point that science cannot be value-free could be defended with a much more thorough discussion than can be presented in this chapter. Those desiring to read deeply on the question should consult Putnam (1981), Kuhn (1970), and Rorty (1979). The myth of a value-free science is so deeply embedded in modern culture, however, that philosophical argument, alone, is unlikely to displace it.

What would be so terrible about allowing the myth to perpetuate? It would be unwise because persistence of the myth has some unfortunate consequences. Almost all science, in one way or another, places power at the control of humans—power that they did not possess before. Some kinds of power have awesome potential for destruction. If we believe that science should not be guided by public discussion about values, we hand over part of the control of our future to those who shape science. The power delivered by science will continue to be controlled by those who command major societal institutions. *Science that pretends to be value-free will serve the values of those who rule "the establishment."* Scientists inevitably are affected in their inquiries by the values of those who have the money to hire them and provide the materials for them to practice their science. The value-laden subservience of science to the establishment is camouflaged under the myth of value-free science.

Environmentalists are often perceived as being antiscience; please be clear, I am not against science. I am not calling for outright prohibition of scientific inquiry and technological development, not even for their stringent regulation. We need the knowledge science can develop to be able to deal more effectively with our many social problems, including environmental problems. External controls on science would restrict freedom of speech and inquiry, and they could inhibit valuable inquiry. I believe that the best way to control science and technology is for all of us to learn how to think about them in a much larger socioeconomic-political context that gives adequate consideration to their value implications; doing this will help us decide how, when, and where to support science. This admonition is especially urgent for scientists and public officials. Scientists (that is, the science establishment) need to develop some of their own internal controls by careful discussion about the values they serve.

Scientists are only human and, as with most of us, have given little systematic thought to values and value structures. When they make a new discovery, or develop a new insight or technology, it is quite natural for them to emphasize the good it might do for people and to give less thought to its potential harm. They desire praise and esteem, just like everyone else. Even if aware of harm, as well as good, they will have difficulty balancing the bad against the good. Others, who may be less biased in perspective, need to enter the discussion about the long-term effects of utilization of a discovery.

Science needs to be thought about, planned for, supported, and controlled in a context of full awareness of a societal value hierarchy. I restate the syllogism I used in Chapter 2 to derive a value ranking for a sustainable society:

I can imagine a biocommunity thriving well without any human members but I cannot imagine human society thriving without a well-functioning biocommunity. Similarly, I can imagine human society functioning well without a given individual but I cannot imagine an individual thriving without a well-functioning biocommunity and a well-functioning human community. Therefore, individuals desiring quality of life must give top priority to protection and preservation of their biocommunity (their ecosystem). Second priority must go to preservation and protection of the good functioning of their social community. Only when people are careful to protect the viability of their two communities is it acceptable for individuals to pursue quality of life according to their own personal desires.

The obligation to think deeply about and protect these values lies just as firmly on the shoulders of scientists as it does on the rest of us. I believe we will eventually come to recognize that adhering to the myth of a value-free science is one of the most dangerous characteristics of modern thought. If we fail to give deep thought to these matters, we will continue to be the victims of misdirections of science and technology. If a mishap involves powerful technologies, such as nuclear energy, our biosphere could be gravely injured.

I endorse the four principles that Hans Jonas (1984) suggested for developing new ethical controls of technology: 1) we must develop better ways of predicting the long-range effects of technology; 2) give greater priority to prophecies of doom than to those predicting bliss so that we will more prudently face multiplying unknowns; 3) never place humankind's existence, or basic humanity, at risk; and 4) recognize our duty to ensure a decent future to our posterity.

Searching For A Sustainable Value Structure By Reasoning To Ground

Even though values are fairly stable psychological structures that typically are strongly reinforced by such societal institutions as churches, schools, and government, they do change with time. We learn and relearn our values. If our lives are not working well, or if our society is not working well, we get the message that we need to rethink our value structure. The arguments presented in the previous chapters show that we have built a society that cannot sustain itself. As we try to envision a society that can sustain itself we will find it helpful to begin by reasoning to ground about values.

The Core Values

When we look for the core of something, we seek the very essence of its being. For human values, then, we are seeking the very core of human nature, something that is biologically based that will become the centerpiece of a new, yet old, value structure that constitutes the underpinnings for a sustainable society.

Clearly, the most central value for every person is an instinctive desire to preserve one's own life. This value is so fundamental that it is hardly ever asserted as a value; it is simply taken for granted. In this basic sense, we are all selfish, and we want our lives to be good. We want not simply to have life but to have a high quality of life. Realizing a high quality of life involves many choices, and in the working out of this realization, a value structure becomes elaborated.

Distinguishing a personal value structure from a societal value structure is useful. Some humans narrowly focus on the preservation and enhancement of their own lives with little or no consideration for others; this is typical of some criminals, for example. Such value structures cannot be tolerated in civil society. Most people reject total selfishness and develop a value structure that shows consideration for others. Narrowly selfish value systems of individuals are not analyzed in this book; our task is to devise a value system that people hold for their society, not an individual value system.

I argued earlier that a society could not elevate the preservation of every human life to be a central concern. Because every person does eventually die, it would be a societal value that could never be realized. Instead, we should assign top priority to preserving the viability of our ecosystem and second priority to nourishing the good-functioning of our society. When those two systems are working well, all creatures would have a decent opportunity to live a healthy and satisfying life.

Why would it not make sense for a society to elevate some other value above preserving the viability of its ecosystem? In actuality, this is what modern industrial society has done by giving top priority to economic growth and development. I have argued that giving priority to economic values will produce a society that cannot sustain itself and I will argue in Chapter 17 that it would not be a desirable place to live even if it could sustain itself.

Some people have suggested that *Homo sapiens* is such a destructive species that the world would be better off if humans no longer existed. Therefore, why not give top priority to destroying ourselves? Ecosystems work just fine without humans so why do we not accelerate the process by which our species becomes extinct? This might be called an extreme bio-centric position—the good-functioning of the biosphere is the supreme value. This suggestion leads to a logical trap. It is evident that only humans can know, state, and pursue values; other species have preferences but only humans value. Nature does not value, it is indifferent to whether our, or any other, species becomes extinct. Biologists and paleontologists estimate that only a small percentage of the species that have ever existed are still alive today (I have seen estimates of 2 percent and 10 percent). Most of those exterminations occurred before humans came on the scene. If humans say that their own species should die, they are asserting this value only to other humans—the rest of nature does not know and cannot care. In a strictly logical sense, then, because only humans value, all value systems are anthropocentric, which does not mean that humans must value their own unlimited growth and their own enjoyment no matter what consequences it has for the biosphere. Our wisest choice is to select our reproduction rate and our lifestyles so that a viable ecosystem is preserved to nourish a good quality of life for us, other creatures, and our posterity.

But, what does it mean to experience high quality in living? The following definition was developed by my graduate seminar on "Quality of Life." It addresses quality of life as experienced by individuals.

Quality in living is experienced only by individuals and is *necessarily* subjective. Objective conditions may contribute to or detract from the experience of quality but human reactions are not automatic to physical conditions; the experience occurs only subjectively. Personal reports of experiences of quality are much better indicators of these subjective experiences that physical measures of physical conditions. (We should carefully distinguish environmental conditions that can be measured with objective indicators from the experience of quality that can only be measured with subjective indicators.)

Quality is not a constant state but a variable ranging from high quality to low quality. Persons usually experience some combination of high and low quality; they seldom experience only one extreme or the other.

Persons have a high quality of life when they experience the following:

1. A sense of happiness but not simply a momentary happiness; rather a long-run sense of joy in living.

2. A sense of physical well-being; usually this means good health but the sense of physical well-being can be realized by persons having lost certain capacities.

3. A sense of completeness or fullness of life; a sense that one is on the way to achieving, or has achieved, what one aspires to become as a person.

4. A sense of zestful anticipation of life's unfolding drama, greeting each day with hope and confidence that living it will be good.

Persons have a low quality of life when they experience the following:

1. A sense of hopelessness and despair; mornings are greeted with fear and dread. A sense that one is buffeted by fate and has lost control of his life.

2. A sense of having failed to live up to one's image of oneself; that one's life has been a failure.

3. A sense of poor physical well-being; illness, injury, hunger, discomfort.

4. A pervading sense of unhappiness.

Quality of life judgments are more valid if they are considered carefully. A multitude of facets of life should be considered, not simply a few, and it should not be simply a hasty "top-of-the-head" reaction.

We should carefully distinguish quality of life judgments that are individual (personal) and subjective, from prescriptions for a good society. Individual experiences with the quality of this or that aspect of life do not translate directly into policy even though they are important informational inputs to policymakers. Ecosystem and social system values must be served in policymaking as well as quality of life values.

We want a society and an environment that will allow people, as individuals, to work out their own quality of life. But there is a heavy responsibility on individuals to make the best of their situation and to take personal actions to achieve quality in living. We should be cautious about making the inference that a person living in what most people would assess as favorable conditions will experience high quality; or, conversely, that a person living in what most would assess as poor conditions will experience low quality. Yet, policymakers frequently make such inferences (when they report that per capita income has risen, or fallen, for example).

Designing a good society or nurturing a good environment are related but different goals. Good societies or good environments are those in which people have a better chance of finding quality of life, but we should not confuse them with quality of life. The following writers give two prescriptions for a society (and environment) that provide a favorable setting in which people could seek and find a high quality of life. They should be thought of as societal value systems, not personal value systems.

Medard Gable, Director of the Regeneration Project at Emmaus, Pennsylvania, has identified the following values for a desirable future world:

The future world that is most attractive embodies nine qualities:

1. It is *abundant*. The life-threatening scarcities of resources needed to meet humanity's basic needs are no longer with us, having been overcome through use of abundant local resources.

2. It is *regenerative*. These abundant resources are produced and distributed in regenerative ways, revitalizing the natural systems—just as regenerative agriculture rebuilds the soil.

3. It is *dependable*. These abundances are dependable in supply, quality, and quantity, and free from disruption.

4. It is *safe*. All products and services are safe; they don't endanger workers, consumers, or the environment.

5. It is *appropriate*. The primary determinants and limits on what is done in a given area are the culture, geography, local economy, and basic human needs of people living there.

6. It is *equitable*. Resources, products, and services are available for use by all in a fair manner.

7. It is *flexible*. The myriad of systems involved in our products and services are open to change, growth, creativity, and experimentation.

8. It is *efficient*. Production and distribution are as efficient as possible—with efficiency defined for the long term.

9. It is *open minded and locally controlled*.

The preceding goals are all interrelated; they describe a vision of the future that I see as most beneficial to the sustained vitality of the local community, the United States, and the entire world. (*Environmental Action*, Visions, 1985, p. 15)

Murray Bookchin, writing in the same issue, expresses values and beliefs that supplement Gable:

We need a "localist *morality*" as well as a "globalist" one, along with a sense of responsibility to act and think on these two scales. . . . We desperately need a decentralized, revitalized sense of community. We need to recreate communities as vital public spheres in which people can recover control over their destinies. . . . To me, "participatory community" embraces technics as well as people, nature as well as humanly scaled societies. It connotes a shared concept of interaction. Finally, we need a participatory politics to achieve these decentralistic, eco-technical and moral goals. I use the word "politics" in the Hellenic sense of the word to mean the managing of the *polis* or community not in its Euro-American sense of parliamentary and electoral manipulation. Politics is educative, not simply oriented toward winning elections; it is a curriculum for creating real citizenship, not simply a form of public mobilization on the first Tuesday in November. It presupposes a sense of civic virtue, a vital community life and a rich, creative, and supportive collectivity (p. 8).

These statements defining a high quality of life, the well-functioning ecosystem, and the good society in which quality of life might be achieved, are useful goals to pursue. They do not, however, specify the way that society should be structured. The quest to find that structure is the main concern of this book; it is best begun by inquiring more systematically into the values we want this society to foster. The structure that evolves will be my preferred value structure for the society in which I would like to live–I hope it becomes your preferred structure as well. Discussion of such structures is an urgent concern for modern society for it is only as we agree on a structure that it becomes society's structure. I have sketched my proposed structure in Figure 4.1; the reader is invited to consult it as the discussion proceeds. Imaging the figure, however, is not essential to following the discussion.

If a value structure is thoroughly elaborated, it could produce a long and complex list of instrumental values, which could confuse and bore the reader. Instead, I develop a parsimonious structure for identifying those instrumental values that most clearly support a small set of core values. The linkages between instrumental values and core values are not shown in figure 4.1; rather, I discuss them here. Also, note in Figure 4.1 that the instrumental values are surrounded by a fourth ring listing societal systems that serve instrumental and core values; discussion of those systems is interspersed with discussion of the instrumental values to which they are most clearly connected.

As shown in Figure 4.1, life in a viable ecosystem is the centerpiece of a value structure for a sustainable society. It is surrounded by four core values: a high quality of life, security, compassion, and justice. This group of core values is surrounded by a third ring of instrumental values. We value such things as freedom, equality, peace, and order not for themselves alone, but because they are likely to lead to the realization of the core values. These

Figure 4.1
A Proposed Value Structure for a
Sustainable Society

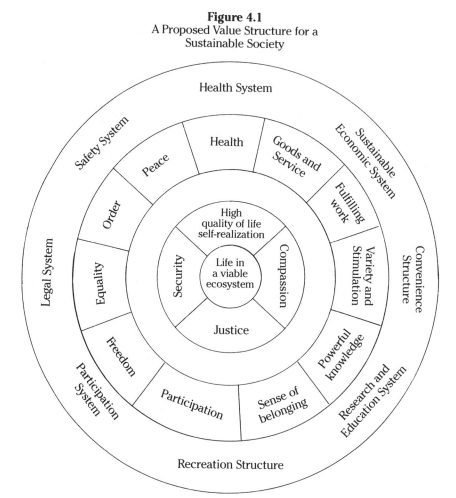

core values are not a means to anything else. When we reason back to one of these values we have reasoned to ground. I have discussed ecosystem viability and quality of life; these two values also are given considerable additional attention throughout the book.

Every society must presume that individuals are innately selfish, but if everyone was allowed to act in a totally selfish way, society could not work; therefore, every society modifies that basic selfishness by the values it promotes. If selfishness is modified only slightly and society encourages competition and aggressiveness, it will be a strikingly different society from one that emphasizes compassion and empathy. Modern industrial societies that use the free market for making many societal decisions have given great emphasis to selfishness, competition, and maximization of material wealth (hereafter, Society A). Many people are now questioning whether such a society is sustainable and are proposing a new, yet old, value structure that gives greater emphasis to cooperation, justice, compassion, and empathy; I believe it is more sustainable (hereafter, Society B). I occasionally refer to these two prototypical societal models.

I turn now to consideration of *security, justice,* and *compassion. Security* is a core value in every society because it is a key condition for survival. Many people link their personal security to the security of their nation – almost invariably, national security is given high national priority. Nations justify the most atrocious actions toward others because they are presumed to promote national security. Security institutions such as police, fire, and armed forces are accepted as governmental obligations in every society.

Security is linked to familiarity and continuity. The well-known resistance of most people to changes in occupation, residence, nationality, spouse, beliefs, and so on is linked to their need for security. When events force people into such changes, most find the transition to be traumatic.

Should a society make sure no one is poor, everyone who wants a job can have a job, everyone who is sick is cared for, all victims (of crime or disasters) are compensated? Most contemporary societies generate considerable discussion and disagreement regarding how much emphasis society should give to providing security to individuals. Persons who emphasize becoming wealthy typically do not want society to provide personal security to individuals; instead, they are likely to believe that people should be competitive, aggressive, and take risks. Ironically, in modern America, these same people are likely to assign high priority to armaments and national defense (for security).

Justice is a core value that is so pervasive and enters human thought and behavior so early in life, that one could almost call it part of the "wiring diagram" of the human brain. As soon as children are old enough to form intelligible sentences, they begin to invoke justice as a value. Every parent is familiar with a child's complaint that a playmate has acted unfairly. Experiencing an injustice is similar to experiencing an injury; it is that close to our inner being.

This personal value is translated into a societal value through empathy with the plight of others. When someone else suffers an injustice, we personally feel pain; we feel compassion for them and desire that their injustice be removed. In this sense, justice and compassion are linked. Our sense of justice can be so strong that we assert it as a value even when it is against our personal self-interest. It is a basic emotive glue for a sustainable society. It is the central value in every legal system in every civil society.

Persons who assert aggressive, competitive, selfishness as their values (as in Society A) tend to be interested in justice primarily for themselves. For most people, however, their sense of justice generalizes to desiring it for everyone; when we do so we are asserting the value of justice not only for our own life, but also for all life. (For an exceptionally thoughtful discussion of environmental justice see Wenz, 1988.)

The Eco-Justice Project

Some religious organizations have been devoting special attention to ecological justice. The Eco-Justice Project, and Network of the Center for Religion, Ethics & Social Policy, and the Genesee Area Campus Ministries (in New York State) recently (1985) joined forces with the Eco-Justice Task Force of the National Council of Churches to form an Eco-Justice Working Group. William E. Gibson, coordinator of the Eco-Justice Project located at Cornell University in Ithaca, New York, describes eco-justice in this way:

The *eco* in eco-justice stands for ecological wholeness—the life-sustaining soundness and integrity of natural systems with continuing balance and harmony between human creatures and the nonhuman creation. The second part of the hyphenated word, justice, is to be understood in the broad sense of "common good," the well-being of a whole society, with each member contributing to and sharing in the good of the whole.

Sometimes I have said, "the 'eco' is for ecology and 'justice' is for Jeremiah." This to stress what the prophet stressed: that the right-to-participate of the powerless, the vulnerable, and the needy has to be vigorously upheld in any just society. Jeremiah and the other prophets always understood God as both Creator and Deliverer—the Creator still caring for the creation and the Deliverer setting things right for the victims of those who draw greedily and carelessly upon the sustenance intended by the creator for all.

Eco-Justice means both ecological wholeness and social justice—never one without the other. Eco-Justice extends the concept of justice to the earth as well as to its people, to the nonhuman as well as the human. It is the well-being of all humankind on a thriving earth.

... In the effort to overcome want and insecurity, industry, and agriculture have employed a violent technology that undermines the long-term capacity of the earth to provide the material basis for the human economy. Justice in our time cannot be justice even in the sense of meeting basic human needs unless it is understood as eco-justice.

Eco-justice signifies understandings and responsibilities necessary to human survival. The concept loses power, however, unless it transcends the concern of men and women to save themselves. Its power reflects nature's claim—and the Creator's claim—upon the human creature to respect the integrity and honor the intrinsic worth of the whole created order.

Eco-Justice as a Project or a Working Group focuses upon long-neglected issues of environmental health and wholeness. At the same time, it gives concerned attention to issues of human well-being that now must be addressed with recognition of human dependence upon nature and the resource base. The economy, we now see, cannot be just if it is not sustainable.

Source: *The Egg,* vol. 5, no. 2 (Summer 1985), published at Anabel Taylor Hall, Cornell University, Ithaca, New York 15853.

Feelings of *empathy and compassion* are essentially an expression of love for one's own self that generalizes to others. It is just as fundamental to human nature as aggression and competition; and it serves as a much better core value for society. Being compassionate is a central value that links people in more peaceful, more cooperative, and more durable relationships (Society B). For most of us, it is a part of our life force. We can learn to express compassion readily and openly; a sustainable society would reinforce that learning. In the most highly developed "civil" societies, love and compassion generalize to humans in other communities, in other countries, to other species, and to future generations.

For most of the centuries of human civilization, people expected that their learning and development would lead to a better future for their children. The planet is now called upon to support so many humans, who are so resource consumptive, that there is good reason to fear that our own activities are detrimental to the welfare of future generations. In 1983, the U.N. General Assembly established a World Commission on Environment and Development. This commission issued a report in April 1987 titled, *Our Common Future,* which contained these poignant phrases:

. . . our children will inherit the losses. We borrow environmental capital from future generations with no intention or prospect of repaying. They may damn us for our spendthrift ways, but they can never collect on our debt to them. We act as we do because we can get away with it: future generations do not vote; they have no political or financial power; they cannot challenge our decisions.

Compassion (read, love) is the core value that links are lives to the destiny of future generations. Iroquois Indians teach that we should always act so that the seventh generation to come will also have a good life. The completely selfish person could ask, "What has posterity ever done for me?" Heilbroner (1975) has written a classic essay dealing with that rhetorical question. He grounds our sense of obligation in a built-in love of what is honorable and noble; no normal man could willingly assume the responsibility for the extinction of mankind. This topic has received considerable attention from philosophers (Partridge, 1981, has edited a collection of relevant essays; see also Jonas, 1974; Sikora and Barry, 1978; Shrader-Frechette, 1981; and Scherer & Attiq, 1983). Jonas admonishes:

> Act so that the effects of your action are compatible with the permanence of genuine human life. . . . Act so that the effects of your action are not destructive of the future possibility of such life. . . . Do not compromise the conditions for an indefinite continuation of humanity on earth. . . . In your present choices, include the future wholeness of Man among the objects of your will. (Jonas, 1974, p. 13)

Our love for ourselves becomes a moral obligation to our own well-being which generalizes, through love, to the welfare of other people, other species, and posterity. We recognize that our own welfare and quality of life cannot be separated from the welfare of other creatures—from the good-functioning of the biocommunity.

Partridge says that our obligation to future generations rests on a need for self-transcendence:

> Accordingly, if one feels no concern for the quality of life of his successors, he is not only lacking a moral sense but is also seriously impoverishing his life. He is, that is to say, not only to be *blamed*; he is also to be *pitied*. . . . By claiming that there is a basic human need for "self-transcendence," I am proposing that, as a result of the psychodevelopmental sources of the self and the fundamental dynamics of social experience, well-functioning human beings identify with and seek to further, the well-being, preservation, and endurance of communities, locations, causes, artifacts, institutions, ideals, and so on, that are outside themselves and that they hope will flourish beyond their own lifetimes. . . . [T]his claim has a reverse side to it, namely, that individuals who lack a sense of self-transcendence are acutely impoverished in that they lack significant, fundamental, and widespread capacities and features of human moral and social experience. Such individuals are said to be *alienated*, both from themselves and from their communities. (Partridge, 1981, p. 204, emphasis in the original)

Partridge, in philosophically justifying the need for self-transcendence, draws on Erich Fromm's description of "love" as the emotion/value transcending the reach from one's own self to another self:

> There is only one passion which satisfies man's need to unite himself with the world, and to acquire at the same time a sense of integrity and individuality, and

this is *love*. *Love is union* with somebody, or something, outside oneself, *under the condition of retaining the separateness and integrity of one's own self.* It is an experience of sharing, of communion, which permits the full unfolding of one's own inner activity. The experience of love does away with the necessity of illusions. . . . [T]he reality of active sharing and loving permits me to transcend my individualized existence, and at the same time to experience myself as the bearer of the active powers which constitute the act of loving. (Fromm, 1955, p. 36-37, 41-42, quoted in Partridge, 1981, p. 212, emphasis in the original)

Our earlier discussion showed that for most of the centuries that humans have lived on this planet, their value-system oriented them to living in harmony with nature. For many people, this love for nature and other species links back to life; we realize that our own lives can only be sustained within a viable ecosystem that supports many thousands of other forms of life. Our own well-being is not separable from the well-being of our biocommunity. More than 2,000 years ago, Chinese philosophers of the *Tao* perceived a basic unity and harmony between humans and their biocommunity; they characterized these communities as having, "an intrinsic principle of ultimate creativity. . . . consists in an unlimited . . . expression of life forms and life processes in a state of universal harmony and in a process of universal transformation." (Cheng, 1986, p. 357)

One of my students, who read an earlier draft of these pages, pointed to an anomaly that she felt I must discuss. In my proposed value structure, I recognize that everyone values his own life. I also emphasize compassion for other humans, and other species, as a core value. I appear to advocate the sanctity of life; yet, at the conclusion of chapter 2, I said that sanctity of life cannot be an operative value in a sustainable society. How do I resolve this seeming inconsistency?

This is, indeed, a troublesome dilemma that has bothered environmentalists for a long time. They value a healthy and good life for all creatures; yet, they also know that if any one species reproduces rapidly, it will have the effect of inflicting death on other species and impairing the good-functioning as well as the carrying capacity of the ecosystem. They also know that all lives are terminated by death. Furthermore, if there were no death, there could be no life because all creatures, in one way or another, live by the death of other creatures. Only plants make their own food while all animals live, directly or indirectly, from the food provided by plants. Decayed plant and animal tissues also are the major food for plants. Biocommunities are characterized by a constant cycling of life and death. Life is the very essence of a biocommunity but no one life is elevated as supremely valuable and no one species can be allowed to reproduce wildly.

I have no doubt that nearly every one of us, no matter how loving and compassionate toward others, values his own life intensely. Most of us would refuse to sacrifice our own life for that of a plant, an animal, a person in another country, or a person yet unborn. A society that asked its members to make such a sacrifice simply would not be successful. If the human species refuses to limit its own reproduction, however, nature will do it for us, at great cost in pain and death not only for ourselves but for many other creatures as well. Perhaps that is our destiny.

Humans in a good and wise society will anticipate this problem and will voluntarily limit their reproduction. In a well-functioning biocommunity that also serves as their home, humans will practice self-control against excess; this kind of balance is called *homeostasis*. People in such a society will simultaneously value life and limitation on reproduction. They will recognize the naturalness of death. They will see death, not as an unmixed evil but as a part of community life. Native Americans have understood this duality; they loved and respected the animals that they killed to maintain their own lives. They did not try to prolong human life when that life no longer had quality or when it interfered with the ability of the community to survive.

The operative central value for such a society, then, is the healthy functioning of its ecosystem, its biocommunity. Wenz (1988) argues (pp. 300-304) that humans living moral lives in harmony with their biocommunity will follow the principle of "avoiding process harm." By this principle, it is immoral to interfere in the good-functioning of any natural organization that involves interdependent processes or activities. Evolution has moved from simpler to greater complexity and a greater variety of species are nurtured in ecosystems. Human development of agriculture and industry has reversed that process, however. We have simplified ecosystems to make them serve uniquely human needs. This drives other species from their niches and makes ecosystems more vulnerable. We are flattening the biotic pyramid. By what right do we do these things?

To summarize, the central/core value for a sustainable society is life in a viable ecosystem. This means life for all creatures, not just humans; the inherent value resides in the biocommunity. The humans living in the biocommunity also practice love and compassion for other humans and other creatures, all of whom "have a good of their own." This love for life, this love for the biocommunity, does not treat life as sacred; it also recognizes that death is essential to the health of the community. In the well-functioning biocommunity all members live off the lives (deaths) of other members; their relations are interdependent and symbiotic. By practicing self-restraint in reproduction and consumption, human societies living in harmony with their biocommunities can avoid much of the pain, turmoil, and death that will follow human excess.

What do we do if our societies do run to excess in reproduction and consumption? What kind of values do we follow to get out of that predicament? I would still pursue ecosystem integrity as the highest societal value. It would be folly to give priority to serving human needs over the needs of the biocommunity (although that may well be what will happen). If that were to be done for any extended period, it would severely injure the biosphere and further reduce its carrying capacity. If we wait that long to wake up, we may have to turn to draconian measures, such as China's one child per family policy, in order to save the biocommunity. I would not like to live in that kind of society, but the alternative to such measures could well be massive starvation—of all creatures—and severe deterioration of the ecosystem.

I turn now to a discussion of the two outer rings of Figure 4.1—the instrumental values and the societal processes that support them.

Instrumental Values

Sense of Belonging. Almost without exception, humans need to belong to a community where they are valued and loved. This gives them a sense of security and identity. In return, they empathize with and extend compassion to others in their community; this feeling often extends to all the creatures in the biocommunity. Having a community that provides a sense of belonging serves the more fundamental values of security, love, justice, and a high quality of life.

Participation. People need, and the community should provide, a meaningful personal role for their participation. Asserting the right to participate is another way of saying that it is only fair for people to be given an equal chance to determine their future. Participation means something more than working to win elections; it may be directed to all aspects of community living. It is educative and provides opportunities for people to realize all they are capable of being; in that sense, it is a vehicle for self-realization and quality of life. Participation also is important for achieving a sense of belonging and freedom.

A Participation System. A sense of belonging and participation does not just happen; it is achieved within an institutionalized participation system that includes not only the opportunity to vote but also a network of organizations to link individuals with each other and to the community's formal and informal decisional processes. People need to know what others are thinking and doing so that their own role may be defined meaningfully.

Freedom. We all desire freedom and would like to do anything we wanted at any time; but completely unrestrained freedom cannot be a societal value because it does not generalize. If everyone insisted on doing whatever he wanted, no one would have freedom. Because of this, freedom cannot be a core value. We should strive for freedom within the restraint of the simultaneous achievement of such other values as security, sense of belonging, participation, justice, equality, and order. Freedom of speech and assembly are crucial to a democratic civil society and need to be protected. Freedom is an important aspect of self-realization; it implies lack of unnecessary restraints and the provision of meaningful opportunities.

Equality. Why is *justice* listed as a core value in Figure 4-1 but *equality* is designated an instrumental value? Many people confuse the two terms and ask for equality when they really mean justice. By keeping the terms distinct, we can recognize that sometimes treating people equally, even though they have made unequal contributions, is perceived as being unjust. We value equality instrumentally as it leads to a sense of justice. Curiously, we cannot say that justice is instrumental to some other value; we value justice for its own sake. It is the very essence of being human. Equality is an important instrumental value as humans strive to realize the values of belonging, participation, freedom and justice. Equality also is essential within a participation system and a legal system, mainly because it is the *just* way to proceed.

Order. Order is essential for self-realization and quality of life. We need to carry out daily functions in a somewhat orderly fashion; otherwise, they would become so burdensome as to consume most of our time. We achieve order in our personal lives only if the behavior of others also is somewhat ordered—if the socioeconomic-political system proceeds in an orderly fashion. I did not list *honesty* in Figure 4.1, not because it is unimportant, but because it is instrumental to order. Imagine how chaotic life would be if we never knew if other people were telling the truth. Life would also be awkward and embarrassing if we were completely honest in every reaction to a situation. Somehow we find a workable level of honesty. Societies differ in the emphasis they place on being orderly which, in turn, influences the amount of freedom available and also the level of peace achieved in a community. Order should be seen as another of those values that is balanced out in the simultaneous achievement of such other values as freedom and peace.

The Legal System. The values of participation, freedom, equality, peace and order are supported in civil society by a legal system. All governments utilize legal systems; the trick is to so design and maintain them that they do, in fact, support the values identified above rather than subvert them.

Peace. We seek peace mainly because it enhances quality of life. It means not only the absence of injury but also the absence of the threat of injury. In a civil society, peace is achieved only in the context of a social order and a well-functioning safety system, including national defense. Peace also is important for realizing good health.

The Safety System. Safety systems include such institutions as police forces, investigative systems, fire protection systems, and national defense systems. They are essential to the realization of order, peace, and health; as such they are crucial to the realization of high quality of life. They are intertwined with the legal system and the health system.

Health. By definition we must have good health to maintain life; good health is essential to the realization of high quality of life. A well-functioning health system and safety system are important for maintaining good health but a community that provides ample goods and services as well as peace and order also is important to health.

The Health System. This system is composed of medical practitioners and their supporting institutions such as hospitals, laboratories, research institutions, pharmaceutical companies, and the makers of prosthetic devices. The educational system also is important for guiding individuals to wellness understanding and behavior. These systems must be supported by well-functioning economic and safety systems. The relationship of such a system to good health and quality of life is obvious.

Goods and Services. Even though it is obvious that material goods and services are needed in order to lead a high quality of life, we should not automatically assume that material wealth leads to it. A strong emphasis in a society on industrial production, material goods, and maximum consumption in order to boost employment is misplaced. Such an emphasis injures ecosystems and human health; it deprives future generations of necessary natural resources; it cannot be sustained over the long haul; and gives insufficient attention to other important values. An intense struggle for material gain can diminish rather than enhance self-realization. A wise value structure for a sustainable society, that focuses on realization of the core values we have identified, assigns economic values a more subdued and appropriate role than they receive in modern industrial societies.

Fulfilling Work. Most of us believe that work should be personally fulfilling, as well as provide income. Placing an exceptionally high emphasis on economic values in industrial societies leads to defining work as employment that must be kept high to stimulate the economy. I believe this is an inappropriate definition of the meaning of work; work should provide opportunities for development, creativity, beauty, sense of belonging, variety, and good health; it should be a primary vehicle for self-realization. Opportunities for fulfilling work should not be limited to those who hold jobs; we should treasure and reward with money and societal recognition those who make work contributions that are not now paid a wage (for example, housework, mothering, voluntary services, meals on wheels, and so forth).

A Sustainable Economic System. I intentionally did not call this a productive economic system or one that maximizes wealth. An economic system must serve a variety of values; it must:

1. Preserve and enhance a well-functioning ecosystem.

2. Provide humans with goods and services—necessities for a good life.

3. Provide opportunities for fulfilling work—self-realization.

4. Achieve and maintain economic justice.

5. Utilize resources at a sustainable rate—justice for future generations.

Variety and Stimulation. A self-actualizing fulfilling life has a good deal of variety and stimulation which people need to experience high quality of life. A society that allows people to seek quality of life in a multitude of ways, as long as the ecosystem and social system are preserved, will have variety and stimulation. A well-functioning research and education system, a recreation structure, a convenience structure, and an economic system are needed to provide the variety and stimulation.

The Convenience Structure. The phrase "convenience structure" is a shorthand way of talking about the ease with which important daily tasks can be completed in one's everyday life. It is the physical structure (roads, parking, distances) as well as the sociobehavioral patterns that impact us in our daily routine. Barriers to the completion of tasks are frustrating and diminish the fulfillment one gets from work and other daily tasks; they also diminish self-realization and quality of life. A well-functioning community is a convenient community; achieving convenience should be an important goal in community planning.

The Recreation Structure. A good recreational structure provides ample and varied opportunities so that people can recover from the wear and tear of their daily work. People express their recreational needs in various ways. A good society will encourage wide variation, but with forethought so that ecosystem preservation and justice are served as well. Some highly consumptive activities (dirt bike racing, for example) not only waste resources but seriously injure the ecosystem. A wise recreational structure serves such values as a variety and stimulation, self-realization, health, knowledge, participation, and a sense of belonging.

Powerful Knowledge. Learning and knowing are major resources that humans use to manage their lives, realize quality of life, and protect their ecosystem. I use the phrase "powerful knowledge" to emphasize that this knowledge needs to be both broad and deep. Powerful knowledge is the key to self-realization; it alerts individuals to the ways that other important values are served, or are failing to be served, by individual and societal actions. While specialized knowledge is clearly useful, it is equally important for people to obtain and utilize integrative knowledge. A civil society that is merely a collection of experts cannot function well. In order to make wise policy and build a good society, people need to learn how to integrate many strands of knowledge.

The Research and Education System. Powerful knowledge must be supported by a research structure that foresees problems and develops new knowledge as well as an educational structure that transmits that knowledge to the people in the community. Research and education also provide variety and stimulation; their very nature is stimulative. This system provides important support in training people for fulfilling work. Learning also typically occurs on the job and is one of its most fulfilling aspects. In addition, the research and education system is very important for self-realization and quality of life.

Generalizations About This Value Structure

To summarize points about this value structure:

1. It is a very intricate structure in that each component is linked to many other components of the structure. A society that hopes to be sustainable would strive to implement the whole package.

2. The reciprocal relationships between many of the components give the whole structure balance and stability. It is unlikely that any one value would become so emphasized as to unbalance the structure and deny realization of other important values.

3. The core values are a limited set; all other values are instrumental. This limitation makes it easier to avoid confusing instrumental values with core values and to perceive which instrumentalities are needed to realize a given value.

4. It is a parsimonious structure, easily learned and applied; one need not deal with a set that has hundreds of components.

5. It is a set based on nature; it allows, even encourages, people to be their natural selves.

A basic objective in every civil society is to develop an ethical system so well thought through, and so widely accepted, that morally correct decisions are made routinely—almost as if on autopilot (Bookchin, 1986). We want not simply moral rules but a moral system for judging rules.

Why Is A Compassionate Society (Society B) More Sustainable?

The emphasis on compassion instead of competition in Society B would make it a more balanced society, less likely to go to extremes. There would be built-in value constraints on the efforts of some to dominate others. Additionally, there would be constraints on technology due to the more subdued role of economic values (as compared to Society A) and an elevated valuation on justice, compassion, and the preservation of nature. Overall, there would be a greater recognition of the need to preserve the common good and also a greater understanding of what leads to quality of life. There would be better recognition of the full range of societal systems that need to be kept in good working order so as to provide quality of life.

In public discourse, we often hear that we must continue to grow economically and exploit natural resources vigorously (particularly to produce more energy) in order to maintain a high standard of wealth – or else we will go back to the horse-and-buggy days and freeze in our homes. People who pose such a choice assume the selfish, aggressive, competitive value structure of Society A. Within the value structure of Society B, it is unlikely that such a choice would be posed. Ways can be found to limit population, husband resources, protect the ecosystem and yet find richness and quality in living that is not dependent upon high consumption. Life in Society B really is attractive, making it easier for us to abandon the beliefs, values, and social structure of Society A.

The Need To Learn New Ways Of Thinking

Throughout this chapter, I call for new ways of thinking. I have not discussed how to do this; that is the focus of the next chapter on social learning. Most of my admonitions with respect to values lead us in the direction of clarifying our values. Why is that useful? A clarified value structure provides a citizenry with a solid basis for policy analysis. Many policy disagreements are based on confusing presuppositions about values and beliefs that have not been clearly illuminated by reasoning to ground. We often dispute over instrumental values that are assumed to be fundamental values. Economic growth, for example, is clearly an instrumental value, but our public discourse often elevates it to an end value. We need to learn how to dissect policy proposals for their implicit values and examine them to see if they are the values that are really important. This examination may lead us to ask what other values in our structure are not being served by the proposed policy.

Fundamental values often lie implicit but unanalyzed in policy proposals. Bringing them to public attention can clarify discussion and may even lead some individuals to change policy positions. Over the past decade or so, many modern industrial societies have been learning how to conduct environmental impact analyses; more recently some have been using social impact analyses. It is ironic that the value components of these analyses, which are the most fundamental parts, typically are ignored, evaded or assumed. Maybe the time has come for us to learn *value impact analysis.* Since knowing means and ends, and clarifying the relationships between them, is the central component in all impact analyses, using this value structure to differentiate means from ends would make it a useful intellectual tool in all environmental and social impact studies.

Is it possible to use value analysis/clarification to resolve most value conflicts? Even though I believe this process has excellent potential for clarifying values and their hierarchical relationships, and thus has high potential for resolution of value conflicts, I also recognize that people are emotionally committed to their values—perhaps so strongly committed that they are impervious to argument or new information. For example, both sides in the current controversy over whether the state should allow abortions proclaim life as their major value; but they have quite different definitions of life. The controversy focuses more on the definition than on the value. Neither side seems very open to new information or to rational argument. Such emotionally laden controversies are unlikely to be resolved by value analysis. The controversy would be easier to resolve if our society realized that its core value should be the good-functioning of the biocommunity, the integrity of its ecosystem; pursuit of that value requires us to limit human reproduction (but not necessarily by abortion).

Thinking about this proposed value structure can stimulate persons to reexamine their own value system. Comparing this structure against their own constitutes a kind of test of the validity of their personal value structure. Even if individuals decide that they prefer an amended structure, the encounter with this proposed structure will have clarified their understanding. Value analysis is a learning process.

Summary: An Agenda For New Ways Of Thinking

Succinctly summarizing this new way of thinking:

1. We must learn how to become conscious of our ways of knowing. That is the first step toward fully using our reason to deal with our problems. As we do so, we will come to recognize the key role that society plays in knowledge development, in development of beliefs about how the world works, and in value clarification (or obfuscation). Recognizing that our beliefs and values are culturally derived frees us to reexamine them to see if they can be revised to serve us better.

2. We must recognize that the structure of inquiry by which we come to know facts about the way the world works is fundamentally the same as the way we come to know values. This recognition will enable us to reason together to arrive at our social values in the same way we reason together to arrive at an understanding of how the world works physically.

3. We must learn that beliefs are crucial for shaping values and values are crucial for shaping beliefs. In effect, we must study them as a package.

4. We must recognize that we are deluding ourselves if we believe that science is value free. Believing that it is value free has the effect of delivering the control of science into the hands of those holding power and wealth. We need to pay close attention to the values that science serves and redirect its thrust so that it serves society's highest values.

5. We must learn to reason together in public debate about our values and use the resulting understanding to control the direction and speed of scientific development, as well as the direction of society.

6. We must learn how to use our values to take control of technological development and deployment; otherwise, our technological binge will carry us all to destruction. Risk-taking with a rifle may seem to be similar to risk-taking with nuclear weapons/power, or with recombinant DNA, but the quantitative difference in the power employed makes an enormous difference in the need for social control. Society, and the biosphere, can tolerate a few accidents with a rifle but we cannot tolerate nuclear or bioengineering catastrophes.

7. We must learn that ecosystems are so complicated and interconnected that almost no action is isolated. Our motto should be, "We can never do merely one thing." We should continually ask, "And then what?"

8. We must learn to think holistically, systemically, integratively. In effect, we must downgrade the emphasis of modern science and society on minute dissection and analysis, on extreme division of labor and specialization, on abdication of responsibility behind a facade of, "That's not my job." I do not deny the usefulness of specialized analyses and cause-effect thinking, but they need to be incorporated into a larger holistic and organic thought structure.

9. We must learn that we are *not* exempt from nature but must learn to live in harmony with it rather than try to dominate it.

10. We must learn to live lightly on the earth: avoid manipulating or thoughtlessly interfering with nature's systems and cycles; avoid depletion or despoilation of soils, waters, air, plantlife, wildlife, etc.; avoid consuming more than necessary; avoid casting our wastes into the biosphere.

11. We must learn to understand the meaning of these related concepts: *exponential growth, carrying capacity, overshoot,* and *dieback.* As we do so, we will recognize that we must limit growth in human population, must accept social control of family size.

12. We must learn to empathize with, and extend our compassion to people in other lands, to other species, and to future generations in order to preserve the integrity of the ecosphere and the survival of us all.

We will not fully preserve our own lives, and live them with high quality, until we transform our ways of thinking, our society, and our civilization. We have two vital human qualities to assist us in that mammoth task, our reason and our compassion. Changing our way of thinking is the first step to changing a civilization.

An Inquiry Into Social Learning

Chapter 5

Understanding ... is to see the way things are put together, and to see why and how they work as they do. ... The urge to understand ourselves and the world we live in is one of the noblest and most powerful of human motives; but often when we ask for the bread of understanding we are fed with the stone of knowledge. (From the Spring 1987 issue of *UniS*, a new Journal for Discovering Universal Qualities, Box 6615, Bridgewater, N.J. 08807.

In a New Context, the primary educator as well as the primary lawgiver and the primary healer would be the natural world itself. The integral earth community would be a self-educating community within the context of a self-educating universe. (T. Berry, 1987, p. 79).

Learning is a central concern in our civilization. Every societal institution—family, school, church, business firm, government, voluntary organization—devotes considerable time and effort toward learning. We have mostly studied the way individuals learn and have paid little attention to social learning. What is social learning? How does it work? Why is it important? How can it be nourished?

Social learning is not easy to define because it takes place in several different ways. Also, we cannot specify a physical entity, such as the brain of a mammal, as the main site of social learning. Thomas Berry's phrase, "a self-educating community," comes close to what I mean; but even that phrase needs explanation. In a community, we learn from each other and from nature. It is meaningful to speak of a "learning community" or that "a community makes up its mind about something"—even though a community does not have a mind.

Some examples may breathe life into the concept: I wrote these words while the Iran Contra Hearings were taking place before the U.S. Congress. The hearings were purposely staged for social learning. The media-hype coverage of the hearings brought the entire nation into the learning process and the country literally "made up its mind" about policies relevant to the topic.

I was living in Norway in the early 1970s while the nation prepared for a referendum on whether it should join the European Common Market. Every adult seemed concerned; the debate was intense in personal conversations and in the media because it involved the very definition of the Norwegian way of life. The country was overt and self-conscious about its social learning experience. By a slight majority, the Norwegians decided not to join. Despite the strong views of those who had lost, the nation soon settled down to acceptance of the decision. In 1988, Canada went through a similar national debate about whether it should consummate a free-trade agreement with the United States.

Social learning may not be very visible while it is underway but a look backward may disclose that social learning has occurred because a dominant institution or practice is replaced by another. For example, over the last century civilization has abandoned slavery and colonialism even though they had once been dominant institutions. Their previous widespread acceptance has been displaced by widespread disapproval.

Learning Is Ubiquitous In Nature

Verbal communication between persons facilitates social learning, but words are by no means necessary. "Higher" animals such as dogs or chimpanzees have a large potential for learning; but all creatures, even one-celled bacteria, have a rudimentary capacity for learning. For more than 2,000 million years (nearly one-half the time earth has existed), microbes were the only living things on this planet. They learned to organize themselves, to cooperate in symbiotic relations, to differentiate, to adapt to changing conditions.

The view of evolution as chronic bloody competition among individuals and species, a popular distortion of Darwin's notion of "survival of the fittest," dissolves before a new view of continual cooperation, strong interaction, and mutual dependence among life forms. Life did not take over the globe by combat, but by networking. Life forms multiplied and complexified by coopting others, not just by killing them. . . . In their first two billion years on earth, prokaryotes (microbes) continuously transformed the earth's surface and atmosphere. They invented all of life's essential miniaturized chemical systems—achievements that so far humanity has not approached. This ancient high *bio*technology led to the development of fermentation, photosynthesis, oxygen breathing, and the removal of nitrogen gas from the air. It also led to worldwide crises of starvation, pollution, and extinction long before the dawn of larger forms of life. (Margulis and Sagan, 1986, pp. 14-15)

The calamities mentioned by Margulis and Sagan illustrate the way communities learn by natural selection in evolution. For many millions of years microbes lived without oxygen; nearly all the oxygen on earth was bound with hydrogen in water. Certain bacteria evolved the capability to photo-synthesize their own food; this process released free oxygen; as these bacteria multiplied, the percentage of oxygen in the atmosphere increased steadily from 0.0001 to 21 in a relatively short time. However, this new gas was toxic to most existing life forms; it created the greatest pollution crisis the earth has ever endured; there were mass extinctions.

This life-threatening event did not extinguish all life, however. Some microbes continued to learn and evolve; "the cyanobacteria invented a metabolic system that *required* the very substance that had been a deadly poison. . . . Cyanobacteria now had both photosynthesis which generated oxygen and respiration which consumed it. . . . Given only sunlight, a few salts always present in natural waters, and atmospheric carbon dioxide, they could make everything they needed: nucleic acids, proteins, vitamins, and the machinery for making them" (Margulis and Sagan, 1986, pp. 108-110). These new life forms were more efficient utilizers of energy; it turns out that their development was prerequisite to the later evolution of plants and animals.

As this episode portrays, evolutionary learning is slow but effective. The selective process did not "know" where an evolutionary experiment would eventually lead; nature merely selected what worked. Evolutionary learning is accelerated by death as well as by great challenges that are traumatic to individuals and systems.

The science of cybernetics suggested to Gregory Bateson how living systems and organisms (even those with no brain such as bacteria) can learn and be self-organizing. He characterized learning as the core act of mentation. It is a series of acts including a perception. With each act there is a perception of the difference between the effect and the expected effect (a sense of good fit), followed by a next step of correction in the direction of more accurate perception or action–this is the way automatic control devices work. A learning system always operates with energy and information. The energy makes it able to move and the information is a set of signals that enable the movement to be orderly. Information often consists of pulses of energy, but the energy is not the relevant part. It is the *difference* between the pulse and the quiet instant following it that constitutes the information. That which causes orderly motion is the circular flow of information in loops (often called feedback loops), where inputs are influenced by previous outputs–where the smallest bit of information enables a more accurate movement in correction.

In this sense, just as for the Greeks, the same "mind" is evident in the self-corrective balancing of a climax ecosystem as in the speculations of a philosopher. We are wholly natural and nature is wholly kin to us. Just as for the Greeks, the design of every organism is the evidence of "mind," so for us, the mental process of evolution in the dance of random mutation, comparison with the selecting environment, deselection of the random changes that do not fit in an orderly way within the whole, is a fully cybernetic process of information flow and self-correction. It is "of the same kind" as every mind, every mental process.

As the organism, so the ecosystem. As the ecosystem, so the biosphere. If I am alive, with self-regulating flows of energy, so is the ecosystem. If I have mind, which operates by guessing and correcting, by trial and refinement, so does the biosphere. And, as the Buddha said, if we sit quietly for a long time we will come to notice that there is no "substance" which is a "self" in me. There is a process. And I share this process with all creatures and with the whole earth. (Cashman, 1987, p. 32)

The evolutionary story tells us that social learning involves more than winning or losing; organisms found that cooperation was even more effective. For example, trillions of bacteria in the microcosm learned how to cooperate and perform services for each other. The biological term for this is *symbiosis.*

The trip from greedy gluttony, from instant satisfaction to long-term mutualism, has been made many times in the microcosm. Indeed, it does not even take foresight or intelligence to make it: the brutal destroyers always end up destroying themselves—automatically leaving those who get along better with others to inherit the world. ... While destructive species may come and go, cooperation itself increases through time. People may expand, plundering and pillaging the Amazon, ignoring most of the biosphere, but the history of cells says we cannot keep it up for long. To survive even a small fraction of the time of the symbiotic settlers of the oceans and earth, people will have to change ... we will have to dampen our aggressive instincts, limit our rapacious growth, and become far more conciliatory if we are to survive, in the long term, with the rest of the biosphere. (Margulis and Sagan, 1986, p. 248)

Human Learning

If learning works so beautifully for all living creatures, even the biosphere, how could it happen that humans have so tragically degraded our biosphere? Of all creatures, we should have the greatest advantage in learning, yet, we seem unable to understand the great damage we are inflicting on other creatures, the biosphere, and ultimately ourselves. Are we like mindless creatures that do not understand their "lack of fit" and suffer elimination from nature's systems? It is not inconceivable that nature will solve the problem of human degradation of the biosphere by extinction of our species. Does social learning have to be so tragically painful? Can we humans be foresighted enough to learn better and faster so as to avoid severe trauma and death? An important first step is to try to understand more about social learning.

The great advantage possessed by humans is our capability of participating in our own learning; we are able to undergo enormous transformation within a single lifetime. "From the early human phase until now the brain has tripled in size, growing at the approximate rate of 100 percent every million years." (Margulis and Sagan, 1986, p. 253) We can perceive our past, assess the present, and imagine the future. We can perceive our progress in learning and make plans for further progress. This powerful capability of human individuals for learning is the underpinning for social learning. Understanding how social learning can extricate us from our societal predicament with our environment is the main concern of this chapter, in fact, for the whole book.

Social learning in human communities seems to advance by three processes: accumulation of knowledge, technological challenge, and elaborate forms of communication. Our specific species, *Homo sapiens sapiens*, developed about 100,000 years ago but it was not until the advent of civilization fewer than 10,000 years ago that knowledge began to accumulate. Keeping written records, that began about 5,000 years ago, allowed swift accumulation. The accumulated learning offered to children today, and carefully spoon-fed to them between fifteen and twenty-five years, is vastly greater than that offered a young person 10,000 years ago.

People had probably figured out the rudimentary laws of nature thousands of years ago, but the last two centuries of scientific inquiry have accumulated knowledge at such a swiftly accelerating rate that it is now estimated to be doubling every ten years. The findings are recorded so that they may be copied, stored, and disseminated across community boundaries and beyond generations. The worldwide collective enterprise of science constitutes a regulated group perception that far transcends the activity within one person's brain. This accumulating store of knowledge surely is social learning.

As we saw in the previous chapter, society's traditional beliefs and values provide a structure that shapes the way people perceive and learn; helping them to make the learning task manageable. Learning the structure offered by one's society is often spoken of as "socialization" and begins from the day of birth. The meaning we give to what we observe is culturally defined by a society's belief structure; in this book, I call that structure a *social paradigm*. Social paradigms are not only important for social learning but also are dominant cultural influences. They are so important that I devote the entire next chapter to them.

Another way society tries to make our learning task manageable is to encourage people to choose a specialty. We often feel competent to challenge accepted knowledge, or to contribute new knowledge, only within our specialty. The remainder of our daily working knowledge is handed to us by our culture. In this sense, then, we do have a group mind. Most of what we "know" and use to make most of our decisions is derived from that group mind. This group mind, as with individual minds, is not a static thing; it learns new knowledge, new concepts, new worldviews (paradigms)–that is social learning.

Science begets technology that awards power to those that control it. Much of our social learning is a response to the challenge of new technology. We all know how the use of the automobile has taught us to restructure our cities, change our work patterns, amend our moral norms, and waste our resources. Computers are now teaching us new work and play patterns. My university is on the verge of requiring all students to become computer literate. Within a decade, nearly all business transactions in developed countries probably will be conducted via computers. The pace of technically induced change is so swift that many people find it dizzying. Technology is the product of our social learning, but it has become the greatest prod to new social learning.

Many people have come to believe that technical change is unstoppable, that they are mere pawns in a process they do not understand, that there is no point in trying to direct the course of technology. In Chapters 12 and 13, I address the problem of social control of technology. My focus in this chapter is on the way science and technology induce social learning.

Written language, mathematics, and our new capability for electronic storage and transmittal of information have greatly accelerated communication and learning. Our global communication linkages have become a collective social nervous system that in many respects is able to learn and think. Its importance for future social learning may be as great as development of the first nervous systems in living organisms.

In addition to communicating with words, we are all familiar with body language, visual images, sounds, smells, touch, and taste. Tonight's news carried the story that organisms possess a sixth sense–the ability to sense magnetic fields. Scientists were even able to identify the specific cell in the nervous system of a sea slug that could sense a magnetic field, enabling it to learn a simple maze. Evidence suggests that simple organisms with none of our five senses, and without a nervous system, are able to communicate with each other and communicate with the earth. (Margulis and Sagan, 1986, passim)

Although science has not identified a specific mechanism for extrasensory perception, nevertheless, many people believe that it is possible. Can we participate in each other's thoughts, even at long distances, despite the absence of verbal communication? Are we all participants in some kind of cosmic group mind? If that were so, it would help explain some puzzling instances of swift social learning. Science can only say it does not know much about our ability to participate in each other's thoughts; it cannot either accept or reject the idea.

Structurally Nourishing Social Learning

Admonishing people to improve their learning will do little good; we will have to design social structures that nourish it. Developing a learning society will require many years of criticism, social experimentation, failed experiences, and a great deal of thinking and discussion. Harmon (1988, preface) writes, "All my life I have heard the admonition, 'Don't just talk; get out there and do something!' The problem is that in times like these we are all too likely to do what turns out to be the wrong thing. If it is to represent the best advice for such uncertain times, the maxim should probably be turned around: 'Don't just do something; get out there and talk.' "

Argyris and Schon (1978) distinguished between single loop learning, which is experientially based incremental learning, and double loop learning, in which the learner becomes aware of the assumptions and values on which it is based and is capable of major shifts of reference. Bateson (1975) advanced a similar idea in his theory of *deuterolearning*, in which learners are encouraged to become conscious of the context of their learning–even of the context of this context. Trist (1980) believes we are moving into a turbulent environment that will require developing the capability for double loop learning at a social as well as personal level.

The reader will recognize that the ideas just reviewed are similar to those expressed in the previous chapter where I analyzed many dysfunctional aspects of our present ways of knowing and valuing. I also offered an agenda of new ways of thinking that would facilitate adaptive social learning.

In this chapter, I discuss some social-structural characteristics that will nourish social learning. The ancient Greeks had a word for a society that nourished personal and social learning–they called it *paideia*. They believed that promotion of self-development and lifelong learning should be the central project of a society–this is feasible in a society that has found ways to readily provide sufficient goods and services. "*Paideia* meant the task of making life itself an art form, with the *person* the work of art. In theory, at least, the achievement of the human whole–and the wholly human–took precedence over every specialized activity or narrower purpose." (Harman, 1988, p. 147) As nature's limits force us to turn away from massive production and consumption, we should make achieving *paideia* the central project in our society. Activities such as research, education, self-discovery, artistic expression, participation in public discourse and public affairs, imaging, and choosing a better future, are nonpolluting, nonstultifying, and fulfilling. Unlimited numbers of people, not needed for other sorts of work, can be absorbed in this central project.

A Learning Society Utilizes A Wealth Of Information

How can a society structurally nourish social learning? Easy sharing of information, unfettered by legal or financial barriers, is crucial. People must value and protect the openness of a society if they wish it to be a learning society.

Although a wealth of information is obviously desirable, the human organism can receive and process only a limited amount. The totality of messages thrust at us each day via the media (especially advertising) makes most of us suffer information overload. What do we do then? People adopt a variety of coping strategies. One possibility is to shut out most messages, the strategy used by a surprising proportion of people. Ignorance about public affairs, as revealed by surveys of the public, is shocking to most believers in democracy.

We all use selective perception to cope with the overwhelming number of messages that afflict us. A message has to trigger a predisposition within us in order to be picked up and incorporated into our thinking and behavior. An easy selection strategy is to attend to those messages that intrude most prominently and not worry about messages that do not. We may call those using this strategy *passive information processors*. The media gatekeepers (those who decide which stories to highlight, which to give less prominence, and which not to present at all) have a powerful influence on passive information processors. The gatekeepers determine levels of public concern and the mix of items on the public agenda. Advertisers who obtrusively repeat their advertisements are likely to receive the attention of the passive information processors. Because wealthy people tend to control both media and advertising, passive information processors will mainly receive messages that encourage them to support the established system; obviously this will have a conserving effect on a society.

A small proportion of people, the *active searchers*, pursue a more focused searching strategy. They do not wait for messages to be thrust at them, but communicate with special people or subscribe to special publications in order to get the information they want. Joining an organization of people with similar interests brings in a host of new messages; most of these organizations have special publications presenting information relevant to the interests of the group. Persons with identifiable interests, organizations too, may join information exchange networks; people active in such an exchange are said to be *networking*. As you might imagine, I am part of an "environmental network." Hundreds of environmentally relevant messages cross my desk every day—far more than I can possibly attend to. Because few of these are picked up by the media, the average person would not be aware of them. Networks are an essential resource for social learning and change. Active searchers also can subscribe to a wide variety of newsletters, abstracts, and data services that scan and bring to the attention of their subscribers those pieces of information they believe will interest them.

The new information technologies (television, computers, data bases, and so forth) reduce the time and expense of sharing information. Electronic technologies also facilitate lateral sharing: from person to person, group to group, institution to institution. When I visited China in 1985, I frequently heard complaints against vertical bureaucratic hierarchies. Messages and information travelled up to the top of one hierarchy before being passed across and down another; messages seldom were passed across laterally from bureaucracy to bureaucracy. This procedure not only took unreasonable time but also introduced additional distortion. Such hierarchical structures are a barrier to social learning.

Imaginative solutions can be found to this hierarchical rigidity. One of the most interesting was advanced by Andrews (1984), who proposes establishing information routing groups (IRGs) in big organizations. An IRG is an interlocking network in which each member understands his sphere of operation and its interactions with adjacent IRGs. IRGs would be designed to facilitate lateral communication using personal computers.

Persons who actively seek information that has some real depth have an especially troublesome problem of information overload. I have suffered with this problem for years and still have no satisfactory solution to it. I have taught myself to scan quickly and discard hundreds of messages that are of only peripheral relevance to what I am doing. Among those I would *like* to read, I actually read no more than 10 percent. Among those I feel I *must* read, I do not manage more than one-half. The most frustrating aspect is that some of the seemingly highly relevant messages turn out to be fairly useless; yet, I could not know this until I read them. Meanwhile, something that I might find really valuable does not get read, either because I do not know about it or because I cannot find time to read it.

In the previous chapter, I made the point that raw information (data) is not useful to us until we give it meaning—only then does it become something we know. In modern society with its heavy load of information, we are likely to become information-rich and knowledge-poor. As society gets more complex and bureaucratic, we confront a continuous race between "knowledge demand" (the growth of complexity) and "knowledge supply" (the growth of learning). If complexity outraces the ability of a society to learn, it will be knowledge-poor and its capacity to make skillful decisions will be commensurately diminished (Elgin, 1981, Appendix III).

The learning curve, or knowledge-supply curve, of a society or complex organization, is subject to biological, mechanical, and time limitations. It rises rapidly at first and then begins to bend over and flatten out; in other words, learning capacity soon reaches a saturation level (see Figure 5.1). If the society continues to become more complex (especially the organizations that managers are charged to administer), the knowledge-demand curve that rose slowly at first will rise steeply with increasing scale (again, see Figure 5.1). This curve rises steeply because knowledge demand is a multiplicative factor of size of bureaucracy, complexity of bureaucracy, and interdependence of its units. At point B of Figure 5.1, knowledge demand has far exceeded knowledge supply. As we lose the race with mounting complexity, overly simple solutions probably will be applied to increasingly complex problems and the overall performance of institutions will drop; ironically, this will exacerbate the problem of mounting complexity. Almost any legislator or top administrator would agree with former Senator Adlai Stevenson, "We're frantically trying to keep our noses above water, racing from one problem to the next."

Figure 5.1

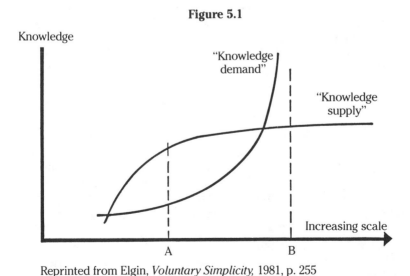

Reprinted from Elgin, *Voluntary Simplicity,* 1981, p. 255

A society that hopes to maximize its capability for social learning must find some way to develop a better information abstracting, organizing, and sifting capability so that those under great pressure to learn swiftly will have some hope of trying to keep up with knowledge demand.

A Learning Society Finds Better Ways To Disseminate And Utilize Information

Our society is remarkably effective at generating new information, but not at meaningfully disseminating it. Too much of it is underutilized because it is not getting to the people who need it in a form they can use. The number of publications each year (new books, articles, etc.) is staggering; I cannot even read all the notices that cross my desk. Much information that would be very useful for social learning gets buried in that flood of publications. Good books struggle to get noticed in a flood of mediocre ones. We need better methods for abstracting, coding, and retrieving information so it can be directed to potential users, who also need to be identified.

Computer search and retrieval systems are now being used, but they turn up much more noise than genuine knowledge. Theodor Nelson, 1981, claims he has a better system called *hypertext*. Documents and books would be stored in computers with gigantic memory; eventually they would become a world electronic library. Users would be able to link up any part of a document they were reading with any other document in the system. If one were reading a document that contained a citation to another document, one could point to the citation on the screen and the other document would be brought up for immediate comparison with what one had just been reading. One could gather, read, and compare several treatments of the same subject without leaving one's chair.

Eric Drexler discusses the advantages and disadvantages of hypertext and imagines how it would work much more effectively with "nanotechnologies" (tiny but enormously fast and clever computers, discussed in Chapter 12 herein). "Better indices will make information easier to find. Better critical discussion will weed out nonsense and help sound ideas thrive. Better presentation of wholes will highlight the holes in our knowledge." (Drexler, 1986, p. 230) Drexler has elaborated the way hypertext could be developed using nanotechnologies in a new paper, "Technologies of Danger and Wisdom," 1987.

Many people who need information do not know that it exists or where to find it. A targeted dissemination service (patterned after the Agricultural Extension Service in the United States), that could direct relevant knowledge to people who normally would not know how to get it, would encourage social learning. In the past, such services have mainly disseminated information that helped people to be more productive, efficient, and competitive. Similar services could just as well disseminate information that would help people to understand ecosystemic processes, resource stocks and depletion rates, how to anticipate consequences elsewhere in the system for initiatives currently being planned, how to think systemically and integratively, how to think creatively, and so forth. These services would be especially helpful to adults who failed to learn this perspective in their formal schooling. The learning society will encourage lifelong learning.

A Learning Society Emphasizes Integrative And Probabilistic Thinking

I have previously raised the point that people in modern society are overspecialized and overemphasize linear mechanistic thinking. People need to be trained how to think integratively, to expand their frame of reference to include the biosphere as a total system, and to think in long time frames—in other words, to think holistically. The schools should routinely teach students how to think holistically and should help them learn about (see their world as) complex sets of interactive systems—students need to be rewarded and socially encouraged for doing so. Schools should encourage this mode of thought just as much as they encourage specialization. The larger society needs to reverse its current adulation of experts and should use social respect and monetary rewards to encourage integrative holistic thinking. Integrative thinking can be in error, just as any other kind of thinking; therefore, we need to develop a societal tradition of constructive criticism of this style of thinking.

A sustainable society also should encourage probabilistic thinking. This contrasts with mechanistic thinking that emphasizes searching for "concrete" facts, precisely measured, that unambiguously link cause with effect. For example, those who promote chemical lawn sprays (or nuclear power or smoking) frequently claim that there are no cases unambiguously "proving" that their agent *caused* injury or death to a person. If the mechanistic-like linkage cannot be established, they claim there is insufficient basis for public policy and action.

The main reason for supplanting mechanistic thinking with probabilistic thinking is that the natural world really is not a mechanism. Most scientific knowledge is based on probabilities. This statement holds even for the most exact of sciences—physics. As physicists have searched deeper and deeper into the atom, they have had to abandon their former insistence on exact mechanistic connections and have turned to viewing their subject probabilistically (Capra, 1982). People must learn to distinguish high probabilities from low probabilities and use those as guides to action. They should abandon the idea that certainty is possible and must be achieved before action can be taken. They should learn the meaning of statistical significance—that is, the use of statistical evidence to estimate the probability that an observed relationship could have occurred by chance alone. If that probability is very low (five or fewer chances out of 100), it is probable that the observed relationship holds in the real world. Such knowledge can serve as a basis for action.

Probabilistic thinking is more helpful than mechanistic thinking for seeing the world as a complex set of interactive systems. Mechanisms provide leverage, they connote power and domination by humans over nature. Probabilistic thinking is more attuned to inquiring, to trying to understand how complex systems operate. Persons who recognize they are part of nature, and who desire to live in harmony with the whole of nature, are most likely to use probabilistic thinking. Persons who desire to dominate nature are most likely to approach their task in a mechanistic mode, without much concern for the delicate systems being interfered with. If mechanistic thinking dominates in our society, it will be to the detriment of all of us.

A Learning Society Emphasizes Values As Much As Facts

I tried to show earlier that many of the unsustainable characteristics of modern society (destruction of nature, depletion of resources, lack of security, etc.) can be traced to inadequate understanding of values. Most of society's institutions (organized religions, schools, business firms, governments) operate on the presumption that we already know which values are best; that the main task is to reaffirm and uphold them. Most social institutions are especially conservative in this respect. If we are to understand our values deeply, and design a sustainable society, we must learn how to reexamine the value structure that was handed to us when we were socialized into society. This is a social as well as an individual task; we need to learn how to seriously discuss our societal values with each other and think our way through to a deep understanding. People, and their institutions (especially their schools), should help each other to learn how to think deeply about their values.

I pointed out in the previous chapter that many people in modern society believe personal values are not a proper subject for examination or discussion. They seem unsure where values come from and accord equal deference to any value. This feeling creates a "value vacuum" that prevents reasoned discussion about values. I argued that this societal posture toward values is not viable. We will never develop a sustainable society, a learning society, if we cannot talk about the things most important to us. I believe it is possible to reason to ground about values, and that this mode of reasoning can be applied not only to our personal values but also to societal values. We should develop a tradition of having deep public discussions about values. Value analysis is the core of public discourse and is absolutely essential to effective social learning.

Lawrence Kohlberg has for some years advanced the idea that moral reasoning develops in stages—that people advance to higher stages of moral reasoning as their cognitive ability develops. In his later work (1984), he refers to the stages of moral reasoning as stages of justice reasoning and calls his formulation a "theory of the development of justice reasoning." The highest stage of moral reasoning he calls the "postconventional level." Two researchers, working at the German Armed Forces Institute for Social Research, have developed a research instrument for measuring Kohlberg's stages of moral reasoning (Kohr and Rader, 1984). As their work has progressed, they found that those who have achieved the postconventional level of moral reasoning are most likely to think ecologically (Kohr, 1985). If the preliminary conclusion from this line of research withstands further examination, it suggests that growing ecological awareness, heightened moral consciousness, and the ability to carefully examine and discuss values will probably develop simultaneously.

A Learning Society Is Critical Of Science And Technology

Our modern civilization accords a worshipful deference toward science and expertise. Obviously, scientific knowledge has brought great benefits to humans, but it is unhealthy for the biosphere, and for society, for us to be so worshipfully uncritical of what science does and where it is leading us. Society needs to find structural ways to critically evaluate the role science plays in society. Some suggestions for systematically and regularly doing this are presented in Chapter 12.

News media recently reported three stories that illustrate how difficult it will be to obtain critical public review of scientific initiatives. Jeremy Rifkin is the foremost advocate in the United States for rigorously controlling recombinant DNA projects, often called bioengineering (see Chapter 9 for fuller discussion of this technology). In 1984, he filed suit in federal court to require bioengineers at the University of Pennsylvania to file an environmental impact statement assessing the consequences of a proposed research project. In a second case, Rifkin requested the Environmental Protection Agency (EPA) to delay the use of an "ice-minus" bacteria engineered to make plants more resistant to frost. The suits were denied in both cases.

Rifkin had acted out of his deeply felt concern that some newly invented creature would be released into the environment and have devastating unforeseen consequences. There was little support demanding the review he requested. Cell and microbial biologists recognize the danger Rifkin points to, so they, in cooperation with the U.S. Government, have established rigorous review procedures that they believe will forestall the release of potentially dangerous creatures. Among the few people who are aware of the power of recombinant DNA technology, most seem to be satisfied that the scientific reviews will provide safety.

Unfortunately, the controls set up in the United States are not applicable in other nations where this research also is being vigorously pursued. DNA research is the new rage; there are large and small fortunes to be made and many people want a part of the action. Currently, more than 60 percent of graduate students in biology elect to study that specialty. That brings us to the third story reported on television early in July 1987. An entrepreneurial scientist in Alabama equipped a mobile laboratory that he is taking from town to town demonstrating to high school biology teachers and students how recombinant DNA research can be done. The film clip showed fourteen-year-old students snipping a portion of a strand of DNA from one bacteria, combining it with a strand from another, and transplanting it into a third. The students were ecstatically exuberant at experiencing this new-found power. Now they were part of the forefront of science, the potentials were limitless, and wasn't science wonderful? There was not a hint from entrepreneur, teacher, students, or reporter that they were playing with a dangerous power. Can you imagine controls on DNA research ever being effective if the technique is taught and encouraged in hundreds of thousands of high school biology laboratories?

For the time being, the public surely will go along with the thrust of the establishment toward support for ever greater scientific and technological development. Considerable social learning must occur before society will be disposed to review scientific initiatives. Regretfully, that learning probably will not take place until some new technology turns out to have strongly adverse consequences for the biosphere and society. By then, the technology will probably be so well established, with so many vested interests, that it will be extremely difficult to change course. We are caught in a potentially tragic catch-22: we cannot develop effective controls on science until social learning has progressed much further; yet, we cannot learn the need for the controls until a technoscientific catastrophe occurs.

Our regard for science and our belief in its power has frequently led us to call on that power to solve difficult policy questions. It is common to redefine policy questions so that the decisional task is mainly one of using science to get the "facts" while the value components of the issue are swept under the rug or assumed away. The implicit values embedded in the organization that poses the issue, and funds the science, are typically carried forward into the policy without any serious examination of their validity. These organizations typically give great emphasis to the immediate material gains and either ignore or seriously discount potential long-term negative effects.

To illustrate, nearly all public decisions to regulate for environmental protection involve judgments about risk and values. No matter what decision is taken (to regulate or not to regulate, or how much to regulate), someone benefits and someone else gets hurt. Important values are being traded off in these regulatory decisions; thus, they are prime candidates for value impact analysis. The risk component of the decision also is heavily value laden. The risk to health has to be weighed (valued) against risk of loss of money, for example.

Elected officials hate to make these decisions because their own careers are at risk. Therefore, they typically delegate the decisions to the bureaucracies in the executive branch. Bureaucracies also do not wish to be at risk, so they redefine the problem as a scientific one of estimating probabilities of injury; values are not visibly analyzed. Once the probability estimates are made (by their nature they cannot be made precisely), they are entered into a risk-benefit analysis in which values (such as health) are converted into monetary terms and a risk-benefit ratio is calculated that, presumably, provides a scientific answer to the policy question. This procedure will not survive careful, critical scrutiny.* However, the aura of science and the smokescreen of "objective" mathematics usually are sufficient to finesse the issue—resulting in the values of the establishment being preserved.

Citizens should not allow decisions to be "stolen" using a smokescreen of science; they should reserve to themselves and their elected representatives the right to weigh their value choices. There is no easy answer to the problems I have posed in this section. We must become much more critical of science and technology; yet, the current worship of science is so great that those who do criticize are likely to be dismissed. Langdon Winner (1986) reports that his engineering colleagues gazed at him incredulously when he asked them what were the central values in their discipline. They simply had not thought about it. Those who constructively criticize science are likely to be taken seriously only when science stumbles badly and injures many people. That is a painful way for society to learn.

A Learning Society Combines Theory With Practice

In public debate over an environmental policy issue it is not uncommon to hear something like this: "Yes, I see your point that we should protect

---

*See Winner (1986, Ch. 8), for an especially cogent critique.

the environment but we have to be *practical.*" What does *practical* mean in this context? Usually it means that protection of the environment can wait and that the urgent thing right now is that people need to make a living. "Being practical" seems to carry one of four connotations: first, that we should operate in a short time frame and attend to immediately urgent matters; second, that we should put aside theory and try this or that in an aimless manner until one seems to work; third, that being practical is more valuable than being theoretical; and fourth, in order to win support, we must make arguments that emphasize the "buzz phrase" that everyone is using in public discourse at the time. The buzz phrase was *competitiveness* during the time I wrote this book. If one wished to garner support for a project, it was effective to claim that the action would make us competitive. Obviously, being "practical" in these senses is not very conducive to social learning; it is not even very practical for solving problems.

It seems obvious that practice should be embedded in a theoretical understanding of the way a system works. Decisions made in the context of a larger theory are more practical because they are cast in a longer time frame. Actions that appear to be practical in the short run often turn out to be counterproductive in the long run. Our search for practicality deceives us because it is easier to perceive the obvious costs of action than the hidden costs of inaction. With respect to the environment, usually the most practical actions are those that will make for a flourishing ecosystem and a sustainable society. If an option does not withstand careful scrutiny in this larger theoretical context, it is not a practical solution. A society that maximizes social learning will perceive as practical only those options that are likely to make for long-run sustainability. (For an illustration of this point, see the example in the section that follows.)

A Learning Society Is Consciously Anticipatory

The need for being anticipatory is well-expressed in two ecological principles that I have enunciated previously: "We can never do merely one thing" and, "We should always ask 'And then what?' " Almost every policy will have unintended consequences because we are connected into a complex set of interactive systems. Human experience with chemical pesticides illustrates the consequences of failing to use these two principles.

A great effort was launched to invent chemicals to kill pests after World War II. These pesticides that so easily killed the age-old nuisances of insects and weeds were first heralded as great boons to humankind; no serious effort was made to foresee unintended consequences. It was first thought that the lungs of mammals would screen out the injurious chemicals. We did not know that they also were absorbed through the skin (although insects absorbed them that way). Another route to our bodies was discovered when lower animals ate the toxified insects; the chemicals in the insects were absorbed, passed up the food chain and entered our bodies anyway. Because many such chemicals "bioaccumulate" as they are passed up the food chain, the doses of toxics entering our bodies were far larger and did more damage than originally thought possible. They have slowly found expression in humans as disease and genetic mutations.

DDT, the first widely used powerful insecticide, was absorbed by birds that ate poisoned insects. It bioaccumulated in their tissues and altered the cell structure of their egg shells. Egg shells became so fragile that the birds could not sit on their nests without breaking the eggs. It took many years before scientists realized the connection between the insecticide and the inability of birds to reproduce. Rachel Carson alerted us in 1961 by her warnings of a "silent spring." An alarmed public demanded action, but DDT was not banned until 1972 due to the stalling lobbying tactics of the chemical industry. Chemical manufacturers were allowed to continue making the product for sale abroad. We still failed to ask, "And then what?" Now we find that DDT levels, that had been falling for fifteen years, are rising again in the United States. Apparently DDT is returning to the country in food products imported from abroad; some is volatilized and returns via winds; also migrating birds eat toxic insects when abroad and bring DDT back to the United States. We can never do merely one thing.

Meanwhile, chemical manufacturers invented a variety of other pesticides that presumably were less injurious to humans and the ecosystem. As we developed experience with them, many have proven so injurious that they also have been banned. We still do not insist that chemicals be thoroughly tested for effects before being spread in the biosphere. We seem to follow the maxim, "Innocent until proven guilty." That maxim makes sense for persons accused of crime but it makes no sense when some people wish to make money at the possible expense of injury to the biosphere. To nourish life in a viable ecosystem we must reverse the maxim to, "Guilty until proven innocent." Only chemicals that pass that screening should be licensed for use.

The chemical industry continues to boom, and pesticides are now applied copiously to home lawns and forests, as well as to farm crops. A recent study by the National Research Council disclosed that a great number of pest species have become resistant to chemicals. Prior to World War II, only seven species of insects and mites were known to be resistant to chemicals. Now the council estimates 447 species can survive chemical attacks. (Gradual adaptation toward chemical resistance also applies to plants that have been treated with herbicides.) They recommend use of pesticides only intermittently (that is, every other year) in a program that also uses natural predators and crop rotation. We are learning that chemicals that work today will probably not work tomorrow if used continually.

Now we are also discovering that certain people are chemically sensitive and have violent reactions when they come into contact with chemical sprays; however, chemicals are so ubiquitous in modern society that it is extremely difficult to avoid chemical contact. Society, today, has a far greater problem than it started with forty years ago: chemical poisons are everywhere and circulate endlessly in our air, water, and land; the health of humans and other creatures is being impaired; yet, we are not successfully eliminating the pests. Pesticides are now heavily and thoughtlessly applied worldwide. The chemical industry is so large and well entrenched that banning chemicals will create severe economic hardship; thus, any serious move to reverse course on use of chemicals will be bitterly resisted. This tragic situation developed because we were structurally incapable of being anticipatory; because we chose short-term "practicality" over long-term sustainability. Eventually, enough people will recognize the insanity of using this technology and its use can be terminated. It is hard to imagine more painfully expensive social learning.

It is equally important for people to develop a personal awareness of the need to be anticipatory. This fundamental principle of thinking and learning should become so ingrained that people will apply it everyday to their own decision making. Our educational system should make it a policy to help people adopt this style of thinking as a habit.

A Learning Society Believes That Change Is Possible

How do we convince people that social change is possible? This barrier is probably the most difficult to surmount of all the societal characteristics discussed in this section. The system seems so big and so impervious to individual inputs that many people do not see any point in pushing for change; they are likely, in fact, not to think about it at all.

People today seem to have two postures toward change that are ironically contrastive. On the one hand, they perceive that their world changes swiftly. I spoke to a church study group in 1984 and asked how many of those present (about forty) believed that the way they currently go about their daily lives would be much the same in the year 2000 as it is today. Not one person believed it would be much the same. On the other hand, people seem to believe that the socioeconomic-political system is so strongly entrenched that there is little hope that it can be influenced or changed by citizen action. They are cognizant of their own weakness; many have a fear of being wrong, especially of appearing foolish if they take a stand in public.

People's sense of the possibility of system change is influenced by the attitudes and behavior of the people around them. If many people believe that things are so bad that change must take place, more of them are likely to believe that change is possible. Conversely, if most people perceive that present conditions are good, even merely tolerable, the possibility of change will seem more distant. Seeing that some people are able to bring about a change, even a small one, also is important; it provides some evidence that further change is possible. Citizen action success stories are rare, but they do exist. A group of women living at Love Canal in Niagara Falls, New York were so upset about injuries to the health of their families from toxic wastes that they formed and sustained the Love Canal Homeowners' Association. It was so successful in winning people to its cause that local, state, and national governments adopted new policies with respect to the handling of abandoned toxic waste sites as well as future disposal of toxic wastes. If people rally in support of an appeal for a proposed change, everyone begins to sense that change is possible.

This same phenomenon could be seen with respect to Gorbachev's *glasnost* policy in the U.S.S.R. When he opened the socioeconomic-political system to criticism by the public, formerly reticent citizens spoke up quickly and launched many public protests. Their desire and expectations for change far exceeded the capability of the system to absorb change.

If the structural changes recommended in this chapter, and those to follow, were to be put into effect, it would help to build a sense that change is possible. Having an open information system that is easy to use, for example, supports a sense that change can be attained. Developing institutions for helping people anticipate the future would stimulate a feeling that the future could be shaped and make life better. Building a society that is designed to learn would help to empower people with the belief that change is possible.

A Learning Society Examines Outcomes To Learn From Them

The following scenario is quite common in modern society: a social problem is identified and there is clamor for change; the political system responds with a new policy (laws, regulations, agreements); the new policy is implemented (usually only halfway); and we never find out if the new policy worked. Often the government does not even try to find out. All learning is dependent on perceiving what works (fits) and what does not work. A special subfield of social science called *evaluation research* has developed the capability to infer validly whether a policy has succeeded;

yet, it is seldom used in public policymaking. Officials seem reluctant to spend money on such research or to use it when they make decisions.

Social interventions can be set up in a quasi-experimental fashion. For example, a proposed new policy could be tried out in one or two places, while the remainder of the system is left to function in the previous fashion, thereby producing a result that is more interpretable for policy learning. America's fifty states constitute something of an experimental laboratory. A good example is provided by the policy, adopted in some states, requiring a small monetary deposit on beverage containers in order to encourage their recycling (which Oregon was first to adopt). This experiment was studied carefully and its success became social learning for other states. Now this practice has been adopted in fifteen or twenty states; eventually, the U. S. Congress will use this learning to make a nationwide policy on container deposits. Nations also study each other's policy experiments to make their own national policy.

Some experiments are neater than others. Beverage containers are visible and localized; the policy takes effect in a specific area and the results can be observed there. Acid rain control presents a much more difficult experimental problem. If one state enforced stricter controls on sulfur dioxide emissions (as did New York), while others kept their old emission standards, the results would be much less interpretable. Because air moves freely across state lines, only a continentwide policy will have much impact in controlling acid rain.

As another example, Congress lowered speed limits to fifty-five miles per hour in the 1970s to conserve petroleum, but we soon discovered that deaths from accidents also dropped; therefore, the fifty-five mile per hour limit was retained when the oil crisis eased. In 1987, those valuing speeding prevailed over those valuing safety and states were allowed to raise the limit to sixty-five and deaths from accidents rose again.

Social learning will not be very effective if we do not systematically evaluate the outcomes of policies to aid our learning; and then use that learning in conjunction with our values to make better policy.

A Learning Society Develops Institutions To Foster Systemic And Futures Thinking

People and institutions in modern society are not familiar with, and seldom use, systemic and futures thinking. We could establish special institutions where futures and systemic thinking is consciously practiced and methods are further refined for doing it well (in Chapter 14, I propose a Council for Long-Range Societal Guidance to do so). Systemic and futures thinking modes could be developed and applied by a variety of institutions, such as private consulting firms, governmental bureaus, business corporations, professional organizations, and citizens organizations. The results of these institutional efforts should be shared so that the public not only can learn about the specific problems under investigation but also about the practice and utility of these thinking styles. Eventually, the public could learn to demand this kind of thinking in the planning and decisionmaking of their governments and other social institutions. This mode of thinking would be a key component of a society programmed to learn.

A Learning Society Institutionalizes A Practice Of Analyzing Future Impacts

The environmental problems that modern society faces have been significantly worsened in many instances when humans have acted without sufficient forethought about all the consequences that would follow from the action. Burying toxic wastes in landfills is an example from the recent past. Cutting down tropical forests for agricultural cultivation is a current example. The U. S. Congress recognized this problem when it passed the National Environmental Policy Act of 1969, the major innovation of which was a requirement that an environmental impact statement be written, filed, and accepted before any project using federal funds could get underway. Although those statements have not always been well-prepared, the basic idea of taking a look before we leap is sound and has begun to be institutionalized in an ever-widening circle of public initiatives. Environmental Impact Analysis has been adopted by more than 100 countries as well as by many state and local governments. Even the new initiatives of some private firms are subjected to similar reviews.

The basic idea of looking carefully before charging ahead has also been extended to foreseeing impacts on society. Many projects that physically impact Mother Earth have substantial consequences for society as well; social impact assessment has been designed to assess them. The idea has also been extended to proposed laws, programs, and new institutions. Social impact statements have not been used very much, probably because no law requires them, but they have been the subject of methodological research by sociologists (Finsterbusch and Wolf, 1981; Finsterbusch, 1982; McAllister, 1982; Murdock, et al., 1982; Branch et al., 1984 and Rickson, Western, and Burdge, 1986).

The reader will recall that I advocate extending impact analysis to values; I would like to see value impact analyses adopted as a standard practice for major social, technological, and scientific initiatives (I discuss this further in Chapter 14). The methods and lore for value impact analysis need to be developed much further. This task will undoubtedly be more difficult than environmental impact analysis, because values, and the way they function, are much less visible than physical things such as soil, water, and plants. Yet, values are centrally important to everything we do; developing ways of understanding how our actions could impinge on them is crucial to making our society into an effective learning society.

A Learning Society Reorients Education Toward Social Learning

Educational institutions need to be reoriented toward helping students to learn the ways of thinking identified earlier in this and previous chapters: systemic thinking, futures thinking, integrative thinking, probabilistic thinking, creative thinking, values analysis, moral reasoning. Reforming the U. S. educational system will not be easy. With each passing year, some major figure or institution puts forward a proposal for educational reform; yet, the schools continue to perform poorly. It is obvious that schools have the potential to strongly influence the ability of the entire society to learn. It will require the thinking of many creative minds to transform them into strong positive forces for social learning. Here, I can only bring to your attention a few points on which the schools need to be reformed.

Many recent studies of the U. S. educational system point to deficiencies—not to small deficiencies, but to a real crisis. The Carnegie Corporation Task Force on Teaching as a Profession issued a report in May 1986 titled, "A Nation Prepared: Teachers for the 21st Century" which identified a crisis with the schools and pointed out that the problem lies most deeply with teachers. It declared that bolstering the teaching profession is the key to transforming factory-like public schools into places where children will learn to think for themselves. Our goal should be to train "our young people how to be flexible, how to be creative, how to think on the job." Clearly most teachers in most schools simply do not know how to do that. The task force called for abolishing undergraduate education degrees, creating a board to certify top teachers, and pegging teacher's salaries, in part, to their student's performance.

MacCready, who headed the team that built the Gossamer series of light-weight airplanes—one of which flew the English Channel, powered only by human energy—believes that his team succeeded where others failed because they had the ability to break away from old patterns of thinking. He would have creative rational thinking skills routinely taught and tested in the school system; he believes these skills can be taught to people of any intelligence quotient or educational background. Even though he has been exceptionally successful in technological achievement, he doubts that technology alone will ensure our future; "One thing seems clear, getting through this stressful time is not going to depend on any technological advance–an improved antimissile system, better computers and communications, free energy, or a cure for cancer. It will depend on using our brains to get along better with one another and our globe. Our brains are the problem, but they also represent the only possible solution." (quoted in Parrish, 1985, p. 41)

Another set of innovative educators identify student self-esteem as a critical component of creative thinking. Young people who do not believe in themselves, in their ability to have successful experiences, are unlikely to think creatively, or integratively, and cannot imaginatively face the future. These educators (see *New Options*, April 1986) note that parents have been unable to instill this sense of self-esteem in most children. They believe that a sense of self-esteem can be learned in the guise of play and that development programs for four-year-olds should be established in the public schools.

A Learning Society Supports Research

It is well-accepted in modern society that research is just as crucial to learning as is education. However, we must distinguish various kinds of research if we desire to understand its role in social learning. Modern institutions have learned that scientific research "pays off" in greater wealth and power for the sponsor; thus, it is conducted in and supported by private business as well as government. It positions the institution to be a more successful competitor. In fact, the fear of falling behind and losing out in competition with other institutions is a major spur to scientific research. Unfortunately, research that is undertaken to get a pay-off also typically provides ever greater leverage for humans to dominate nature. Learning that is oriented mainly toward dominating nature for a quick pay-off does not devote sufficient attention to the long-range consequences of such domination and does not qualify as good social learning.

Private business has a clear incentive to sponsor physical science and engineering research that may lead to greater wealth and power. However, little incentive exists for such institutions to sponsor social science research that is more likely to help society with its social learning. That means that most social science research must be supported by the government. Alas, governments under pressure to reduce public expenditures usually choose to cut back first on social science research; when that happens, the private sector seldom steps in to fill the gap. If a society wants to nourish social learning, it must expect its government to support research into societal processes and institutions.

Our capability for enhancing social learning by doing research is hampered by specialization into disciplines. Undoubtedly, specialization helps persons develop mastery of an area of inquiry, but society must guard against allowing disciplines to become fiefdoms where an inner group defends its boundaries against intruders. Society currently rewards specialization and turf defense. Instead, we need to reward interdisciplinary research; we need more *inter*disciplinary focus within *multi*disciplinary research. We need more scientists who are willing to interpret their research in a larger context – to see new relationships, formulate broader hypotheses, and struggle for grander conclusions. Most social learning occurs when knowledge from more than one traditional discipline is synthesized and applied in new ways. In the struggle to encourage interdisciplinary research, universities are likely to be a serious obstacle because they are structured to conserve the status quo. It is ironic that institutions dedicated to learning can become barriers to social learning.

A Learning Society Maintains Openness And Encourages Citizen Participation

The importance of an open society was discussed above. The desire of a people for an open society, and a determination to maintain it, is, itself, something that a society must learn. This learning is best enshrined and protected in a basic social contract, such as a constitution. But simply having it written in a constitution is not sufficient to guarantee openness. The belief expressed by the phrase "eternal vigilance is the price of liberty" is still true. As I explained earlier, an open society would encourage lateral communication so that connections between citizens are frequent. People should try to maintain a public aura of openness that welcomes ideas. It is especially important for them to be skeptically critical of all dogmas, scientific as well as religious, because dogmas interfere with the openness that learning requires.

A strong tradition favoring citizen participation, and vigorous encouragement of it, is an important aspect of openness that stimulates social learning. Citizen participants are in constant contact with other citizens, with officials, and with the world of ideas – those stimuli are bound to foster social learning. In trying to influence policy, citizens are challenged to dig up information, to be creative and present ideas of their own, thereby becoming an innovative part of the social learning process. A group called "Choosing Our Future," which operates in the San Francisco Bay area, is utilizing television and other electronic technology to support a public dialogue similar to traditional New England town meetings. It designs institutional processes specifically to encourage citizen participation and social learning.

In a society that possesses a strong tradition for citizen participation, governments may go out of their way to encourage citizen input into public decisions. Over the past forty years in the United States, a pattern has developed that encourages citizens to participate in *bureaucratic decisions.* Governments literally reached out to citizens and asked them to become involved. The opportunity for citizens to try to influence their elected officials is guaranteed in the First Amendment to the U. S. Constitution, but there is no similar formal right of citizens to influence bureaucratic decisions. This movement for citizen participation in bureaucratic policymaking reached its peak in the 1970s, especially under the Carter Administration, in what has been called the "participatory epoch." The Reagan Administration was much less encouraging of citizen participation. For example, the federal government under Carter hired hundreds of "Citizen Participation Coordinators"; under Reagan, however, most of those positions have been phased out.

It cannot be inferred that this outreach by government to invite citizens into the decision process will result in better or more democratic government. Such other traditional democracies as Great Britain, Canada, and Australia, with whom we share a common political heritage, do not similarly reach out to citizens to invite them into bureaucratic decisionmaking. It seems obvious, however, that regular citizen interaction with bureaucratic officials–troublesome and time consuming as it may be for both sides–is bound to encourage social learning in a society.

Postscript

All the suggestions made in this chapter for helping a society to learn have probably sounded reasonable to the reader, because reason is essential to communication and learning. But, is reason sufficient to achieve the necessary social learning? Will learning take place swiftly enough to save our biosphere from severe injury, to prevent great human suffering, to make our society sustainable? Is it not evident that most people are not listening, that they have not really awakened to the need to foster social learning?

Psychologist Douglas Hofstadter (1985) suggests that awakening and learning are not merely a function of conscious reasoning, but also take place at a deeper subconscious level that is not apparent even to ourselves and certainly is not visible to others. He quotes William James on waking up in the morning; James describes how we are torn between the desire to lie in bed and the resolve to get up. For a time, the stuggle paralyzes activity, but at some moment the inhibitory ideas cease and we suddenly find that we have gotten up—a decision was made at a subconscious level.

Hofstadter believes the waking of an individual has its parallel in the collective waking of a nation. A critical point is reached when the number of concerned citizens rises above a threshold, producing a quick turnaround at the national level. No one can predict how or when that will happen, but we know it does happen. These "phase transitions" happen in physical systems (for example, schools of fish, brains), groups, mobs, and countries (Iran is an example from recent history). Most of us, at some deep level, have a sense that our present social system is not working well, that we cannot keep going the way we have been. That feeling, which constitutes a kind of social learning as well as a readiness for further learning, may be tapped some day by a massive confluence of thoughts or events. The whole society may quite suddenly turn around in a bewildering phase transition. Whatever social learning we have absorbed before such an event occurs will help if a time of great social turbulence comes upon us.

Stories About The Way The World Works: Belief Paradigms

Chapter 6

A ll societies develop a story about the way the world works. Margaret Mead once remarked that she never found a primal people who lacked a cosmic story. "Humans will have their cosmic stories as surely as they will have their food and drink." Stories also are necessary to our own definition of ourselves. "Each human enters the world and awakens to a simple truth: 'I must find my own story within this great epic of being.' " (Swimme, 1987, p. 83)

The stories that dominate interpretation of reality in each society might be thought of as sets of cultural lenses: they provide the structure for social learning. When reality changes, those lenses distort some aspects of reality and may lead observers to completely ignore other aspects. Reality does change, as discussed in Chapter 2 with respect to such phenomena as population growth, economic growth, technological development, and environmental degradation. Our interpretations of reality also change as we learn more from science and experience. Perceiving humans to be harmonious participants in biocommunities is a new, yet old, example of such a reinterpretation.

Curiously, in the modern age of the ascendancy of science, we are not very conscious of our story. We believe we have only facts as revealed by science and that we no longer have a need for a story. We fail to perceive that we do use a story that assumes the primacy of the human who has the right to dominate nature for his own benefit. This story ignores the dependency of humans on the continued good functioning of the ecosphere and the intimate connection between the human story and the cosmic story. We are failing as a society because our story is far out of step with reality.

The particular belief structure (cultural lens) dominating a society, then, may be made up partly of images that reflect reality and partly of images that are myth. As reality grows further and further out of synchrony with dominating images, the gap between reality and image creates a tension leading some people to change their beliefs while others stubbornly continue to believe in what they have always believed.

Thomas Berry said it eloquently:

It's all a question of story. We are in trouble just now because we do not have a good story. We are in between stories. The old story—the account of how the world came

115

to be and how we fit into it—is not functioning properly, and we have not learned the New Story. The old story sustained us for a long period of time. It shaped our emotional attitudes, provided us with life purpose, energized action. It consecrated suffering, integrated knowledge, guided education. We awoke in the morning and knew where we were. We could answer the questions of our children. We could identify crime, punish criminals. Everything was taken care of because the story was there. It did not make men good, it did not take away the pains and stupidities of life, or make for unfailing warmth in human association. But it did provide a context in which life could function in a meaningful manner. (T. Berry, 1978, p. 1)

Wanda Urbanska (1986) studied the generation of young Americans coming of age in the 1980s; she characterizes them as having nothing to feel secure about (as having no story), thus they turn inward, seeking security in their own singular persona. They live in the present because their future seems so ominous. They focus on their homes, careers, hobbies; many don't know how to relax.

As some people think about the deficiencies of the old story, they begin to develop new ideas that depart in crucial ways from it. This process eventually takes the form of a new belief structure, some might call it an ideology or worldview, that challenges the old. This new worldview may be hidden from public consciousness for a time. At the public level, many people continue to engage in the old political rhetoric which is embedded in the old story—probably because everybody understands the old story and it is the "only game in town," even though their inner thoughts already are moving on to a new story and a new politics.

Scholars who study worldviews usually call them *paradigms*. Fritjof Capra, in a special issue of *ReVISION* devoted to paradigm thinking offered this definition of a *paradigm* (vol. 9, no. 1, 1986 p. 14): *A constellation of concepts, values, perceptions, and practices shared by a community, which forms a particular vision of reality and a collective mood that is the basis of the way the community organizes itself.* A belief paradigm that is dominant in a given society could be called its *dominant social paradigm* (DSP). *A DSP may be defined as a society's dominant belief structure that organizes the way people perceive and interpret the functioning of the world around them.* From time to time, dominant paradigms are challenged so fundamentally that they give way to new paradigms; this process is called *paradigm shift*. A defining characteristic of paradigms is that changes of paradigms occur in discontinuous revolutionary breaks, which distinguishes paradigm shifts from more gradual kinds of social change.

Many scholars who study beliefs and values, myself included, perceive that the United States, indeed the entire Western World, is currently undergoing paradigm shift. Our politics has, so far, failed to recognize that this shift is underway. Perhaps that is because paradigms are so fundamental a part of social architecture that most people take them for granted. Perlmutter and Trist discuss this characteristic as part of their definition of the concept:

Paradigms are the "logics" or "mental models" that underlie the missions, systems of governance, organizational character and structures (including socio-technical systems) which are the parameters of the social architecture of institutions. ... A paradigm expresses a self-consistent world view, a social construction of reality widely shared and taken for granted by the members of a society, most of whom are aware only to a limited extent of the underlying logic, which is implicit rather than explicit in what they feel and think and in the courses of action they undertake. A paradigm provides, as it were, the medium in which they exist and tends to become explicit only when the need for a new overall perspective arises through increasing dysfunction in the prevailing paradigm. (Perlmutter and Trist, 1986, pp. 2-3)

The perspective I take here on social change is akin to Toynbee's model of cultural dynamics. He has studied the history of civilizations and perceives that they rise and fall in eras, as people abandon inadequate old social structures and struggle to develop new and more adequate structures. A social paradigm incorporates beliefs about how the world works physically, socially, economically, and politically. Cotgrove has done some of the clearest thinking about paradigms:

A paradigm is dominant not in the statistical sense of being held by most people, but in the sense that it is the paradigm held by dominant groups in industrial societies; and in the sense that it serves to legitimate and justify the institutions and practices of a market economy. ... it is the taken-for-granted common-sensical view which usually determines the outcome of debates on environmental issues.

Paradigms then provide maps of what the world is believed to be like. They constitute guidelines for identifying and solving problems. Above all, paradigms provide the framework of meaning within which "facts" and experiences acquire significance and can be interpreted.

Paradigms are not only beliefs about what the world is like and guides to action; they also serve the purpose of legitimating or justifying courses of action. That is to say, they function as ideologies. Hence, conflicts over what constitutes the paradigm by which action should be guided and judged to be reasonable is itself a part of the political process. The struggle to universalize a paradigm is part of a struggle for power.

The protagonists face each other in a spirit of exasperation, talking past each other with mutual incomprehension. It is a dialogue of the blind talking to the deaf. Nor can the debate be settled by appeals to the facts. We need to grasp the implicit cultural meanings which underlie the dialogue.

It is because protagonists to the debate approach issues from different cultural contexts, which generate different and conflicting implicit meanings, that there is mutual exasperation and charges and counter charges of irrationality and unreason. What is sensible from one point of view is nonsense from another. It is the implicit, self-evident, taken-for-granted character of paradigms which clogs the channels of communication. (Cotgrove, 1982, pp. 26-27, 33, 82, 88)

Empirical Evidence Of Paradigm Differences And Paradigm Shift

I participated in a three-nation (England, Germany, and the United States) comparative study of environmental beliefs and values. One of our purposes was to study the contrasting beliefs and values of the currently dominant social paradigm and challenging paradigms. (Cotgrove was our collaborator in England and scholars at the Science Center in Berlin comprised the German team.) Questionnaires were filled out by a random sample of the public and also by samples of the following elites: environmentalists, business leaders, labor leaders, and elected and appointed officials. The study was conducted in all three nations in 1980 and was repeated in 1982. The questions were drafted through extensive pretests and were as close to identical as possible in each of the three countries as well as across the years (see Milbrath, 1984, for details). Mail questionnaires limit the number of questions one can ask and still expect a good return; therefore, we could not exhaustively examine beliefs and values surrounding the DSP and challenging paradigms. We do believe we have identified those that are centrally important.

That study provided solid evidence that a new paradigm is emerging, one which differs significantly from the dominant social paradigm. This emerging paradigm is being developed by environmentally oriented thinkers who constitute a kind of vanguard. They advocate a new set of beliefs and values that people have begun referring to as a *new environmental paradigm* (NEP) (Dunlap and Van Liere, 1978). Naturally, those who believe in the wisdom of the DSP wish to defend it; they have become a rearguard in opposition to the challenging vanguard with its NEP. The labels *rearguard* and *vanguard* are my own. I do not intend for these labels to be pejorative. The outstanding characteristic of the rearguard is that these people are defenders of the DSP and the outstanding characteristic of the vanguard is that these people are trying to bring about a new society with a new paradigm. The people in the rearguard and vanguard may not perceive themselves as such; if so, it is beside the point that, from the perspective of the social analyst, they are playing out these roles as social change is worked through.*

Major Differences Between DSP And NEP

Many of the beliefs and values that differentiate the two paradigms have been considered in earlier chapters. The discussion that follows is based substantially on findings from the three-nation study and focuses on differ-

---

*Perlmutter & Trist (1986), discuss three paradigms and link them to twelve major societal institutions.

Table 6.1
Contrasts between Competing Paradigms

New Environmental Paradigm	Dominant Social Paradigm

1. High valuation on nature
  a. nature for its own sake—worshipful love of nature
  b. wholistic—relationship between humans and nature
  c. environmental protection over economic growth
	1. Lower valuation on nature
	  a. use of nature to produce goods
	  b. human domination of nature
	  c. economic growth over environmental protection

2. Generalized compassion toward
 	
  a. other species
 	
  b. other peoples
  c. other generations
	2. Compassion only for those near and dear
	  a. exploitation of other species for human needs
	  b. lack of concern for other people
	  c. concern for this generation only

3. Careful plans and actions to avoid risk
  a. science and technology not always good
  b. halt to further development of nuclear power
  c. development and use of soft technology
  d. government regulation to protect nature and humans
	3. Risk acceptable in order to maximize wealth
	  a. science and technology a great boon to humans
	  b. swift development of nuclear power
	  c. emphasis on hard technology
	  d. deemphasis on regulation—use of the market—individual responsibility for risk

4. Limits to growth
  a. resource shortages
  b. increased needs of an exploding population
  c. conservation
	4. No limits to growth
	  a. no resource shortages
	  b. no problem with population
	
	  c. production and consumption

5. Completely new society
  a. serious damage by humans to nature and themselves
  b. openness and participation
  c. emphasis on public goods
  d. cooperation
  e. simple lifestyles
  f. emphasis on worker satisfaction
	5. Present society okay
	  a. no serious damage to nature by humans
	  b. hierarchy and efficiency
	  c. emphasis on market
	  d. competition
	  e. complex and fast lifestyles
	  f. emphasis on jobs for economic needs

6. New politics
  a. consultation and participation
  b. emphasis on foresight and planning
  c. willingness to use direct action
  d. new party structure along a new axis
	6. Old politics
	  a. determination by experts
	  b. emphasis on market control
	  c. opposition to direct action—use of normal channels
	  d. left-right party axis—argument over ownership of means of production

Source: Adapted slightly from Milbrath, 1984, p. 22.

ences in specific beliefs that separate adherents to the two paradigms. In order to provide a handy referent, the contrasting beliefs are summarized in Table 6.1; the discussion follows the order in the table.

Valuation On Nature

Everyone values nature, of course, but some people mainly desire to exploit it for goods and services that can be consumed while others lovingly desire to preserve nature. Our study showed that most environmentalists value it for its own sake; many of them have an almost worshipful love for it. The defenders of the DSP believe we should emphasize using nature to produce material goods. One of our questions asked if people would prefer to live in a society that emphasizes "preserving nature for its own sake" as contrasted to a society that "uses nature to produce the goods we use." Those opposing views were separated by a seven-point scale; respondents could select any one of the points between the two poles. Public respondents in all three countries were fairly evenly distributed across all seven categories. This distribution shows that people were not experiencing great inner conflict between these two values; if they had been, there would have been high percentages choosing the middle categories. The data showed, rather, that people held widely differing beliefs about the proper relationship between humans and nature. Nearly all of the environmentalists believed that nature should be preserved for its own sake. Business leaders were spread quite broadly across the scale but, more than any other group, leaned toward using nature to produce material goods.

Economic Growth Versus Environmental Protection

Economic growth is highly valued in nearly all countries; it is a key belief in the DSP. Most people also value a safe and clean environment. Because the vigorous pursuit of one of these values may diminish realization of the other, it is important to know how people would trade them off when they select an emphasis for the society in which they would like to live. One of the items in our study asked respondents to choose on a seven-point scale whether they would prefer to live in a society that "emphasizes environmental protection *over* economic growth" or a society that "emphasizes economic growth *over* environmental protection." This item turned out to be one of the most powerful in our study. It taps a central belief and value difference between the vanguard proponents of the NEP and the rearguard defenders of the DSP. Table 6.2 reports the distributions of the various samples from each of the three countries on this item for both 1980 and 1982. The great deal of information in this table should be studied carefully.

Table 6.2
A Society that Emphasizes:
(Percentage in each response category)

	Environmental protection over economic growth								Economic growth over environmental protection						Economic growth over environmental protection	
	3		2		1		0		1		2		3		Mean*	Mean
	1980	1982	1980	1982	1980	1982	1980	1982	1980	1982	1980	1982	1980	1982	1980	1982
United States																
general public	26	21	18	21	18	17	19	20	8	9	5	7	6	5	2.99	3.17
environmentalists	52	45	24	27	8	12	12	12	1	1	1	3	2	1	1.97	2.12
business leaders	8	5	5	7	16	8	30	30	28	29	6	14	8	7	4.16	4.40
labor leaders	20	23	18	14	13	17	29	26	5	12	8	3	6	5	3.26	3.16
appointed officials	12	5	15	19	21	17	34	34	11	17	7	8	1	0	3.45	3.62
elected officials	5	13	19	17	23	17	32	36	12	9	5	4	4	4	3.58	3.40
media gatekeepers	11	–	15	–	24	–	29	–	17	–	1	–	4	–	3.48	–
England																
general public	29	28	19	24	18	11	22	16	6	8	2	6	4	7	2.80	2.95
conservation society (same as environmentalists in 1982)	66	57	25	27	3	9	5	4	1	2	0	1	0	0	1.48	1.70
nature conservationists	44	–	23	–	15	–	13	–	2	–	2	–	2	–	2.18	–
business leaders	13	10	15	18	20	17	37	35	10	12	3	6	3	3	3.38	3.49
labor leaders	28	–	14	–	19	–	26	–	5	–	4	–	4	–	2.95	–
public officials	13	14	19	19	25	22	27	29	11	8	3	6	3	1	3.23	3.22
Germany																
general public	38	31	13	13	9	11	18	13	6	8	5	12	11	12	2.99	3.38
environmentalists	56	61	20	14	5	5	6	5	4	3	3	7	6	5	2.16	2.14
business leaders	17	7	14	14	11	14	31	29	15	19	9	15	4	2	3.55	3.92
public officials	26	13	22	22	9	18	28	28	9	10	4	6	1	3	2.88	3.28

Source: Reprinted from Milbrath, 1984, p. 121.

*The scale shown at the top of the table was converted to a linear scale 1-7 running from left to right for computation of the means.

One would suppose from attending to the media and listening to political discourse that most people prefer economic growth over environmental protection. That is what the DSP tells them and the idea is constantly reinforced by leaders in business and government. The main summary message from Table 6.2, however, is that the public in the United States chose environmental protection over economic growth by a ratio of 3-to-1 in both 1980 and 1982. The public in Germany also selected environmental protection over economic growth by a ratio of 3-to-1 in 1980 and 2-to-1 in 1982; in England the ratio was nearly 5-to-1. These high ratios in favor of environmental protection were not simply an artifact of question wording or of the use of a mail questionnaire. A 7,010-respondent personal interview study, conducted nationwide by the Harris Poll Organization for the U. S. Department of Agriculture in the fall of 1979, used a question nearly identical to the one in the three-nation study; the Department of Agriculture study found a ratio in favor of environmental protection of 2-to-1.

A similar question was asked in nine European countries in the Eurobarometer study in 1982. In this study, respondents were asked if they would favor protection of the environment even if it caused companies to raise their prices. They also were asked if environmental protection should take priority even if it risks holding back economic growth. The percentages selecting protection of the environment as more important than prices and more important that economic growth are set forth in Table 6.3. Substantial majorities of the public in eight out of nine European countries preferred environmental protection to continued economic growth. Only in Ireland did less than a majority favor environmental protection in this trade-off. In

Table 6.3
Opinions on Environmental and Economic Trade-Offs

	Environment More Important than Prices	Environment More Important than Growth
Belgium	55%	60%
Britain	59	55
Denmark	81	81
France	66	62
Ireland	34	33
Italy	67	73
Luxembourg	72	67
Netherlands	74	59
West Germany	58	73
European Community	63	65

Source: Eurobarometer 18.

1984, the Gallup Poll put the identical question used in the Eurobarometer study to an American public sample; 61 percent chose environmental protection over economic growth, 28 percent chose economic growth over environmental protection, and 11 percent had no opinion.

The emphasis on environmental protection over economic growth also holds in some Pacific countries. In a public opinion survey conducted in Japan in 1982, respondents were asked to choose among three options: giving priority to the environment, giving priority to economic growth, or the belief that environmental conservation and economic growth are compatible. The data reported in Table 6.4 (from Watanuki, 1982) show that most people in Japan would like to have both economic growth and environmental conservation (our studies in the United States also show that people will choose both values if that option is offered). The table also shows that among those people choosing one of the values as having priority, two or three times as many chose environmental conservation as chose economic growth. The higher the education, the greater the tendency to choose environmental conservation.

Smaller-scale companion studies to our three-nation study were conducted in Australia and Taiwan. In Australia, environmental protection was chosen over economic growth by a ratio of 6-to-1; in Taiwan, the ratio was

Table 6.4
Environmental Conservation and Economic Growth

	Priority on Environment	Priority on Economic Growth	Compatibility of Environment Conservation and Economic Growth	DK, NA	Total
Average	28	11	41	20	100% (N = 2, 486)
Years of Age					
20-29	27	9	44	21	100% (N = 354)
30-39	27	9	48	16	100% (N = 613)
40-49	29	12	41	18	100% (N = 560)
50-59	32	14	38	16	100% (N = 447)
Older than 60	25	13	31	31	100% (N = 452)
Years of Education					
Fewer than 9	26	15	34	25	100% (N = 978)
9-12	28	9	45	17	100% (N = 978)
More than 12	32	8	48	12	100% (N = 343)

Source: Naikaku Soridaijin Kambo Kohoshitsu (Public Relations Section, Prime Minister's Chamber), *Kogai nikansuru Yoron Chosa* (Public Opinion Survey on Pollution), March 1982, p. 116. Reprinted from Watanuki, 1982

2.5-to-1.* The evidence from studies in thirteen nations suggests that our finding—that most people prefer environmental protection to economic growth— cannot be dismissed as a specious finding from a flawed study; rather, it is clear that widespread acceptance of a key belief in the NEP has already occurred in many countries. To my knowledge, these are the only countries in which the trade-off between economic growth and environmental protection has been posed for the public; it is significant that the public ranked economic growth above environmental protection in only one of the thirteen countries studied.

Returning to Table 6.2, we can see that the trade-off between environmental protection and economic growth much more strongly differentiates groups within countries than it shows differences across countries. Environmentalists in all three countries were clearly in the vanguard, being nearly unanimous in favor of environmental protection over economic growth. It might be supposed that business leaders would be nearly unanimous in the opposite direction, but that was not the case. Business leaders in all three countries, and in both waves of the study, tended to be undecided when they were requested to choose between these two values; high percentages took neutral, or near neutral positions. Labor leaders in England especially favored environmental protection, but labor leaders in the United States also leaned in that direction. A substantial number of the public officials took a neutral position but, overall, they favored environmental protection over economic growth. This trade-off seemed to be somewhat more divisive among the public in Germany than in England and the United States.

Even though the public favors environmental protection over economic growth, large differences between the rearguard and vanguard within countries signals that this will be an issue of strong political contention for many years to come.

Generalized Compassion

In Chapter 4, we identified compassion as a core value of a sustainable society. Compassion is a universal human trait; nearly everyone feels compassion for those near and dear. But, people differ in their readiness to extend compassion to those more remote from themselves. Our study showed that environmentalists, more than nonenvironmentalists, extended their compassion to other species, to people in remote communities and countries, and to future generations.

---

*Because these were collaborators, we have access to the raw data for these, as yet unpublished, studies.

The competitive market system of capitalism urges us to look out for ourselves first and to strive mightily to defeat our competitors; it also pressures people to produce and to consume at high rates and to give little thought to resources for the future. Most participants in our three-nation study did not strongly endorse those beliefs. For example, we asked if people would prefer to live in a "country that saves its resources to benefit future generations" or one "that uses its resources to benefit the present generation." In all three countries, a strong emphasis was placed on saving resources for future generations. It was stronger among environmentalists, but even business leaders mostly favored saving for future generations. (Milbrath, 1984, p. 122)

How Should We Handle Risk?

Modern industrial society encourages people to extract finite resources from the earth, to manufacture products, to consume them, and to discard wastes at each of the stages. Presumably they are making a better life for themselves. Unfortunately, we are learning the bitter lesson that as people pursue these activities, they injure the biosphere and place themselves, and other people (even future generations), in serious risk of physical injury. After repeated exposure to a variety of hazards, people are beginning to demand that their government adopt policies to protect them from risk and/or to help alleviate the losses that they suffer as a result of exposure to hazard or injury.

Nearly all activities involve some level of risk, but what is an appropriate and tolerable level? How much risk should we accept in a good society? In our three-nation study we asked several questions about risk. We discovered wide differences of opinion as to how much risk should be accepted. Environmentalists strongly emphasized that society should avoid physical risks in the production of wealth, whereas business leaders believed that accepting risks was necessary to make wealth. When we examined the data to try to discover why some people were risk acceptors and others were risk avoiders, a connection between a person's posture toward risk and that person's perception of what is happening in the world was apparent. Those wishing to avoid risk perceived exceptionally high danger from nuclear power, for example. Risk avoiders also had much less faith in science and technology and were, in fact, fearful that technology was running away with society. Risk avoiders also perceived higher levels of damage to nature from the actions of humans. The strong defenders of the DSP believed that science and technology had been a great boon to humans and they had a deep and firm faith that swift and continual technological development will continue to be good for humans.

People see two sides to technology; they like the good things that technology can do but they also perceive the risks to humans and the biosphere from the use of technology—Chernobyl and Bhopal being horrible examples. Most of the public has great faith in science and technology. Most environmentalists do not reject technology; they use it everyday. But they have greater reservations than most people about being caught up in and dependent on technology; they would proceed with more caution on new deployment of it. In the 1982 study, most of the American public (66 percent) strongly endorsed the policy that governmental regulation should be used to protect the environment; this policy was even more strongly endorsed by environmentalists (79 percent), but less so by business leaders (47 percent).

Environmentalists often propose "soft" or "appropriate" technologies. Compared to the large projects favored in modern society, these technologies are smaller in scale, less complicated to build, less difficult to control, less wasteful of resources, less expensive, less damaging to the biosphere, and, because they decentralize their operations, they are less likely to result in massive disruptive breakdowns. The development of solar energy as an alternative to nuclear energy illustrates the difference in perspective. Supporters of the DSP, in contrast, generally encourage such "hard tech" and "high tech" development as nuclear power. Our questions about whether we should use nuclear power showed some of the largest differences between DSP supporters and NEP supporters.

Limits To Growth

I extensively discussed in Chapter 2 why growth in population and economic activity cannot be unlimited. This argument has been heard repeatedly throughout the past two decades. Yet, the controversy rages in most countries as to whether there truly are "limits to growth." Former President Reagan, in his Second Inaugural Address in Juanuary 1985, declared that "there are no limits to growth." Probably each of you have your own answer to that question. The public's views on the question are important because they shape policy. If people anticipate that there are limits and make adjustments in advance, the consequences will be less severe than if they make no preparations and suddenly overshoot the carrying capacity of the ecosphere.

Most NEP advocates believe there are limits to growth, whereas most DSP defenders believe there are no limits. This is a crucial difference between the two, so we studied it carefully in our three-nation study. One of the items asked respondents the extent of their agreement or disagreement with this statement: "There are limits to growth beyond which our industrialized society cannot expand." Figure 6.1 presents the mean scores on that

Figure 6.1

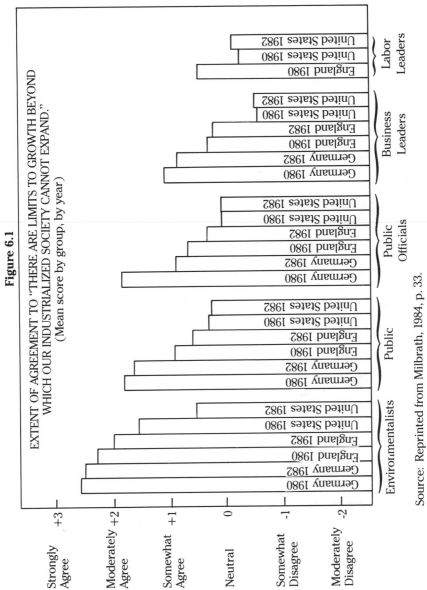

EXTENT OF AGREEMENT TO "THERE ARE LIMITS TO GROWTH BEYOND WHICH OUR INDUSTRIALIZED SOCIETY CANNOT EXPAND."
(Mean score by group, by year)

Source: Reprinted from Milbrath, 1984, p. 33.

item, sample by sample, for each of the three countries, for 1980 and 1982. It can be seen from the figure that the German people are much more accepting of the idea of limits to growth than people in the United States, about one-half of whom deny limits. The English generally were more accepting than the Americans, but less accepting than the Germans.

The U. S. respondents were widely distributed across the scale; this means that there was strong disagreement about this question in the United States. In Germany, however, 78 percent of the public were in the top three categories of agreement and 95 percent of the environmentalists were in those categories.* The reader can also see in Figure 6.1 that environmentalists were most likely to accept limits to growth and business leaders were most likely to deny limits, with the public falling in between.

Clearly, the people in these three countries are not ready to take concerted action to adjust to a world of limits. The idea simply has not been accepted at a sufficiently basic level, nor is it sufficiently widespread, to allow us to make appropriate preparations. If anything, many people in developed countries have moved even further away from seeing such a need during the intervening years since we made our study.

Do We Need A New Society?

Everyone would agree that plenty of things are wrong with modern industrial society. The disagreement is centered on whether those flaws are fundamental to the nature of that society or whether it is basically sound and only needs minor adjustment and technical development. As expected, DSP defenders take the latter view. Those who believe society needs fundamental change have not agreed upon a design for a superior society, but there are some common elements in their critique of the old society and in their beliefs about what would be better. The contrasts are sketched in Table 6.1.

Humans Seriously Damaging Nature?

Modern industrial society inflicts some damage on nature. DSP defenders tend to believe that damage is not serious and can be taken care of by technical fixes, whereas NEP advocates believe the damage to be so serious that we must drastically alter our way of doing things. Our study shows

---

*Figure 6.1 shows U. S. environmentalists with a lower mean in 1982 than in 1980; this lowering occurred because we deliberately added a higher ratio of "nature conservationists" to our 1982 environmentalist sample than were in the 1980 sample. We checked specific individuals and discovered they had not changed their views from 1980 to 1982.

that in all three countries the great proportion of the public (75 percent or more) agree with this statement: "Mankind is severely abusing the environment." This item correlated higher with an "environmentalism scale" than any other item in the study. Environmentalists tended to agree strongly whereas the responses of business leaders were spread widely across the scale; even among them, a majority agreed.

Openness And Participation

Among people supporting the DSP, it is widely believed that hierarchical structures, particularly in business firms, lead to efficiency as well as to speedy and wise decisions. The environmental vanguard objects to that way of making decisions because it so often has been used to exploit and injure nature and/or to dominate weak and poor people; they prefer citizen participation in decisions. An item in the three-nation study asked people to choose, on a seven-point scale, the extent of their preference for "a society which is willing to put up with some delay in order to let more people have a say in the big decisions" versus "a society which is willing to let a few people make the big decisions in order to get things done more quickly." In all three countries, a strong preference was expressed for letting more people have a say even if it causes delays; but there was a very strong emphasis on this by the environmentalists, in contrast to the business leaders, who were more likely to want a few people to make the big decisions.

Public Goods Versus The Market

It is a central tenet of the DSP that the supply and demand market can be relied upon to allocate resources so as to maximize the public good. DSP adherents typically advocate low taxes, very few public enterprises, very little regulation, no public planning. Many believe that this system will adequately provide public goods as well as private goods. (A public good is available to everyone; examples are public parks, clean air and water, a public road system.) A contrasting view insists that we must rely on foresight and planning in order to maximize public goods. Our study showed that people generally preferred to use foresight and planning to provide public goods but they were reluctant to give this task over to a central plan made by the government. This probably reflects a greater distrust of government than it does of planning. Once again, we found the familiar pattern that the environmentalist vanguard leaned distinctly in the direction of foresight and planning, as contrasted to the business leaders who distinctively leaned in favor of using the market to provide public goods.

Competition Versus Cooperation

Biological communities, and human-social communities are similar in the sense that both work best if there is a moderate proportion of both cooperation and competition. Just as humans compete with each other, as well as with other organisms in the biocommunity, plants and animals compete for energy, nutrients, territory, security—even love. But plants and animals also *cooperate* in a network of symbiotic relationships. Human communities also work best if restrained competition is balanced by cooperation; finding that balance usually is not easy.

People in modern society hold quite different views about what proportion of societal activity should be cooperative versus competitive; inevitably this becomes a question for political contention. DSP defenders assert that society benefits when people compete vigorously with one another. NEP challengers assert that society works best if people cooperate rather than compete. When we asked respondents which emphasis they preferred for society, environmentalists displayed a sweepingly strong preference for cooperation, in contrast to business leaders, who showed a nearly equally strong preference for competition. This was one of the most sharply divisive items in our study. Among the public in the United States, approximately twice as many preferred cooperation as preferred competition.

Lifestyles

Beliefs about the proper structure for society carry over to preferences for lifestyles. DSP adherents are likely to prefer complex and fast lifestyles; typically they consume more energy and are more destructive of the biosphere. (My favorite horrible example is summertime snowmobile racing on grassy meadows.) Modern day advertising urges people to "live life in the fast lane." NEP advocates urge simple lifestyles and more contemplative pleasures because they believe that exuberant lifestyles will injure the biosphere and diminish quality of life.

One of the items in our study asked if an effective long-range solution to environmental problems depends upon "changing our lifestyle" or "developing better technology"; these extremes were separated by a seven-point scale. Environmentalists strongly and nearly unanimously believed that an effective long-range solution would require a change in lifestyles, whereas business leaders equally strongly and unanimously favored developing better technology. This item also was highly correlated with the environmentalism scale. The U. S. public had no consensus; responses were widely distributed across the scale.

The Meaning of Work

DSP defenders, with their emphasis on economic values and material wealth, tend to view employment as a means to obtain material goods; they also believe in organizing work so as to maximize productive output. The challengers to the DSP argue that work ought to be satisfying in its own right; that fulfillment in the work setting is an important component of high quality of life. If necessary, maximizing productive output should be secondary to pursuing personal fulfillment.

An item in the three-nation study asked for the extent of preference people had for "a society which emphasizes work which is humanly satisfying" versus "a society where work is controlled mainly by economic needs." In all three countries, for both years, a strong preference was shown for a society that emphasizes work that is humanly satisfying. The by-now-familiar differences between elites appeared again: environmental leaders urged work which is humanly satisfying while business leaders' responses spread across the scale with a substantial proportion subordinating work to economic needs.

A New Politics

Persons who advocate a new belief and value structure for a new society inevitably urge a new politics. From the NEP perspective, politics should be consultative and participatory, whereas defenders of the DSP believe that many political decisions should mainly be determined by experts. Public respondents in our study overwhelmingly preferred a society that is consultative and participatory even if it slows down decisions and action. We also saw that NEP advocates believe that foresight and planning are essential to protection of the biosphere and to the provision of public goods, while the DSP defenders believe that market mechanisms are adequate for those tasks.

Our data show that most people in the environmental vanguard are willing to use direct action tactics such as demonstrations, protests, and boycotts, in order to get their message across to public decisionmakers. DSP defenders nearly unanimously, and quite strongly, are opposed to direct action. They believe that all participants should use normal channels for communication and persuasion (which the DSP adherents believe are adequate but NEP supporters believe are inaccessible to persons holding their point of view).

Politics, for many decades now, has been conceptualized as being arrayed along a left-right axis. The main dispute between political parties on the left and the right is over the ownership of the means of production and over who can best manage the economy. The advocates of the NEP feel that capitalism versus socialism should no longer be the major dispute in modern industrial democracies; many NEP supporters refuse to participate in that political argument. From their perspective, both left and right desire to dominate nature to maximize economic output; both have despoiled nature to pursue that end. They believe that it is much more useful to devote their political energy to finding a more satisfactory relationship between humans and nature. As they push strongly for a new and more satisfactory relationship between humans and nature, they are likely to be opposed by the rearguard defenders of the DSP. These clashing viewpoints have the *potential* to shift party contesting away from the old left-right axis to a new axis, where the main dispute is over the proper relationship between humans and nature. The motto of the German "Die Grunen" (green) party is: "We are neither left nor right, we are in front."

The Elmwood Institute: A Greenhouse for
New Ecological Visions
P.O. Box 5805, Berkeley, California 94705

The Elmwood Institute was founded to facilitate the cultural shift from a mechanistic and patriarchal worldview to a holistic and ecological view. Its purpose is to nurture new ecological visions–based on awareness of the fundamental interdependence of all phenomena and of the embeddedness of individuals and societies in the cyclical processes of nature–and to apply these visions to the solutions of current social, economic, environmental, and political problems.

It is comprised of a range of theorists as well as strategists and activists for social change who have in common an interest in those contemporary phenomena that indicate a cultural and political shift from a mechanistic and patriarchal world view to one informed by holistic, ecological, and postpatriarchal concepts and values.

We are often told that our institute is unique because we are not concerned with a specific issue or problem but rather with changing the worldview and value systems that underlie *all* our problems. To do so we organize small, informal gatherings of innovative thinkers and artists, policymakers and grass roots organizers. . . . At our Elmwood gatherings, unique dialogues take place, new ideas are conceived, and strategies for change are born.

The Institute sponsored a five-day symposium at Big Sur, California, November 29-December 4, 1985 on "Critical Questions about New Paradigm Thinking." A full issue of *ReVISION* for Summer/Fall 1986 (vol. 9, no. 1) is devoted to reporting those discussions.

Summary On Paradigm Contesting

This brief review of the differences between the DSP and the NEP, which is based on reasonably solid evidence—not mere speculation—has shown us several things that are useful for our inquiry:

1. A very different paradigm than the one that now dominates societal institutions and thought could become dominant. In fact, the public already has adopted many beliefs that would comprise a New Environmental Paradigm; as usual, the people are ahead of most of the leaders of our dominant institutions.

2. These beliefs and values form into structures. The better educated a person is, and the more active, the more tightly the beliefs and values are organized into an internally consistent structure. Our evidence is that those adhering to the NEP tend to subscribe to most of the beliefs on the left side of Table 6.1; while those adhering to the DSP likewise subscribe to most of the beliefs on the right side. We also find highly significant correlations among these belief items.

3. These belief structures are highly contrastive; if each were a dominant paradigm, but in different countries, they would create quite distinctive societies that would have very different impacts on the biosphere. We really are seeing the possibility for a new society.

4. These disagreements between the DSP and the NEP are so fundamental, and the positions of the vanguard and rearguard are held so fervently, that we can expect sharp political contesting on these matters for several decades. The contestants are struggling not simply to see who can win elections but, more fundamentally, to see which paradigm will shape and dominate social learning.

5. Even though the DSP is currently dominant, giving the rearguard the upper hand in the contest, a sufficient number of people doubt the wisdom of many DSP beliefs to make it reasonable to suppose that the DSP could be supplanted.

Had our minds not been conditioned by reiteration of the dominating themes of the DSP in public discourse, we would not have been surprised at the fairly widespread acceptance of NEP ideas. Most of them are a reaffirmation of traditional values (love of nature, living frugally, love of animals and plants, compassion for other people, cooperation). These ideas and values also are similar to the central beliefs of the major religions of the world. It is only within the past seventy years or so that DSP values (wealth, eager consumption, economic growth, big projects, domination of nature, competition) took such a dominating position in our national belief structure. Many people have not abandoned traditional NEP beliefs and values despite the inroads of the DSP; thus, there is a mental readiness for NEP ideas that could ease the transition from the DSP to the NEP.

The three-nation study showed that about one-half of the American public (as in England and Germany) have not clearly worked out a logically consistent belief structure; many hold beliefs and values that come from both paradigms. They love nature but they also want to use nature to produce goods. They see humans interrelated with nature and wish to avoid blind domination of nature, but they also have widely adopted technology to exploit nature. They want economic growth but they want environmental protection even more.

The emerging "new politics" is likely to be a battle between NEP and DSP adherents for this undecided middle. In the present "old politics," *the policies pursued by both American parties are those of the DSP*. It would be foolhardy to recommend that a political party immediately adopt a whole new program based forthrightly on the NEP; people are not likely to abruptly change the way they perceive and think. A great deal of social learning has to take place before that would be feasible (the paradigm itself needs further development and articulation). But the learning process could be significantly advanced if the leadership of one or more parties began to look at the world through new lenses and to adopt new language in public discourse.

Left-Right And Ecological Thinking

Chapter 7

Please Do Not Label Me!

How does the ecological way of thinking fit with modern political ideologies? This is an important question because we all try to put labels on things in order to think about them. Most of us work with a system of categories provided by society. We elaborate it and modify it occasionally, but the basic structure persists for most of our learning life. We find thinking by categorizing so convenient that we often use it as a shortcut. When we encounter some new concept, we search for an appropriate category in our classificatory system, apply the label, and believe that we now understand it. We may, actually, grossly misunderstand it.

Misunderstanding by mislabelling is an especially vexing problem in thinking about ideologies. Most of us have been taught that the contemporary world has two competing ideological systems: socialism and capitalism. The pattern of seating in the French Parliament places the Socialists on the left side of the chamber and the Capitalists on the right; from this seating pattern we have derived the popular labels "leftist" and "rightist"; "centrists" occupy the middle. These labels have now been applied more or less indiscriminately to political parties, politicians, and ideologies around the world. Journalists, political commentators, politicians, and ordinary people seem to have an uncontrollable urge to apply left-right labels to every new political concept, every belief system, that they encounter. Even if one does not want to be labelled, and refuses to label oneself, the labels are likely to be applied—leading people to misapprehend.

Ecological thinking is like an ideology in that it comprises beliefs about the way the world works and contains values and beliefs about the way society ought to work. Paradigms could be called ideologies. An ecological ideology doesn't fit very well on the left-right dimension, however—as discussed in the previous chapter.

Interestingly, the ecological perspective enhances our insight into the ways that "main-line ideologies" (socialism and capitalism) are similar and different. We will be looking at the unspoken assumptions of these ideologies as well as their spoken assertions. These characteristics are first sketched in tabular form, followed by a discussion of the meaning of the

similarities and differences. The discussion will show how useless and misleading it is to apply left and right labels to this new ecological ideology/
paradigm. We can build on that discussion to decide on some desirable and
not so desirable characteristics that are relevant for the new society we are
trying to develop.

Capitalism And Socialism Compared

Over the past century, the human family has had a good deal of experience with capitalist and socialist systems; this experience provides us with
empirical evidence for judging how they work in practice. My analysis
addresses the practice; I do not address the several versions of the theories. Defenders of these ideologies often claim that the systems found in
the United States or the Soviet Union, the two main protagonists in contemporary ideological conflict, are not pure expressions of their ideological
theories; therefore, we should not judge the theories by the poor way they
are put into practice. Granted, the theories have not been purely expressed
in practice; maybe the people living in those systems discovered that a
pure expression of either ideology does not work well in practice.

For this analysis, I developed two tables to compare the belief structures
lying behind the two ideologies, as practiced. The first shows the belief
similarities between socialism and capitalism, and the second shows their
differences. The tables focus on society and economy and not on political
practices such as the party system or human rights protection found in the
two societies (which really are not integral to the basic theories of the two
systems). The reader will soon see that many more similarities than differences are found. The reader should compare these tables with the DSP
versus NEP contrast set forth in Table 6.1 (page 119).

Similarities Between Capitalism And Socialism

Both capitalists and socialists believe that humans should dominate
nature. They perceive nature as a resource base to be exploited for the
welfare and comfort of humans. If anything, Marxian socialists have been
even more exploitive of nature than capitalists. Marx taught that economics and the economic structure of a society shape everything about a society; capitalists also tend to believe this. The great writers setting forth the
theories for each of these systems did not anticipate in their analyses that
humans would one day encounter a scarcity of natural resources. Even
today, in both kinds of societies, most of the people perceive few resource
limits. The leaders in both kinds of societies have generally perceived that

Table 7.1
Similarities between Capitalism and Socialism

1. Humans should dominate nature
   a. resourcism—no perception of resource limits
   b. perception that nature is not seriously damaged

2. The economy shapes society and is the basis of power
   a. maximization of productivity and wealth should be the dominant object of public policy.
   b. economic growth more important than environmental protection
   c. risks to nature are acceptable in order to maximize wealth
   d. orient jobs for economic need, not worker satisfaction

3. Growth is good and limitless

4. Power is needed
   a. to maintain control—distrust of public participation or direct action
   b. to protect the nation—perceived threats and conflicts
   c. to win recruits to the ideology
   d. emphasis on hierarchy and efficiency to maximize power
   e. large projects are efficient and demonstrate power

5. Science and technology are good and should be emphasized
   a. worshipful love of science and technology
   b. deference to experts—emphasize their development

6. Missionary-like need to convert others—doctrinaire

7. Compassion focused on in-group—resist extending compassion to other species, other people, future generations

Table 7.2
Differences between Socialism and Capitalism

Socialism		Capitalism
1. public ownership	—means of production—	private ownership
2. utilize planning	—coordinate economic acts—	utilize market
3. emphasize cooperation		emphasize competition
4. collective goods		individual goods
5. emphasize equality		emphasize individual differences

damage to nature from human activities has been minor and readily correctable by technological fixes.

The orientation toward domination of nature in both societies finds expression in a policy to maximize economic productivity and wealth. As a matter of fact, the two systems compete to see which one can produce the most in the shortest span of time—maximize throughput. Each system has shown that its leaders believe economic growth to be more important than environmental protection. Each system is willing to risk damage to nature in order to maximize wealth. While leaders in each system believe that workers should derive satisfaction from their jobs, they place higher priority on structuring jobs to serve overall economic needs. In both societies, growth is an unquestioned good—necessary to maximizing wealth and power. Leaders in both societies deny that there are limits to growth.

Each set of leaders claims that it is necessary to maximize power to defend the system against its enemies—each perceives monstrous forces arrayed against it. Each uses this accretion of power to maintain internal control of the system it leads. Neither set of leaders desires widespread citizen participation in public decisionmaking. Direct citizen action to make demands on the system (demonstrations and other organized protests) is unwelcome in both societies. Each system uses the power it possesses to try to win recruits to its ideological perspective. Each system emphasizes hierarchical organization and efficiency to maximize power. Each system builds large projects not only because they are perceived to be efficient but also to demonstrate power and technological proficiency. Such systems also are easier to control.

Science and technology are the major means that each society uses to maximize power and wealth. Each displays a virtual worship of science and technology. Each sets aside large appropriations of public funds to develop them further. Strong emphasis is placed on developing expertise, and great deference is shown to experts. Each society has developed a technocracy who make many of the important societal decisions.

If the leaders and practitioners of an ideological system perceive that their system is threatened, they not only will try to maximize power but also will seek new recruits. Each side in this struggle has a missionary-like zeal to convert others to its ideology. Both the Soviet Union and the United States are competing to win Third World recruits. They invest their ideological theories with such powerful emotional commitment that those who believe they have the "true faith" chastise others for not remaining pure. Deviants may be expelled from the group or removed from a position of power.

Ideological struggle divides people and nations into in-groups and out-groups, friends and enemies. Love and compassion is extended to the in-group; animosity and hatred is directed toward the enemy. In neither socialism nor capitalism is much compassion shown for people in other lands, for future generations, or for other species. Both systems are inwardly oriented toward the material welfare of the true believers in the system.

Differences Between Socialism And Capitalism

The most doctrinaire difference between the two systems, the one that is purported to make *all* the difference, is over who should own the means of production. Capitalists believe the means of production should be privately owned, while the socialists believe it should be publicly owned. Each says that its system of ownership will maximize productivity and wealth (and power).

Another crucial difference is that capitalists would, nearly exclusively, use competition in the market place to coordinate economic activities while socialists believe that governmental planning and cooperation are best for coordinating economic activity. Socialists emphasize the collective production and consumption of goods while capitalism is oriented toward production of private goods to be enjoyed privately.

Capitalists believe that rewarding people differentially in a market is the best way to enlist their energy and maximize productivity. Socialists believe that people should be treated equally; at least that equality should be emphasized as a value. Proponents of each system believe their way not only is best but is the *just* way to reward people. The capitalist system leads to great individual differences in wealth and power, especially if wealth can be inherited from generation to generation. Socialists believe that great differences in wealth do not belong in a good society.

This discussion shows that there are many similarities between the socialist and capitalist systems; perhaps more similarities than differences. Each system attempts to maximize productivity and wealth, and they often are compared on that criterion; but is productivity the ultimate value on which they should be judged? Our earlier analysis of values showed that productivity is a means and not an end; furthermore, that a society becomes severely distorted if emphasis is placed on maximizing throughput. Judging the success of the two systems on the basis of productivity misses the point.

It would be more relevant to judge the two societies on their ability to provide high quality of life for their people. Remember, I cautioned that quality of life was an acceptable goal only as long as people structured their society so that it can live in a harmonious and long-term sustainable relationship with nature. On that criterion both systems are performing badly. The leadership in both societies, and their ideologies, do not recognize the importance of ecosystem viability and their ecosystems are swiftly deteriorating.

For the reasons just discussed, labelling ecological thinking as either left or right, or even centrist, makes no sense. The NEP is a challenge to both socialism and capitalism because both encourage ways of life that are destructive of ecosystems and are, over the long run, unsustainable.

What About Ecological Thinking And Democracy?

We often proclaim that we believe in democracy or extol democratic values. Some people think of democracy as an ideology and in certain respects it is. but *ideologies* are usually defined as a set of moral beliefs structuring: 1) the proper relationships between humans, 2) the proper design of social institutions, 3) the proper relationships between individuals and society, and 4) the proper relationships of all of these to the ecosystem that sustains them. Placed against that definition, democracy does not qualify as a full-blown ideology; rather it is a mode for decisionmaking that can be used in capitalist societies, socialist societies, and ecologically sensitive societies.

The essence of democracy is the provision of some regularized societal procedure for consulting the people about the policies and the future direction their society should take. This consultation should control future policy (an election is an example) rather than be merely advisory. Typically, democrats also believe that each person's views should count equally in this consultation. If the consultation is to be meaningful, the people must have access to relevant information and must be able to speak their views freely, without fear of retribution. That is why people in a democracy typically adopt a constitution and a bill of rights.

Historical experience shows that democracy is extremely difficult to implement in practice. Even though the United States has experimented with democracy for 200 years, critics can readily point out many instances where practice departs from democratic norms. Out of about 160 nations in the world today, only approximately twenty-five are real practicing democracies—although many more claim to be democracies. Maintaining democracy requires confidence in the fairness of the system so that leaders who lose an election willingly relinquish power, knowing that they will not be killed or gravely injured for losing, and knowing that their side will have a chance in the future to regain power.

Many people erroneously believe that communist societies are inherently undemocratic—that a necessary linkage exists between communism and dictatorship. It is probably true that those societies we know as communist dictatorships (those under the influence of the Soviet Union and Communist China) use dictatorial methods to preserve their version of doctrinarily pure socialism which might be eroded if people were allowed to freely choose their system. In essence, the leaders of those societies believe they know better than the people themselves what is good for their people and nation. But that is not the whole story. In the recent past, socialists and communists have been elected to power in several places and have not established dictatorships; they also have relinquished power when they lost elections. At the time I write, François Mitterand, an avowed socialist, is the president of France, a country where much of the economy is in private capitalist control. Everyone in France expects that Mitterand will relinquish power if he is defeated in an election. The democratic monarchies of Scandinavia are often labelled as *democratic socialism*, although their economies are actually a mixture of socialism and capitalism. When socialist governments lose an election in those countries, they relinquish power.

Ecological political parties (often called *green* parties) are a new political phenomenon; none has held significant political power, as yet. Grassroots democracy is one of the central pillars of all known green party programs. In fact, the entire environmental movement has a grassroots origin and is largely sustained by grassroots efforts. The green parties are so democratic in orientation that they may weaken their ability to compete with other parties. They so strongly desire to avoid oligarchical leadership that many party theorists have called for rotation in parliamentary office (an elected member would serve a one-half term and then rotate out to allow another member to fill out the term). The German Die Grunen tried this policy for a while but is abandoning it as too weakening of parliamentary influence.

How Do These Ideological Constructs Work In Practice?

What works and what does not work in the ideologies we have examined? This evaluation is based on my reading of history and contemporary practice, as well as on personal experience living in both socialist and capitalist societies. Ultimately, however, such an evaluation must be a personal statement. These experiences have shaped my thinking and my readers are entitled to know my position.

Control Or Ownership?

Ownership makes much less difference than we normally think. What matters is control, and control is only loosely related to ownership. Some examples: At my university, the building that housed the Environmental Studies Center is owned by the state. It contains offices and dormitory rooms; there also are some public lounges. The lounges have been severely vandalized; they are everybody's property and nobody's responsibility. There was an open area outside the suite of offices occupied by The Environmental Studies Center where we had a conference table and a lounge; we had green plants, nonvandalized furniture, and one of the most pleasant indoor settings on campus. We were able to make the area pleasant because we could control access; we knew that any improvement we made would be enjoyed for years. Many people used the area during working hours, but we were able to secure the area at night to keep out vandals or thieves. It was our control, not our ownership, that made the difference. Later, the university closed the center and evicted us from the space. The administrator who ordered this did not own the space but was given control as part of a hierarchical governance structure imposed by the state.

In China, all land is owned by the state. For the first thirty years under communist rule, the state mandated that all land was to be cultivated by agricultural communes where all work and all production was shared. Everyone was equal, everyone had ownership, but no individual had control. Both productivity and maintenance of ecosystem viability lagged. In the early 1980s, the top leadership in China realized that the communes were not working well and abolished them. Peasant families were each given control of a plot of land, even though ownership was retained by the state; they were to deliver a quota of their produce to the state, but could sell any excess that they produced on the free market. Food production has risen markedly. China had 60,000 free markets when I visited there in 1985; the leadership was so pleased with this reform that they planned to have 90,000 by the end of 1986. Giving people some control is an important stimulus to productivity and preservation. Land control by peasants is not complete in China; a peasant who fails to grow crops or who abuses the land can have it taken away and given to someone else, but he now has enough control to make it worth his while to care for the land and to work hard.

In Norway, ownership of land lies with the farmer, but he does not have full control of it (as in the United States). Enjoyment of forests, lakes, and streams is open to the public. A farmer can exclude hikers, skiers, berry pickers and fishermen from his cultivated land but not from pastures, forests, and water bodies. If a farmer wishes to sell his land for development, he must obtain permission from the Ministry of Agriculture. If it is prime agricultural land, permission is likely to be denied. Land is held in stewardship or trust; ultimately it belongs to all the people.

In the United States, the owner of a farm, a house, or a car also has control of it; it can be sold or used as the owner chooses (with some few restrictions). But who controls the giant corporations which dominate so much economic activity in the United States at present? The corporations are owned by the stockholders, but they have no effective control of the firm (except in rare instances where a single stockholder has majority control). Control really lies with the board of directors and the Chief officers; in most instances, these are self-perpetuating oligarchies. The only choice most stockholders can exercise is to sell or to retain their stock.

A poignant story, recently portrayed on television, illustrates that control can sometimes be used destructively. The Pacific Lumber Co., a family firm in northern California had, for many decades, owned 189,000 acres of forested land that contained many prize redwood trees. They harvested trees from this tract but in such a way that the forest's yield was sustainable; they did not clear cut. They felt like stewards of this beautiful natural bounty and were model corporate members of their community. They made the error of selling stock in the firm to the public. They thought they had retained enough shares to secure their control of the business but, in 1985, the firm became the target of an unfriendly takeover. An "investment vehicle," the Maxxam Group, used "junk bond" financing to obtain majority control. They had incurred a large debt to obtain control and looted the firm in order to pay it off—as well as make a handsome profit. They displaced the management, clear-cut the forest, and created a moonscape out of a beautiful sustainable-yield forest. The new owner said he believed in the Golden Rule: "Those who have the gold, get to rule." Even though legal, this was such an immoral deal that both the U. S. Congress and the California legislature are investigating (Anderberg, 1988).

In thinking about ownership and control, reintroducing the distinction between private goods and public goods is useful. A public good is one that can be enjoyed by everyone—a public park, for example. A private good is one that belongs to its owner; others can be excluded from its consumption. Every community has a combination of private and public goods. Some goods are by their nature private (a toothbrush, underwear, a prescription drug, etc.). Many goods can be either private or public (vehicles, books, parks, a communication system). Some goods are inherently public (clean air, a public utility). Every society, whether capitalistic or socialistic, makes distinctions between private and public property. Socialist societies lean heavily toward making goods public. The especially attractive places in the Soviet Union and China, for example, are their public parks, theaters, public buildings, stadiums, and so forth. Capitalist societies lean heavily toward private ownership and control. In the United States, we even allow and encourage private ownership of public utilities, but government keeps control of their actions.

In summary, with respect to ownership and control:

1. No society has public ownership of everything, or private ownership of everything; there is always a mixture.

2. Finding the best mixture of private and public ownership for a society requires long experimentation; the people should be consulted frequently in deciding what works and what does not. Doctrinaire socialism or capitalism interferes with free thought and discussion in making these determinations. My experience with "the middle way" in Scandinavia leads me to believe their mixture is better.

3. Ownership is not closely linked to control. Control is much more important than ownership. We miss the point when we focus on ownership.

4. In deciding what should be kept under public control we should give top priority to ecosystem viability (land, for example, is far too basic to allow private individuals to waste, squander or ruin it). Second priority should be given to societal needs; public access to public goods generally places less stress on a system than private control–a public library rather than each person buying his own books. Only when these two priorities are met is it all right to allow private control of goods.

5. When a society goes too far in allowing private ownership and control, it ruins public goods as well as the enjoyment of private goods. Most observers would agree that the number of private automobiles in Paris has ruined much of what was especially delightful about that city. The freedom that automobiles are supposed to give us is spoiled when we are caught in a traffic jam or cannot find a place to park. Air pollution from automobiles that causes acid rain and injures our health should not be tolerated in any society. Greater use of public transportation may be the only way to restore convenience and enjoyment to urban travel.

Markets Versus Planning

Every society, even the most communistic, has markets; and every society utilizes planning, even the most capitalistic. Ideological contesting has lead us to believe that markets and planning are alternative systems. Actually, they can work side by side within the same system. Even though they may, in certain respects, substitute for each other, they really are complementary; markets inform planning and planning informs markets.

Markets are remarkably effective for providing consumers with a choice of goods and informing producers what consumers want. Occasionally, a society has deliberately tried to suppress markets; such attempts nearly always fail. Black markets quickly arise out of the normal interactions of people. Communistic societies, such as China, have had to introduce free markets to stimulate and direct production. China's current official policy is that markets are not incompatible with socialism.

But markets cannot coordinate and provide everything a society needs. Markets are very poor in providing foresight to deal with future problems. For example, even though society generally recognizes that oil is scarce and that reserves probably will be depleted in no more than a century, the current supply is sufficient to force down the price in order to stimulate greater consumption right now. Markets typically fail to recognize finitude of supply if the current supply is plentiful and cheap. Markets cannot deal responsibly with nature; they have perpetually undervalued the earth. This failure imposes on future generations the costs we blindly refuse to pay. Only collective political decisions can provide focus and direction for the myopic vision of markets.

Markets, with their emphasis on unfettered wealth accumulation, also are power systems. Thought can be purchased to serve private interests. Authors learn that they must please those with money in order to get a platform to expound their message. Wealth can buy a disproportionate say in the political arena by wafting a distorted image of mainstream thought around public decisionmakers. Money is translated into power by choosing the choosers and by shaping the understanding that shapes choices. A society that gives the market free reign also gives power to those who control resources and therefore over opportunities and rewards that shape the other dimensions of cultural life. Money talks not so much by one group exploiting another but by shaping the selective listening of the system that determines which groups will get heard. The power of money injures democracy but it does not kill it. If the market is worth maintaining for other reasons, its role in politics can be restrained by political means. (I am indebted to Schmookler, 1984, pp. 306-18 for some of the thoughts in these two paragraphs.)

A market is not sensitive to what economists call *externalities*. An industrial plant that dumps its pollution on its neighbors and its environment is taking an action that is external to the market; it is a cost that is not calculated in the market price and it helps the plant to compete more effectively. Only government regulation can force the market to include that cost. Clean air is a public good that markets cannot provide.

Markets are also poor substitutes for planning. A community that allows all decisions about land use and development to be determined by the market ends up with a helter-skelter land use pattern that is not pleasing to live with and not functional. Every society has found that it must regulate land use to some extent. Markets are especially weak in providing public goods (nature preserves, parks, stadiums, symphony orchestras, libraries, and so forth). A society that leans heavily on markets is likely to end up with public squalor amidst private affluence.

Planning is ubiquitous in all societies: families plan, enterprises plan, organizations plan, communities plan, governments plan. It is ironic and inconsistent for U. S. business leaders, who prominently use planning in their own corporations, to be so strongly opposed to governmental planning. They probably fear interference more than planning. Planning often fails when we ask too much of it. A planning group is not capable of coordinating all the intricate elements of a modern industrial society. Planners do not possess sufficient information and would not have the capacity to process and utilize it effectively if they did possess it. Planners must presume that they know what people want; very often their presuppositions are wrong. A market could provide the supplementary information the planners lack.

My university recently built and occupied a new campus, which has taken about twenty years. When the project started, the State of New York decided to build a new town adjacent to the campus to provide housing and other services to the university community. The state established a public corporation to plan the town, acquire and develop the land, and see the new community into existence.

The original plan was imaginative. Green space was reserved and other recreational facilities were developed. A mixture of housing would be provided for low-, medium- and high-income families. A special section would be devoted to the elderly. Industry would be placed in an industrial park. A town center would provide shopping and other community services.

What actually happened is quite instructive for our understanding of markets and planning. The plan could not be realized as conceived; market decisions by developers and potential owners reshaped it. The original plan did strongly influence the development pattern, but was far from determinative. The town commercial center never developed. After the first subsidized housing for the poor was built, no further housing of that type was begun. The industrial park is thriving and recently had to be doubled in size. Campus-like office complexes, not in the original plan, have sprouted like mushrooms. Most of the housing has turned out to be single-family detached houses—similar to that found in most American suburbs. Architectural controls still apply to all structures and are hotly debated in meetings of owners. A new town is being built, it has many attractive features, but it is not the town that either the plan or the market, acting alone, would have produced.

We should think of markets and plans as informational services, not as dominating systems. Each provides unique information that every society needs to function well. We should not expect either one to do everything. Each society must work out for itself the particular balance of markets and planning that is most suitable to its circumstances. We will be more open minded and sensible in making these choices if we stop treating planning and markets as ideological norms to be fought over.

Cooperation And Competition

It is obvious to those who observe nature that it is a setting for fierce competition in a contest of "survival of the fittest." That aspect of nature is often used to justify fierce competition in human society; some even argue that competitive vanquishing of a foe is morally correct. Persons making that argument misread both nature and society. Cooperation in symbiotic relationships is just as much a part of nature as competition. Creatures that do not even have a brain learn to cooperate in order to survive.

Competition in which the strong wins has been given a good deal more press than cooperation. But certain superficially weak organisms have survived in the long run by being part of collectives, while the so-called strong ones, never learning the trick of cooperation, have been dumped onto the scrap heap of evolutionary extinction.

If symbiosis is as prevalent and important in the history of life as it seems to be, we must rethink biology from the beginning. Life on earth is not really a game in which some organisms beat others and win. (Margulis and Sagan, 1986, p. 124)

Every society has a mixture of competition and cooperation; a society that emphasizes one to the exclusion of the other probably would not be good. Just as humans are naturally competitive, they also are naturally cooperative.

Competition produces positive benefits for humans when it provides alternatives in the market place and when it eliminates producers who fail to be efficient or who make goods that people do not want. Competition also keeps people "on their toes" and brings out their best talents and efforts. But how do we keep competition from going too far? Competition is destructive when it results in violence or deep injury to the psychological well-being of people. It is destructive when it gives people with ordinary talents a low sense of worth. It is destructive when it shuts down a plant, putting thousands of employees out of work and sending a community into economic decline. It is destructive when a plant externalizes its pollution as it tries to cut costs and remain competitive. It is especially destructive when it results in war or degradation of the ecosphere.

Cooperation has numerous benefits. People can cooperatively complete many tasks that individuals could not hope to complete. All organizations, all firms, all governments depend on internal cooperation to function—even if they externally compete with other organizations or firms. A cooperative atmosphere makes all participants feel useful, even those with lesser talents; it nourishes individual psyches. In a nationwide survey, Americans preferred a society that emphasizes cooperation over one that emphasizes competition by a ratio of 2-to-1. (Milbrath, 1984, p. 38)

History has shown that, when societies try to organize on a strictly cooperative basis, they have difficulty succeeding. Such communities typically emphasize equality, expecting everyone to share in the work and the benefits. Alas, some people reap the benefits but fail to do their share of the work; naturally, the workers perceive that as unjust. People are different, and society needs to take due recognition of those differences. Just as I would not like a society that was totally competitively oriented, I would not like one that was totally cooperatively oriented; finding a balance is better than emphasizing an extreme.

Equality

Equality was just mentioned; it is a value that is prominent in the ideologies of both the Soviet Union and the United States. Chapter 4 showed that we value equality when we perceive it as just, but not when it leads to injustice. If we equally reward unequal contributions, it will probably be perceived as unjust. A society's emphasis on equality needs to be tempered by a recognition of individual differences. On the other side, a strong societal emphasis on individual differences also is likely to be perceived as unjust. When a young person who happens to be born beautiful can attract 100 times the income of one born plain, society has acted unjustly.

Freedom

Many people superficially identify freedom with capitalism and lack of freedom with socialism and communism. As I explain in Chapter 4, no society can allow persons to do anything they wish—all societies restrain freedom in several respects. The extent of freedom in a society is determined most strongly by the circumstances in which it exists. Freedom is much more restrained by overcrowding and shortages than it is by ideology. There is as much political freedom in semisocialistic Scandinavia as there is in capitalistic North America.

Summary

As I set forth my vision for a sustainable society in the following chapters I urge the reader to remember the following:

1. Do not apply an ideological label to my thinking; it does not fit on the left-right dimension.

2. Capitalism and socialism have more similarities than differences. They both seek to dominate nature, to maximize productivity and wealth, to promote growth, to promote science and technology, to maximize power, to convert others to their ideology.

3. Ownership is less important than control. Locating control at the right place, so that societal values are served, requires careful societal learning.

4. Markets and planning are informational systems; we need both of them to make wise decisions about our society. We should be aware of their strengths and weaknesses as we carefully employ them.

5. People are naturally cooperative, as they also are naturally competitive. We should avoid emphasizing either one in our society.

6. Equality is an important value in a society, but the good society also recognizes individual differences. We must learn how to keep them in balance.

7. Political freedoms are just as compatible with socialism as with capitalism. No society provides total freedom; any attempt to do so would be destructive to both ecosystems and social systems.

8. Being open minded in a learning mode is the posture most likely to lead us to a good society.

Summary Of Part I

Summary Of Concepts

A clearer vision of a sustainable society will emerge in Part II if I take a moment to review the major points made so far in this book:

1. Compared to most other species, humans have lived on planet earth for a very brief time, (only 11 minutes of our year-long movie). During most of that time humans lived in harmony with nature; their home was that environment in which they evolved. It is only very recently that our species created an unnatural home for itself as it set out to dominate nature. In that brief period (only two seconds of our year-long movie), we have built a civilization that cannot sustain itself.

2. It is normal in a civilization for theory to lead practice. Within the last century that has changed. Growth in human population, technology, and economic activity is so swift that practice develops without theory. We adopt new technologies and practices with little or no consideration of their impact on our social structures, our behavior, our values, and our quality of life. We need to view technological development as "legislation" that is every bit as governing of societal (and ecosystem) functioning as is political/legal governing. This signals that, not only must we develop more adequate theories about the way the world works, but we also must try to recapture control of our lives by scrutinizing and reining-in the pell-mell dash of technological development.

3. All growth in population and in resource consumption is exponential. Growth in numbers of humans, in their use of resources, and in the wastes they discard into the biosphere, is so steep that it cannot be sustained, We have no choice but to change. If we thought the matter through, we really would not wish to grow. Growth is not an operative value to hold for our society.

4. Inherent weaknesses of markets make it impossible for us to depend on them to allocate resources so that a society can become sustainable; government must play a critical role in achieving sustainability.

5. The parable of the tribes shows us that power has become the principal basis for selection in cultural evolution when there is anarchy among societies.

6. The pursuit of power in the dominator society devastates the environment and deflects humans from their most sublime aspirations.

7. The pursuit of power has enthroned and enforced male domination over women. Patriarchal dominator structures are ubiquitous in contemporary societies, but partnership societies have flourished in the past and could do so again.

8. Becoming aware of the way the dominator society enslaves us is the first step to winning our freedom. A dominator culture is maladaptive and cannot be sustained.

9. Men need to learn the basic insights that feminism offers into the corrosive effects of power and dominance.

10. The basic value (ultimate value) for society, any society, is life in a viable ecosystem.

11. In order to lead a high quality of life, our first priority should be to maintain a viable and flourishing ecosystem. Maintaining a viable and flourishing social system is our next priority. This social system would uphold three other core values: security, compassion, and justice. Only as these priorities are maintained is it possible for a society to encourage individuals to pursue quality of life as they see fit.

12. Sanctity of life cannot be an operative value in a society; carrying capacity is a more meaningful value.

13. We need to become more aware of our ways of knowing. This will disclose that: 1) society plays a key role in knowing, valuing, and believing; 2) beliefs are crucial for shaping values and values are crucial for shaping beliefs; 3) we come to know values in the same way we come to know facts; 4) it is erroneous to believe that values are necessarily subjective; we can reason together to arrive at our social values in the same way we reason together to arrive at an understanding of how the world works physically; in that sense, values can be just as objective as facts; 5) we delude ourselves if we believe that science is value-free; 6) we must learn to reason together about our values and use the resulting understanding to control the speed and direction of scientific and technological development.

14. We must learn that ecosystems are so complicated and interconnected that almost no action is isolated. Our motto should be: "We
can never do merely one thing." We should continually ask "and
then what?"

15. We must learn to think holistically, systemically, and integratively;
we should beware of over-valuing specialization and cause-effect
thinking.

16. Our lifestyles should seek harmony with nature rather than domination of it; we should strive to live lightly on the earth. Carrying capacity must be protected, which means we must limit growth in human
population.

17. Learning to empathize with, and extend our compassion to, people
in other lands, to other species, and to future generations is essential to preservation of the integrity of the ecosphere and to the survival of us all.

18. Social learning will give humans some hope that change can be guided
to encourage the flourishing of the ecosystem within which humans
can enjoy quality in living. We need deliberately to design a society
that encourages social learning by: 1) providing a wealth of information; 2) finding better ways to disseminate and use that information
3) emphasizing integrative and probabilistic thinking; 4) emphasizing values as much as facts; 5) being critical of science and technology; 6) combining theory with practice; 7) being deliberatively
anticipatory; 8) believing that change is possible; 9) examining outcomes to learn from them; 10) developing institutions to foster systemic and futures thinking; 11) reorienting education and research
toward social learning; 12) maintaining openness and encouraging
citizen participation.

19. Every society needs a story that tells its people how their world
works and how they fit into the picture. Modern industrial society
no longer has a valid story; we are between stories. Many people are
casting about trying to find a more satisfactory story; but it will take
some time for a new story to replace the old one. Social analysts call
this process "paradigm shift."

20. A very different paradigm than the one that now dominates societal
thinking could become dominant; the public already has adopted
many beliefs that would make up a "New Environmental Paradigm."
This new paradigm presents a strong contrast to the old paradigm;
the disagreements between the two paradigms are so fundamental,

and the views of their adherents are held so fervently, that we can expect sharp political contesting on these matters for several decades.

21. Left-right ideological thinking is inappropriate for ecological thinking. From the perspective of ecology, capitalism and socialism have more similarities than differences; both kinds of societies destroy the ecosphere and cannot be sustained. Their dispute about ownership of the means of production misses the point: It is control, not ownership, that counts.

22. We should view markets and planning as informational systems that supplement each other, and not as mutually exclusive modes of ordering our relationships. Similarly, the good society balances cooperation and competition in socio-economic relationships.

23. Neither equality nor freedom can be fully realized in a good society. It is unwise to make either a dominant value in an ideology.

Part II

Elaborating A Vision Of A Sustainable Society:

Finding Quality Of Life In A World Of Limits

People have trouble accepting the idea that their way of life cannot continue, especially if they perceive that way of life to be good. It is human nature to deny the evidence pointing to a necessary change. Such denials are likely to persist for many people until the march of events forces the unwelcome reality upon them. Many other people are more open to the idea of change, but do not perceive how the change can come about and cannot envision the new situation.

In the next part of the book, I will try to project an image of how that new society might work. Seeing the direction in which society should and could move might help people to accept the necessity of change and will give them more hope that working for the change will be worth their time and effort.

I aim to design a new society that will function *harmoniously* and *sustainably* within the biocommunity in which it is embedded. I believe the reader will perceive that life in such a society can be of high quality. It also will be designed to fulfill the value structure presented in Chapter 4.

You and I both know that the design will not be complete and will not be perfect; no one person has sufficient wisdom and foresight to do that. I will achieve my purpose if I encourage you to think and to contribute your own good ideas for further improvement. As I proceed, I will *not* feel obligated to show how to achieve, or bring about, each of the conditions I recommend. Of course, some notion as to what could reasonably be achieved will constrain what I recommend, but I choose not to get bogged down with an immediate consideration of "being realistic" when sketching out what a sustainable society would be like.

It is useful to distinguish realism in designing components of a new social structure from realism in trying to figure out how to get to the new society. I will try to be as realistic as possible in designing components that can harmoniously work together in a new social system, but I will not constrain that effort by trying to decide in advance whether any given component can realistically be agreed to in today's political climate. A judgment about a realistic course of action inevitably is based on the situation we perceive ourselves to be in when we make the judgment; hence, it will be time and place bound. Strategies that would make sense today might not be valid tomorrow, and strategies that would appear hopeless today may seem feasible tomorrow. First we must develop a shared vision of the society we want; after we achieve that, it will be easier, and much more meaningful, to be realistic in our choice of action strategies.

Ecosystem Viability

Chapter 8

"It is enough to recognize,
without trying to catalog,
the brilliance of the sunshine;
... the fragrance that the sun draws from the pine
   tree,
... the rough and satisfying feel of sun-warmed
   granite."
                    Charlotte Mauk from "Homecoming"

"Come hither, ye who thirst;
Pure still the brook flows on;
Its waters are not curst. ...
Here's a mossy bank—come sit down:
'Twas Nature that planted this wood,
Unknown to the sins of the town.' "
                    John Clare from "Come Hither"

Life in a viable ecosystem is the top societal value; without ecosystem viability, neither a good society nor a good life is possible. During most of their history, humans sought harmony with nature and seldom injured the ecosystem. "Civilization" changed our perspective; domination of nature became a driving cultural force. We thought this power would give us security. Now we have so much power that we can destroy nearly all of nature in a few hours—but we do not have security. Furthermore, most people have not found the good life. Dominance turned into desecration: polluted air, acid rain, polluted water, toxic poisoning, soil depletion, poisoned soils, extinction of species, dwindling energy supplies, dwindling metallic ore supplies, dwindling water supplies, loss of forests, depletion of fish and other marine life, catastrophic accidents, scars and disfigurement of the landscape, human-induced droughts and deserts, human-induced loss of wetlands—desecration is worldwide and increasingly threatening to all life.

Views Of The Ecosphere

John Livingston (1985, p. 3) identifies three major perspectives in the ways people view the ecosphere: as a system, as a home, and as a being. Those using the *system* metaphor see the Earth as a spaceship, and the various components of the system operate in orderly ways to make for a harmonious and unified whole. The purpose of the system is "life support."

Those using the *home* metaphor see the ecosphere as place. Individual and group attachment to place is fundamental; "not place in the sense of turf but place in the sense of *belonging*. . . . Ecosphere as both spatial and social place for all who live within it, in which all *belong*, in context of which all are defined. Ecosphere as home." (pp. 3-4, emphasis in the original)

This last point is similar to the expression of botanist Hugh Iltis:

> Every basic adaption of the human body. . . . demands for proper functioning access to an environment similar, at least, to one in which those structures evolved through natural selection over the past 100 million years. . . . like the need for love, the need for nature, the need for its diversity and beauty, has a genetic basis. We cannot reject nature from our lives because we cannot change our genes. (Iltis, 1967, p. 887)

Ernest Partridge sees this tight linkage with our natural home as the basis for a moral responsibility to protect nature: "There is a 'paradox of ecological morality' which tells us that mankind's best interests will be obtained by not directly seeking mankind's best interest, but rather by acknowledging and regarding ourselves to be what we are in fact – 'plain citizens' and members of the community of life that created us and which sustains us" (1982, pp. 188-90).

Those advancing the *being* metaphor incorporate the ideas of system and home but go further, suggesting that all earth elements (soil, water, air, plants, animals, minerals) are united into one interdependent entity. Those advancing this metaphor typically speak of it as *Gaia*, the Greek name for the goddess of the Earth (the same one Eisler studied; see Chapter 3).

James Lovelock, a British scientist, first proposed "The Gaia Hypothesis," because he saw the earth as more than just a home, it is a living system and we are part of it. The hypothesis is supported by two streams of recent research. First, the study of genetics and microorganisms shows that all living things are constructed out of the same building blocks. These studies also show, as noted in Chapter 5, that the tiniest of organisms can learn and adapt; they even create their own environment (Margulis and Sagan, 1986).

Second, studies of the atmosphere disclosed that its constituent elements were not following chemical laws. How can the atmosphere maintain so much free oxygen and methane; why do they not unite? Why has atmospheric temperature remained so stable despite Ice Ages or a 30 percent increase in the sun's energy output? Why does the Earth's water not freeze or boil away? Gaia sees the evolution of life so intertwined with the rocks, oceans, and atmosphere that they really are one single process. Lovelock concluded that the earth is "homeostatic"; it follows what he calls "the wisdom of the body."

The fact that the biota monitors portions of the planetary surface is as well established as the fact that our body keeps itself at a constant temperature. Gaia thus may keep atmospheric nitrogen and oxygen, so important to life, from degenerating into nitrates and nitrogen oxides, into salts and laughing gas which could halt the entire system. If there were no constant, worldwide production of new oxygen by photosynthetic organisms, if there were no release of gaseous nitrogen by nitrate and ammonia-breathing bacteria, an inert or poisonous atmosphere would develop. ... On earth the environment has been made and monitored by life as much as life has been made and influenced by the environment. (Margulis and Sagan, 1986, p. 265)

In Lovelock's words:

We now see that the air, the ocean and the soil are much more than a mere environment for life; they are a part of life itself. Thus the air is to life just as is the fur to a cat or the nest to a bird. Not living but something made by living things to protect against an otherwise hostile world. For life on Earth the air is our protection against the cold depths and fierce radiations of space.

There is nothing unusual in the idea of life on Earth interacting with the air, sea and rocks, but it took a view from outside (from space) to glimpse the possibility that this combination might constitute a single, giant, living system and one with the capacity to keep the Earth always at a state most favorable for the life upon it.

If we are "all creatures great and small," from bacteria to whales, part of Gaia, then we are all of us potentially important to her well-being. We knew in our hearts that the destruction of whole ranges of other species was wrong, but now we know why. No longer can we merely regret the passing of one of the great whales or the blue butterfly, nor even the smallpox virus. When we eliminate one of these from the Earth, we may have destroyed a part of ourselves, for we are also a part of Gaia.

There are as many possibilities for comfort as there are for dismay in contemplating the consequences of our membership in this great commonwealth of living things. It may be that one role we can play is as the senses and nervous system for Gaia. Through our eyes she has for the first time seen her very fair face and in our minds become aware of herself. We do indeed belong here. The Earth is more than just a home, it is a living system and we are part of it. (quoted in the album book for *Missa Gaia*, a recording by the Paul Winter Consort)

Our traditional model of perception separates the psyche from the material world; the psyche perceives the world in a unidirectional mechanical process. The Gaia Hypothesis suggests an alternative view of perception that shifts the locus of creativity from the individual to the enveloping world.

> Perception is communication. It is the constant, on-going communication between this organism that I am and the vast organic entity of which I am a part. . . . What is important is that we describe it *as an exchange*, no longer a one-way transfer of random data from an inert world into the human mind but a reciprocal interaction between two living presences—my own body and the vast body of the biosphere. Perhaps the term *communion* is more precise than *communication*—a communication without words. (Abram, 1987, p. 9)

The Gaia Hypothesis has generated vigorous worldwide discussion. Is the Earth a living being? Does it have intelligence? Will it restore itself, perhaps eliminating the offending species (*Homo sapiens*) that refuses to fit harmoniously with the rest of nature? Agreement is slowly emerging that the earth is not a living being with intelligence and disease-fighting powers but most participants in the debate acknowledge that the metaphor opened our eyes to new ways of perceiving the planet and to new lines of research. We now more readily perceive and study planetwide biospheric processes; we see our own well-being linked to the good-functioning of those processes.

How Should The Individual Relate To Nature?

A person's conception of the ecosphere strongly shapes that person's sense of the proper relationship between self and nature. Those who see nature as a support system more willingly dominate and exploit nature than those who see nature as home or see the planet as a living being. Those who hardly ever think about nature are probably most willing of all to allow it to be desecrated. Here are some of the typical ways people think about themselves and nature. Every categorization simplifies reality, thus, I beg the indulgence of those who prefer to describe these groups somewhat differently.

Deep Ecology

Deep ecologists have a deep religious-like feeling for nature and believe that it is valuable in its own right. From this core valuation on nature one could derive a biocentric value structure; behavior that diminished nature would be immoral. (Recall my argument in Chapter 4 that a biocentric value structure is a logical impossibility because only humans value.) The most thorough statement of their perspective was authored recently by Devall and Sessions (1985).

Arne Naess is credited with first using the phrase "the deep long-range ecology movement" (1973). Naess and Sessions, while camping in Death Valley California in April 1984, summed up many years of thinking about deep ecology in the following eight principles; Naess calls it "The Platform of the Deep Ecology Movement":

1. The flourishing of human and nonhuman life on Earth has inherent value. The value of nonhuman lifeforms is independent of the usefulness of the nonhuman world for human purposes.

2. Richness and diversity of life forms are also values in themselves and contribute to the flourishing of human and nonhuman life on Earth.

3. Humans have no right to reduce this richness and diversity except to satisfy vital needs.

4. The flourishing of human life and cultures is compatible with a substantial decrease of human population. The flourishing of nonhuman life requires such a decrease.

5. Present human interference with the nonhuman world is excessive and the situation is rapidly worsening.

6. Policies must therefore be changed. The changes in policies affect basic economic, technological, and ideological structures. The resulting state of affairs would be deeply different from the present and make possible a more joyful experience of the connectedness of all things.

7. The ideological change is mainly that of appreciating *life quality* (dwelling in situations of inherent value) rather than adhering to an increasingly higher standard of living. There will be a profound awareness of the difference between big and great.

8. Those who subscribe to the foregoing points have an obligation, directly or indirectly, to participate in the attempt to implement the necessary changes. (Naess, 1985)

Notice that the deep ecologists call for reduction in the size of the human population, for an emphasis on quality of life, and for a new kind of society with a completely new orientation. The Deep Ecology Platform does not address the problem of how to reduce population. They believe the needs of the ecosphere should come before people—as a matter of fact their leading journal is called *Earth First!*.

Social Ecology—Geocentrism—Life Community—Biocommunity

The *social ecologists* try to balance the needs of humans with the needs of other living creatures. The Earth currently holds a huge human population that is destined to grow still larger. So huge a population will demand further "development" of natural resources just to stay alive. Gorz (1980, p. 20) reminds us that "all production is also destruction." Further damage to the ecosphere is inevitable if we are going to try to keep all those people alive. The social ecologists, however, would sharply turn away from the DSP perspective and seek a deep harmony between human activities and natural systems; their perspective is similar to the NEP described in Chapter 6. Murray Bookchin is the most prominent spokesperson for the social ecology perspective.

Social ecology is largely a philosophy of participation in the broadest sense of the word. In its emphasis on symbiosis as the most important factor in natural evolution, this philosophy sees ecocommunities as participatory communities. The compensatory manner by which animals and plants foster each other's survival, fecundity and well-being surpasses the emphasis conventional evolutionary theory places on their "competition" with each other. ... Social ecology, which tries to plant its feet in nature, begins to raise its head in the municipality that is truly participatory. (Bookchin, 1986, pp. 25, 42)

With its balancing of ecological and social needs, social ecology occupies a middle position between the deep ecologists and the advocates of the DSP. Although social ecologists and deep ecologists hold similar core values, they disagree over means to realize the core values. This conflict is sharpened when the needs of people for survival conflict with the urgency of maintaining a viable ecosystem. Earth firsters argue, for example, that we must close our borders to immigration which, if left unchecked, would push other creatures out of their niches and devastate our ecosystem. Bookchin argues, in contrast, that we should act compassionately to welcome and nourish poor immigrants who cannot find a decent life in their home countries. Bookchin would send relief to starving people who have acted foolishly to wreck their ecosystem (probably by reproducing too swiftly) while the Earth firsters would let nature control human population.

Some environmentalists maintain their base in traditional religion while striving to bring an ecological perspective into spirituality—they seek religious reform. Thomas Berry, a Catholic priest, scholar, and disciple of Tielhard de Chardin, has for several decades explored the moral basis of the relationship between humans and nature. He calls for the human community to turn from its present anthropocentric norm to a *geocentric* norm of reality and value. The earth belongs to itself and to all the component members of the community. Industrial exploitation is plundering the earth, creating a deep cultural pathology. The question of the viability of the human species is intimately connected to the viability of the earth. The life community, the community of all living species, is the greater reality and the greater value.

The primary concern of the human must be to preserve and enhance this larger community. The total extinction of life is not imminent but the Earth's ecosystems may be shattered irreversibly. What is absolutely threatened is the degradation of the planet's more brilliant and satisfying forms of life expression.

A central value word used by our society is "progress." But then we see that human progress has been carried out by desolating the natural world. This degradation of the earth is the very condition of the "progress" presently being made by humans. It is a kind of sacrificial offering. . . . The feeling that even the most trivial modes of human progress are preferable to the survival of the most sublime and even the most sacred aspects of the natural world is so pervasive that the ecologist is at a loss as to how to proceed. The profit of the corporation is the deficit of the earth. . . . Strangely enough, it is our efforts to establish a thoroughly sanitized world that have lead to our toxic world. Our quest for wonderworld is making wasteworld. Our quest for energy is creating entropy on a scale never before witnessed in the historical process. (T. Berry, 1987, pp. 78 and 80)

We are proud of our Western legal tradition, with its insistence on personal rights and the freedom of the human. But that same tradition gives entrepreneurs the right to exploit nature without recognizing any rights of other natural beings. We humans determine the limits that humans are willing to accept. To achieve a viable human-Earth community we need a new legal system to articulate a mutually enhancing human-Earth relationship within the integral functioning of the earth process.

Resourcism

Among adherents of the DSP, nature is viewed as a resource for goods desired by humans. "All of nature is one vast bank of raw materials exclusively earmarked for the human enterprise." (Livingston, 1985, p. 4) The elevation of human purposes above the purposes of nonhuman species, as resourcism does, has been called *speciesism* (Singer, 1977). Whether it is morally defensible to use other species for purposes chosen by humans has been considered in depth by many philosophers (see, especially, Roszak, 1972 and Ehrenfeld, 1978).

The movement we know currently as Environmentalism began in the last century as The Conservation Movement. An identifiable subgroup of contemporary environmentalists can be called *nature conservationists* (Cotgrove, 1982; Milbrath, 1984). This group strongly desires to conserve natural resources: wildlife, forests, lakes and streams, wilderness, and wet-lands. They fight against pollution, habitat destruction, resource depletion, acid rain, toxics, and so forth; but they struggle mainly to preserve those resources for human enjoyment. They adopt a "stewardship" role toward the environment. Commonsense prudence dictates the wise use of resources; people may use them but not use them up. They have a duty to pass natural resources on to future generations. "Of course, the morality is toward future generations of people, not toward nature." (Livingston, 1985, p. 6)

By putting human purposes first, nature conservationists believe in resourcism. Most contemporary nature conservationists also accept the basic structure and thrust of modern industrial society as being valid and effective; they are not asking for a new society as the deep ecologists, social ecologists, and geocentrists demand. Most of them believe that their environmental goals can be achieved by developing better technology and better legislation—better enforced. Because they want to reform society rather than fundamentally change it, their perspective has been labelled "shallow," in contrast to deep, environmentalism. (Naess, 1973 and Devall, 1980)

The Environmental Vanguard

Many contemporary environmentalists do not believe that the integrity of the ecosphere can be restored merely by employing better technology and adopting better legislation; they believe there must be sweeping changes in lifestyles and social structures—that we must build a new society. We may call them *transformationalists*. They constitute a vanguard for a new society. The deep ecologists, social ecologists, and geocentrists all call for a fundamental paradigm shift; in that sense they are all transformationalists. Many of them believe that nonhuman nature has inherent value and that nature should be preserved in all of its rich complexity and diversity. The transformationalists focus on the ecosystem as home rather than the environment as house. They seek regenerative healing of the ecosystem not simply reduction of abuses.

Many in the vanguard are more practically oriented than the deep ecologists, however, in that they also recognize that society must deal with an overpopulated world. For that many people to live as lightly on the Earth as the deep ecologists recommend is simply impossible. Even if we are in a crisis, could any political program advocate letting humans die to preserve the good functioning of the ecosphere? The transformationalists participate in the present society while using education and political action to try to bring about a new society. It is their determination to work for a new society that distinguishes them from the nature conservationists. It is their deep participation in contemporary political struggles tht distinguishes them from the deep ecologists.

On many public policy issues dealing with the environment, the deep ecologists, the social ecologists, the geocentrists, the transformationalists and the nature conservationists work together; they comprise what the media popularly labels as environmentalists. Most of the policies currently on the public agenda are reformist in orientation (toxic waste cleanup, for example); all five subgroups can agree to support them. Efforts to supplant the status quo never get on a public agenda; paradigm shift is never legislated, it takes place at a more fundamental level (although it may later find expression in legislation).

In the mid-1980s, environmentalists (broadly defined) constituted somewhere between 10 and 20 percent of the U. S. population; the proportion who are vigorously active is probably .5 percent. (Milbrath, 1984) Studies from other developed industrial countries suggest that similar proportions of the population are environmentalists there (Milbrath and Cheng, 1985). Even though environmentalists are in the minority, they are not the only ones supporting environmental protection; actually, most of the public supports environmental protection (Milbrath, 1984). The environmental movement has made monumental strides in a few decades (popular support, much new legislation, new practices) despite vigorous opposition from DSP proponents.

DSP Defenders

Environmental values resonate so strongly with something fundamental in human nature that one almost never encounters a person who forthrightly claims to be against environmental protection. We do not even have a name in the English language for antienvironmentalists. Persons and policies turn out to be antienvironmental because they place other values above preservation of the environment. A belief that humans have the right to dominate and use nature for any human purpose is basically antienvironmental, even though most holders of that belief would not think of themselves as being against the environment. A belief that economic growth and the accumulation of material wealth should be the highest public priority is basically antienvironmental. The values and beliefs that lie at the core of the DSP are antienvironmental even though that is not their avowed purpose. Given that environmental destruction was never the intended purpose of DSP policies, it is astonishing how destructive of the environment those policies have been.

Even the greatest devotees of the DSP would shrink from deliberately driving a species to extinction. (The drive to extinguish the small pox virus is a possible exception.) Yet, some estimates are that as many as 15 million species may be extinguished by the turn of the century, largely as a result of DSP policies. Holmes Rolston, III, poignantly makes the case for our moral obligation to endangered species:

> Extinction shuts down the generative process. The wrong that humans are doing, or allowing to happen through carelessness, is stopping the historical flow in which the vitality of life is laid. Every extinction is an incremental decay in stopping life processes—no small thing. Every extinction is a kind of superkilling. It kills forms (species), beyond individuals. It kills "essences" beyond "existences," the "soul" as well as the "body." It kills collectively not just distributively. . . .
> Ought species x to exist? is a single increment in the collective question, "Ought life on Earth to exist?" . . . To kill a species is to shut down a unique story. . . . If . . . it makes any sense to claim that one ought not to kill individuals without justification, it makes more sense to claim that one ought not to superkill the species, without superjustification. (Rolston, 1985, p. 723)

Obviously, the first and most fundamental step for restoring and preserving ecosystem viability is to change our way of thinking, believing, and valuing with respect to the relationship between ourselves and the biosphere. We have gotten into our environmental predicament because of fallacies in our thinking; setting our thinking right is the first step in extricating ourselves. Environmental awareness is growing and will continue to grow. Nature is our most powerful teacher for developing this awareness. Many leaders of less-developed countries resisted the recommendation by the 1972 Stockholm Conference on the Human Environment to begin immediately to preserve their environment. They were later forced by the dire environmental consequences of their policies to recognize the urgency of environmental protection. Individuals and society (especially the media) can also play powerful roles in helping a population to become environmentally aware.

Another curious, but not often appreciated, fact is that once environmental awareness has been achieved by a person, it is seldom lost. Other urgent concerns may displace environmental action but the awareness and concern remain, even though a person is occupied elsewhere. It is like learning to ride a bicycle; once it is mastered, one does not forget how to do it even though decades may pass without riding. It is this characteristic of environmental learning that explains why awareness and concern about the environment remained high in public polls in the United States throughout the 1970s and 1980s—even though the priority for environmentally protective actions ranked lower in the 1980s than in the early 1970s. We are not born with environmental awareness, it must be learned by each person, just as we each learn how to ride a bicycle. As environmental awareness deepens in our culture, young people will be exposed to it routinely in their daily lives. The schools have a very important role to play here; regretfully, they have avoided it up till now.

What Must Be Done To Preserve Ecosystem Viability?

We can assume that people really want a viable ecosphere but what can they do to bring it about? How should they structure their lifestyles, their social relationships, and their institutions to allow ecosystems to restore themselves? Because ecosystems flourished viably before humans ever evolved, perhaps the simplest advice is not to interfere with them. Of course, that is too simple. Too many people are demanding sustenance. If we are not prepared to accept the deaths of billions, our only choice is to continue to manipulate (read, dominate) our ecosystems. We must concentrate, instead, on doing as little damage as possible and use loving husbandry to encourage ecosystems to regenerate themselves.

Humans are unique in being able to hold values; other species have preferences and needs, but only humans can adopt and hold values. Ecosystems also have needs and preferences. We humans should be alert to those needs and preferences and take care to see that they are fulfilled. We may value a viable ecosystem because we believe nature has inherent worth and because we believe ecosystem good functioning is essential to living a high-quality life. We can better keep our ecosystem viable if we take a closer look at ecosystemic needs and preferences.

Ecosystemic Needs And Preferences

Basic requisites for good functioning of ecosystems. Ecosystems must have intact and smoothly functioning *biogeochemical cycles* such as the water cycle, the carbon cycle, and the nitrogen cycle. This is not the place for a thorough description of how these cycles work (curious readers should consult a good environmental science text such as G. T. Miller, 1985; for thorough scientific detail, see Bolin and Cook, 1983). Human interference with these cycles can lead to climate change, water shortages, desertification, loss of fertility, and other calamities. Pollution disturbs the carefully worked out patterns of these cycles. For example, trace metals, like aluminum and cadmium, are normally bound to soil particles; acid rain frees the metals which flow with runoff into lakes and streams, eventually killing fish. The more we learn about these cycles the more incredibly complex they seem; almost every human action has unforeseen consequences.

Ecosystems, like all systems, are powered by some form of *energy*; most of it comes from the sun. Some of this energy is stored in soils; some of it resides in plants or animals that eventually die and decay, becoming available for plant food. If humans interfere with this energy flow by depleting or stripping soils, or by consuming plants for fuel, the productivity of the ecosystem is diminished. Recently, we have been "correcting" nutrient deficiencies in soils by applying fertilizers made from fossil fuels. Replacing lost nutrients works for a while but is less satisfactory than normal ecosystemic processes for soil regeneration. Furthermore, imminent scarcity of fossil fuels will raise fertilizer prices, making it less economical to compensate for lost soil nutrients in that way.

Evolutionary processes have laboriously built up *buffers* that shield living creatures from harm. For example, we have recently become aware of an ozone layer in the stratosphere that shields plants and animals on the surface from injurious ultraviolet solar radiation. Scientists believe that man-made chlorofluorocarbons (used as a propellant in spray cans, as a refrigerant, and in styrofoam) rise to the stratosphere and destroy the ozone layer. A single chlorine atom can, over time, destroy upwards of 10,000 ozone molecules. We now have clear evidence that the shield is being destroyed and no idea how it can be restored. A treaty was signed recently in Montreal, Quebec to reduce the production of chlorofluorocarbons by 50 percent by 1998; as usual, the action was too little, too late.

Good and Bad Extremes. Some ecosystemic characteristics have extremes that are clearly good or bad; they can be distinguished from characteristics where a balance between extremes is needed. For example, soil loss and infertility are clearly detrimental whereas soil maintenance and fertility are clearly beneficial. Similarly, routine (cyclically predictable) climate patterns are beneficial while sudden and severe climate changes are severely detrimental. Natural hazards (earthquakes, volcanic eruptions, floods, big storms) are almost always destructive to ecosystems.

Likewise, when a particular species reproduces rapidly and kills off other species, or crowds them out of their niches, the overall effect on the ecosystem is detrimental. When humans are the victims of this phenomenon, we call it an epidemic and the offending species are called pests. If we are the offending species, we claim that we are successfully flourishing; but the devastation we cause to the ecosystem is probably much greater because we are so powerful. The species who are our victims probably would perceive us as an epidemic and would call us pests if they could use language.

A balance between extremes seems to be best for many ecosystemic characteristics. Too much rainfall and too little rainfall are both detrimental; the most beneficial balance is the one which fits into the normal climate pattern for an area and to which its ecosystem has adapted. The same can be said for sunshine and temperature. We are now discovering that human activities bringing about the greenhouse effect are changing routine climate patterns, resulting in devastating effects on ecosystems and socio-economic systems.

Insufficient nutrition is clearly bad for an ecosystem but ecosystems can suffer from too much nutrition as well. For example, lakes with an excess of nutrition suffer from eutrophication. Algae and seaweed grow so profusely that they go into overshoot and die back; this uses up the oxygen in the water, killing fish and other marine animals. The aquatic ecosystem loses its balance, and declines in health and productivity. Human wastes (phosphate detergents, sewage, fertilizer runoff) generally are the main causes of eutrophication. When humans reduced the flow of these wastes to Lake Erie during the 1970s and 1980s, for example, its ecosystem made significant progress in restoring its balance.

Humans must use agricultural cultivation to feed their enormous population but we should also recognize that agriculture constitutes gross human interference in ecosystems to make them grow plants and animals that humans desire to consume. It is not uncommon for a farm to grow a single crop or raise a single animal species; this is called *monoculture*. Monocultural ecosystems are inherently unbalanced; they are more vulnerable to climate changes, diseases, and pests. Farmers typically use chemical pesticides and fertilizers to maintain monocultural production. Over the long term, however, such practices have unintended consequences (for example, toxic poisoning, or eutrophication) that injure many other species, even those living some distance away. Fertilizers and pesticides put on a farm in Minnesota, for example, can be carried by river to the Great Lakes and eventually injure fish in Lake Ontario, almost 700 miles away. We can never do merely one thing.

An Agenda Of Actions To Restore Ecosystem Viability

Humans inflict so many indignities on ecosystems that the list of changes in practices that might be taken to help restore these systems to viability could go on for pages. Groups of scientists and environmental policy specialists at two recent conferences composed similar lists of priorities for ecosystem restoration. In 1982, the Royal Swedish Academy of Sciences convened a week-long international gathering of scientists to look at environmental priorities for the 1980s. Two years later, the World Resources Institute called another international conference on "The Global Possible: Resources, Development, and the New Century" that included seventy-five leaders from science, business, government, and environmental affairs. Both efforts identified the following ten problems as truly serious and deserving of wide international attention. The list below emphasizes the concern to restore ecosystem viability but also note the prominence of human needs (as contrasted to bio-needs) in the phrasing.

1. *Loss of crop and grazing land* due to desertification, erosion, conversion of land to non-farm uses, and other factors. The United Nations reports that, globally, farm and grazing land is being reduced to zero productivity at the rate of about 20 million hectares a year. (One hectare equals about 2.5 acres.)

2. *Depletion of the world's tropical forests*, which is leading to loss of forest resources, serious watershed damage (erosion, flooding, and siltation), and other adverse consequences. Deforestation is projected to claim a further 100 million hectares of tropical forest by the end of this century.

3. *Mass extinction of species, principally from the global loss of habitat, and the associated loss of genetic resources. One estimate is that more than 1,000 plant and animal species become extinct each year, a rate that is expected to increase. (Species extinction is final, it affects not only our generation but all future generations.)*

4. *Rapid population growth*, burgeoning Third World cities, and ecological refugees. World population will most likely double by the early decades of the next century, and almost half the inhabitants of developing countries will live in cities – many of unmanageable proportions.

5. *Mismanagement and shortages of fresh water resources.* It now seems possible to many researchers that water will be to the 1990s what energy was to the 1970s.

6. *Overfishing, habitat destruction, and pollution in the marine environment.* Twenty-five of the world's most valuable fisheries are seriously depleted today due to overfishing.

7. *Threats to human health* from mismanagement of pesticides and hazardous substances and from waterborne pathogens. Waterborne diseases are responsible for about 80 percent of all illness in the world today.

8. *Climate change* due to the increase in "greenhouse gases" in the atmosphere. The steady build-up of carbon dioxide and other gases in the atmosphere, due principally to fossil fuel burning, is predicted to create a greenhouse effect of rising temperatures and local climate change – the question increasingly is not "if?" but "how much?"

9. *Acid rain* and, more generally, the effects of a complex mix of air pollutants on fisheries, forests, and crops. The "export" of acid rain harms not only natural resources but also constructive relationships among neighboring states in political and economic affairs.

10. *Mismanagement of energy fuels and pressures on energy resources*, including shortages of fuelwood, the poor person's oil. Although market forces and government actions have eased pressures, these vital resources are, undeniably, finite in quantity and disparate in locale. Our energy problems may be forgotten, but they have not gone. (Speth, 1985, p. 12)

All ten of these problems stem from the fact that too many humans are trying in numerous ways to control and dominate nature to make it serve human purposes. Because we are only doing what we have always done—only better and better—we fail to understand the cumulative effects of our actions. In some instances, technologies are developed and applied without sufficient consideration of second, third, and fourth order consequences ("We can never do merely one thing"). In other instances (for example, depletion of fuelwood), poor people are grasping at anything just to stay alive. Although the impact of these problems is great everywhere, it is far more serious in the poor countries than in the rich.

This list is different and broader than similar lists from the early 1970s. Those lists would probably have emphasized air and water pollution, mainly seen as local problems. Pollution is not prominent on the 1985 list, although pollution problems are far from being solved. Too many national leaders, and ordinary people, erroneously believe that pollution is the main environmental problem and it can be solved by regulation and technical fixes. All ten priorities are global in scope and require cooperation across nations for their solution. The United Nations has established the United Nations Environment Programme and other agencies to facilitate international cooperation on these problems (see Caldwell, 1984, for details) but cooperation ultimately depends on recognition of the urgency of these problems by people and their leaders; and on their determination to solve them.

The IUCN and the World Conservation Strategy

The International Union for the Conservation of Nature and Natural Resources (IUCN) obtained sponsorship from the United Nations Environment Programme and financial assistance from the World Wildlife Fund for an intensive three-year effort to draw up a *World Conservation Strategy*. The IUCN mainly comprises physical and life scientists, from many countries, with an interest in environmental matters. The strategy was designed to draw the attention of decisionmakers and the general public to the urgent need for the conservation of the world's land and marine ecosystems. Conservation of these systems must be an integral part of economic and social development. The purpose of the strategy is activation. It is essentially a statement of goals and targets.

Its three main objectives are: 1) to maintain essential ecological processes and life-support systems, 2) to preserve genetic diversity, and 3) to ensure the sustainable utilization of species and ecosystems. (This is clearly a posture of resourcism; but it is embedded in an understanding of ecosystem functioning with an eye to their long-range sustainability.) Even though it seems obvious that all national policies, particularly developmental plans, should strive for these objectives, the IUCN identifies six main obstacles to reaching these objectives:

1. A belief that living resource conservation is a limited sector rather than a process that cuts across, and must be considered by, all sectors.

2. Consequent failure to integrate conservation with development.

3. A development process that is often inflexible and needlessly destructive, due to inadequacies in environmental planning and management.

4. Lack of capacity to conserve, due to inadequate legislation and lack of enforcement.

5. Lack of support for conservation, due to lack of awaremess (other than at the most superficial level) of the benefit of conservation.

6. Inability to deliver conservation-based development where it is most needed, that is, in rural areas of developing countries.

The recommended worldwide strategies focused on restoring and preserving tropical forests, arid lands, biospheric reserves (to preserve genetic resources), and such international commons as the atmosphere, open oceans and Antarctica. Developments in international law and international financial assistance were recommended to achieve these objectives.

Strategies also were suggested for regional planning and management to restore and protect regional river basins and regional seas (such as the Great Lakes in North America or the Mediterranean). Regional arrangements and priorities for sustainable development (e.g. in forestry) also were recommended.

Getting scientists from many countries to agree to the World Conservation Strategy was, itself, something of an accomplishment. It demonstrates that a heterarchial worldwide structure already exists for leadership on environmental policy. However, this structure does not command; it can only educate and recommend. Only when official decisionmakers in a multitude of countries come to perceive the wisdom and necessity of preserving ecosystem viability will real progress be made. (I have relied on Caldwell, 1984, pp. 271-74, for much of this information.)

Ecosystem viability surely ought to be a top priority for governments at all levels. Achieving viability is very difficult because it will require conscious changing of behavior patterns by billions of individuals, firms, and organizations, as well as governments. Every individual takes dozens of actions every day that either contribute to or detract from ecosystem viability. Environmentally protective actions are more probable if a person becomes aware of the urgency of maintaining a viable ecosystem and thoughtfully considers the consequences of routine daily activities (once thought through, they do not need to be reconsidered for each occurrence). If environmental awareness and sense of urgency can continue to increase, people will automatically consider the consequences of private and public decisions.

The context for making decisions also strongly influences their outcome. We are all structurally constrained by our society. Consumers who do not want to buy their products in fancy sealed packages may still have no other choice if that is the only way one can buy the product. For years, I have been frustrated in my desire to recycle such household wastes as paper, metals, and glass. The market for waste paper became so depressed that one could not give it away. The town authorities declined to accept cans and bottles in their recycling program "because it was too much trouble." (Now that space in landfills has become extremely costly, however, the town fathers are talking about instituting a recycling program.)

A few years ago, after an intense lobbying struggle, the State of New York passed a law requiring a 5¢ deposit on each beverage container. Consumers and distributors, after some grumbling, changed their behavior with respect to containers. A thriving container recycling industry has sprung up and roadside beverage container litter has declined markedly. More appropriate behavior followed a change in structure. Ordinary citizens in Japan, Germany, and Norway have learned, with structural encouragement from their governments, to recycle up to one-half of their household wastes. (Brown, 1981, Ch. 8)

New knowledge can simultaneously help to meet human needs and restore damaged ecosystems. Many areas in developing countries have been deforested because poor people have taken the few remaining trees and shrubs for desperately needed fuelwood. The soil then dries out or erodes, and the area may become a wasteland. Reforestation clearly is required, but the expectation that it will take a long time can discourage people and leaders from even trying. Forest scientists have recently discovered a leguminosae tree shrub, *Callindar callothyrsus*, which grows so vigorously that it can be harvested at the end of the first year. It can be harvested annually for up to fifteen years and yield thirty-five to sixty-five cubic meters of wood per hectare. In Java, some 30,000 hectares have been planted with this tree in firewood plantations supporting local communities. (Brown, 1981, pp. 184-85)

An even more promising woody legume is the *Leucaena leucocephala*, or "ipil ipil," now widely planted in the Phillipines. These trees can reach the height of a three-story building in two years and a six-story building in six to eight years. They can produce thirty to fifty cubic meters of wood per hectare annually (a well-managed pine plantation in a temperate climate produces only about ten cubic meters annually). Being leguminous, the tree also fixes nitrogen in the soil, making it ideal for revegetating eroded hillsides in the tropics. Its nitrogen rich leaves also can provide nutritious forage for cattle and an excellent compost for crops. (Brown, 1981, pp. 174-85)

As people introduce new species, they should be cautious that their actions do not result in uncontrolled reproduction that dominates or devastates an ecosystem. For example, the kudzu vine, not native to the Southern United States, has so flourished there that it has become a threat to other members of those ecosystems. Humans have had sufficient experience with the species mentioned in the previous paragraphs as to feel they have not started an epidemic but the reality that we can never do merely one thing should always make us cautious about any significant ecosystem manipulation.

Recombinant DNA enables scientists to create completely new species; presumably, they would be tailor-made to fulfill the needs of humans. This new technology is both inviting and frightening. It holds the promise of dramatically improving the productivity of living creatures to do things that humans want. It could allow humans to reproduce even more rapidly and live in even greater luxury. We should be cautious, however; a completely new species would be less likely to have natural predators that could serve as a biological control; therefore, introducing a new species carries the possibility that it will reproduce in epidemic proportions and destroy much of what humans desire in ecosystems. It could injure or destroy our own species.

Another caution must be given with regard to highly productive new species. Even if they reproduce and grow quickly, they are still dependent on the same resource base (soils, energy, water, minerals) that the preexisting ecosystem utilized. That resource base is finite. Whatever portion of it is expropriated for the new human-serving species will crowd other species into decline or extinction. Propagation of new species will most likely be monocultural; that also weakens ecosystems. We cannot blink away the reality that bioengineering is one of the most arrogant forms of resourcism; that it elevates humans to a higher pedestal than other animals, demanding that other creatures serve us. It is the ultimate expression, to date, of the belief that humans should dominate nature. I discuss bioengineering more fully in the next chapter.

In deciding these questions, we need a truly comprehensive understanding of ecosystem functioning and need to rethink our own values and beliefs relating ourselves to nature. We need a highly refined vision of the human-

Earth community. Protection of the ecosphere should be integrated into every form of our decisionmaking. People must learn how to think integratively; to avoid linear thinking that considers only one consequence—often resulting in injury to the ecosystem.

Our demonstrated ability to gather new knowledge and to ingeniously adjust our practices can go a long way to restore viability to our ecosystems. We must, at the same time, restrain our fertility. Other species will not be able to flourish if we take so much of the Earth's bounty for our own selfish growth and consumption. We cannot go on as though the Earth's living systems were created to serve only humans. Even if we do all those good things, humans will still desecrate ecosystems to some extent. Those who develop a deep awareness of the human relationship to nature will probably carry a concern to minimize human intrusion on nature into their thinking and action. If and when that perspective infuses most of society, it will be on the road to sustainability.

Sustaining Our Food Supply

Chapter 9

Barring devastating war, massive famine, or epidemic disease, human population will certainly increase to 8 or 10 billion in the next fifty years. We must greatly increase food supplies for humans while simultaneously doing our best to preserve and enhance the integrity of ecosystems. Because we also wish to live in a world with wildlife and beautiful nature, we must ensure that ecosystems nourish other species as well. This is a tall order made more difficult by the fact that our civilization is heading in the wrong direction if we wish to achieve a sustainable food supply. Massive social relearning will be required to reverse direction.

Seeing where we stand now and what we must do to sustain our food supply is complicated; it cannot be fully discussed in one chapter. Readers wishing to delve deeper will find many references in the following pages to a rich literature setting forth challenging ideas and practical suggestions for change.

The Historical Context Of Our Food Predicament

As with most species, humans have a natural tendency to keep reproducing until their food supply will support no more of them. In this respect, food has always been scarce for humans. Of course, there have been times of plenty, but usually population increased sufficiently to once more make food scarce. People acquired their food by hunting and gathering throughout 99 percent of the time they have lived on the planet. Schmookler (1984) demonstrates that when we developed the ability to domesticate plants and animals, we reached a major discontinuity in our evolution. The evolutionary principle of selection changed from how well a species adapted to its niche to the ability of a tribe to accrue and wield power. This new era, which we call civilization, has followed cultural rather than biological evolution and has brought an ever-accelerating pace of change.

Cultivation had another consequence just as devastating as the advantage it bestowed in the struggle for power: the loss of topsoil. Most of us do not appreciate soil: how it is created, its composition, how it can be ruined, how it is lost, how it can be regenerated, and its value to us (and all other species). The continuing loss of topsoil through ignorance and greed is one of the greatest tragedies of human history. The current famines in Africa and the crisis in American agriculture are traceable to erroneous cultural practices leading to soil desecration and loss. We must halt these practices and regenerate viable topsoil if we ever hope to feed the billions of humans that will soon be born.

Soil is mainly tiny particles of rock, but it is much more. It incorporates humus from decayed animal and vegetable matter; it contains water and air. It is inhabited by bacteria, microbes, worms, insects, and small mammals. The root structure of plants is especially critical for keeping the soil porous as well as preventing its erosion. Soil is a storehouse of energy that plants utilize for nutrients so they can make the food upon which all animals depend. The chemical processes that take place in soils are extremely complex and still not well understood by scientists. This living membrane of the Earth's crust is the source of nearly all life, even for most of the life in the sea. Jackson says we should think of soil as a placenta or matrix through which the earth mother sustains all life, including humans.

When humans learned to plow, they were on their way to creating our civilization but they also broke that placenta and they have been destroying topsoil ever since.

The plowshare may well have destroyed more options for future generations than the sword. So destructive has the agricultural revolution been that, geologically speaking, it surely stands as the most significant and explosive event to appear on the face of the earth, changing the earth even faster than did the origin of life.... Till agriculture is a global disease, which in a few places has been well managed, but overall has steadily eroded the land. In some areas, such as the U.S., it is advancing at an alarming rate. Unless this disease is checked, the human race will wilt like any other crop. (Jackson, 1980, p.2)

Wendell Berry, farmer, poet, essayist, teacher, explains the meaning of soil beautifully:

The soil is the great connector of lives, the source and destination of all. It is the healer and restorer and resurrector, by which disease passes into health, age into youth, death into life. Without proper care for it we have no community, because without proper care for it we can have no life.

It is alive itself. It is a grave, too, of course. Or a healthy soil is. It is full of dead animals and plants, bodies that have passed through other bodies.... If a healthy soil is full of death it is also full of life: worms, fungi, microorganisms of all kinds, for which, as for us humans, the dead bodies of the once living are a feast.... Given only the health of the soil, nothing that dies is dead for very long. Within this powerful economy, it seems that death occurs only for the good of life.... It is impossible to contemplate the life of the soil for very long without seeing it as analogous to the life of the spirit. No less than the faithful of religion is the good farmer mindful of the persistence of life through death, the passage of energy through changing forms.

And this living topsoil–living in both the biological sense and in the cultural sense, as a metaphor–is the basic element in the technology of farming. (W. Berry, 1977, p. 86)

Despite growing concern for loss or exhaustion of topsoiil, Dregne (1986) reports that little prospect exists for improvement in soil and water conservation in the near future because there are insufficient incentives to induce farmers to take corrective action.

Population And Food Supply

Cultivation gave a distinct edge to humans over other animals. Humans propagated those plant and animal species that were of use to them and crowded other species out of their niches. Human cleverness in making clothing, building houses, using tools to make still more tools, enable people to live in a more diverse set of climates and ecosystems than any other species. Human population grew rapidly. People still experienced famines, and human populations were occasionally decimated by disease and war; but, generally, human population growth continued.

Chapter 2 detailed two powerful events that stimulated a swift spiral of exponential growth in population: 1). the discovery and colonization of North and South America by Europeans, and 2). the surging exuberance by which humans dominated nature during the Industrial Revolution. Our use of fossil energy is now so profligate, not only in industry but also in agriculture, that our reserves are dwindling rapidly. Obviously, farmers consume fossil energy to power their machinery; many people do not know that this energy also is the source of most fertilizers and pesticides. We have insufficient reserves of fossil energy to sustain the growth in population and food supply that we are already destined to experience. We are living on "ghost acreage" that is four parts phantom carrying capacity for every one part real (permanent) carrying capacity (Catton, 1980, p. 46).

Two lessons stand out boldly: 1). we must urgently do everything we can to limit growth in human numbers and actually reduce human numbers as soon as that is practicable, and 2). we must ease ourselves away from our dependence on ghost acreage because we are swiftly depleting the finite quantities of fossil fuel and soil that support our current agricultural productivity. These steps will be very difficult because many people today have the money to demand more food and fuel, even if that should condemn many of our descendants to die of hunger.

People think mostly about their own family and not about world population when they choose whether to have children; therefore, if societal authorities do not guide family planning, the planet almost surely will be overrun with people. My students have had great difficulty accepting the possibility that society (the government) might tell them how many children they can have. They cannot imagine that our people will ever accept a policy of one child per family as is now the official norm in China.

Kenneth Boulding (1964, pp. 135-36) proposed that the government of a nation award two licenses to each woman to have a baby. Although the award of the two licenses would be free and automatic, the licenses could be bought and sold in a governmentally supervised market. (Boulding actually proposed an allocation of 2.1 licenses to each woman; this would make for complicated market exchanges of units of a "tenth" of a license.) The moral basis of the plan is that everyone would be treated equally; yet, there would be no insistence on a standard family size. Some women might choose to have one child, or none at all, and could sell their spare licenses to those who are willing to pay for the privilege of having more than two children. As you can easily imagine, this proposal raises many additional questions. For example, what if a baby dies? Daly (1977), pp. 56-61) proposes that the licenses be given for "survivals" not simply for births; he has a thoughtful discussion of some of these variations and complications, as well as of some of the psychological and philosophical aspects of the proposal.

The Tragic Failures Of Our "Successful" Agriculture

By the usual forms of reckoning, American agriculture has been a smashing success. About 3 percent of the U. S. population grows enough food to more than feed the rest of the people. Ward Sinclair reported in the *Washington Post* on January 5, 1986, that food production was up and more than keeping pace with population growth in most parts of the world, even in such densely populated places as India, Bangladesh, Japan, Indonesia, Taiwan, the Philippines, and China. The success was attributed to science and technology that makes it possible "to grow two stalks of grain where only one grew before." The story does not report that fossil energy inputs (ghost acreage) in the form of fertilizers and pesticides make this increase possible. To be sure, there are new plant varieties, but the energy for increased yields has to come from somewhere; if little additional can be wrung from the soil and sun, it must come from fossil fuels. Rifkin (1985, p. 46) reminds us that living resources can be made to reproduce more quickly, but life support systems cannot be similarly expanded.

What is happening to our life support systems while we stress our ecosystems to produce more food for humans? Tillage agriculture, which focuses on growing one species to the exclusion of all others, exposes the soil to erosion from water and wind. Wendell Berry estimates (1981, p. 42) that in Iowa, growing one bushel of corn entails the loss of five or six bushels of topsoil. Despite fifty years of effort by the U. S. Soil Conservation Service, topsoil is being lost from American farms at an alarming rate. Biomass that is removed from farms to be used for food, fiber, or fuel, rather than being returned to the soil as manure, also constitutes removal of topsoil. Our current topsoil losses far exceed any natural soil regenerative processes in ecosystems. When we lose topsoil at such intolerable rates we are losing our lifeblood. Jacks and Whyte (1939) conducted a detailed study of the relationship between soil fertility and erosion and the rise and decline of civilizations. They concluded that:

Erosion has indeed been one of the most potent factors causing the downfall of former civilizations and empires whose ruined cities now lie amid barren wastes that once were the world's most fertile lands. The deserts of North China, Persia, Mesopotamia and North Africa tell the same story of the gradual exhaustion of the soil as the increasing demands made upon it . . . exceeded its recuperative powers. Soil erosion, then as now, followed soil exhaustion . . . The gutted North-West and the Yellow River (in China) are the outstanding and eternal symbols of the mortality of Civilizations. (quoted in D. Andrews, 1984, pp. 191-92)

Modern China seems to have had difficulty learning from earlier lessons in soil loss. After its violent revolution following World War II, Chairman Mao Tse-Tung urged population growth and China astonished itself by doubling its population from 500 million to one billion in only thirty-five years. Mao believed they could feed all these new people by a program of planting all available land to grain. Forests were slashed down, grasslands were plowed and seeded, and the resulting erosion was spectacular. The reservoirs behind dams silted up in five to seven years instead of the expected fifty years. Rampaging waters flooded the lowland plains when earthen dikes could not contain the rivers. Some of the dikes have now become so high that the river bed is fifteen feet above the surrounding plain. The grain-growing program was a resounding failure because so much land was put to unsuitable use. (Remember, we can never do merely one thing.) Now the Chinese have initiated a huge reforestation program and also are trying to restore their grasslands. The deep scars to the landscape and loss of topsoil will take many decades (even centuries) to heal. The Chinese have been forced to adopt the grim policy of one child per family.

Desertification is spreading in many parts of the world, especially in Africa. Northern Egypt along the Mediterranean was a verdant plain 2,000 or 3,000 years ago. Unwise grazing, especially of goats, consumed nearly all vegetation, and the area became a desert. The recent famines in Ethiopia and the Sudan were the result of drought in an already overstressed ecosystem. Too many people and their cattle, more than the ecosystem could support, crowded wild animals from the land. When the drought came, the people and cattle consumed all the plants, insuring their own death and the creation of a desert.

In the Ethiopian highlands, one billion tons of topsoil erode each year. According to U. N. Food and Agriculture Organization estimates, 16.5 percent of Africa's rainfed cropland will be lost by the year 2000 if conservation measures are not taken. Deserts in Africa are spreading and will most likely continue to spread because human population growth rates in Africa are the highest in the world. The United Nations estimates that the current population of sub-Saharan Africa's forty-two nations, some 434 million, will more than triple to 1.4 billion in the next forty years. There also seems to be little recognition among the people that it is their responsibility to control growth of their numbers. Restoration of topsoil and a viable plant community in these deserts may take centuries. The marvelous variety and plentitude of wildlife in Africa also is seriously threatened by growing numbers of humans who have little understanding how their search for food is destructive of ecosystem viability (Yeager and Miller, 1986). (Statistics in this paragraph taken from El-Ashry, 1986.)

How serious is our dependence on fossil energy? Modern American agriculture has gone furthest in the world in its use of fossil fuels and is the most vulnerable. Between 1949 and 1968 American agriculture increased its use of fertilizer nitrogen by 648 percent; pesticide use increased 168 percent in the same period. Yields on Illinois corn went up from fifty bushels an acre on the average to ninety-five bushels; but this near doubling required a fortyfold increase in energy-intensive fertilizer. (Figures from Jackson, 1980, p. 25.) Nitrogen fertilizer gives the earth the chemicals that plants need but in so doing it kills the natural nitrogen-producing bacteria. Before long, the soil becomes " addicted"; without the chemical fix, yields drop drastically. The technical fix becomes costlier and costlier.

Our food processing and distribution systems also are highly dependent on fossil energy. Every calorie of food energy on the dinner table of a typical American requires ten calories of fossil energy subsidy. What will we do when fossil energy becomes scarce or unavailable? On April 5, 1981, the *Louisville Courier-Journal* ran an advertisement by the Mobil Oil Corporation proudly proclaiming that American farmers require more petroleum products than any other industry. It stated further that it required one gallon of gasoline to grow one bushel of corn. Amish farmers, who use traditional farming methods, horsepower, and little fossil energy, grow a bushel of corn for one-sixth the cost incurred by the modern farmer. (W. Berry,

1981, p. 260) "That we should have an agriculture based as much on petroleum as on the soil—that we need petroleum exactly as much as we need food and must have it *before* we can eat—may sound absurd. It *is* absurd. It is nevertheless true." (W. Berry, 1977, p. 37)

The energy-intensive mode of American farming is not only vulnerable to fossil energy shortages but also to high costs. The capital outlay for equipment and machinery (which also is embodied energy), in addition to land and buildings, is enormous. A family beginning farming today on an average-size farm would need capital of from $500,000 to $1,000,000; it is little wonder that most farmers start by inheritance. Annual outlays for seed, fertilizer, pesticides, and fuel also are substantial. Most farmers have to borrow the money, the interest on which adds to costs. If prices for their products fall, they may be unable to make the payments on their loans and their bank forecloses on the land. Madden (1987, p. 19) quotes Neil Harl of Iowa State University, who states, "unless something is done, or circumstances change, as many as one-third of the farmers nationally will move to insolvency, taking down their lenders, their suppliers and other merchants, and inflicting incalculable damage upon the fabric of rural communities."

The Office of Technology Assessment (OTA), an analytical arm of the U. S. Congress, estimated in a March 1986 report that one million American farms would disappear between then and the year 2,000. In 1986, alone 65,000 farms, one every eight minutes, went out of business. Of course, a farm sold to pay debt is bought by someone else, usually a larger or corporate enterprise, and the land continues to be exploited. The OTA report estimated that in the year 2000, about 50,000 large farms would account for 75 percent of agricultural production in the United States. This turnover constitutes not only the personal loss of livelihood of these families; it also is the loss of a culture, an especially treasured way of life. Family farming lost its cultural hold as the driving force of new technology, competition, and finance thrust old ways aside.

Wendell Berry wrote in the preface to the 1986 second edition of his 1977 book that the situation, which was not good in 1977, is now "catastrophic"; every problem he wrote about then has grown worse. Public concern about the problem also has grown and social learning is slowly taking hold among some farmers, some academics in agricultural colleges, and even in that bastion of conservatism: the U. S. Department of Agriculture.

High energy and chemical inputs to modern agriculture are now taking a costly toll in pollution of the environment. Now that many cities are served by sewage treatment plants, agriculture has become the main polluter of our streams, rivers, and lakes. Water erodes fields and carries not only topsoil but chemical fertilizers and pesticides into our waterways. One effect of this pollution is eutrophication (excess nutrients) of water bodies, which upsets the balance of nature among the species in the water. Another effect is the toxification of marine plants and animals, so that they become ill and are unfit for human consumption. For example, New York State residents are warned not eat fish from Lake Ontario because toxics have bioaccumulated in their body fat. Chemical toxins also sink into the soil and eventually down to the groundwater. Wells that tap into polluted aquifers pull up water that is unfit for human or animal consumption; it may even be unfit for irrigation. The EPA is now investigating groundwater pollution; estimates are that as much as 40 percent of groundwater is contaminated by toxics.

Humans began using irrigation to make semiarid areas suitable for growing crops more than 2,000 years ago. Irrigation is widely practiced in all parts of the world. Using fossil energy to pump water has greatly expanded the use of irrigation, even into areas that normally have sufficient rainfall. Irrigation cannot insure a sufficient food supply for us, however, because little prospect for its further expansion exists. Increasing scarcity and cost of fossil energy is one limitation. Irrigation of land for extended periods also builds up salts in the soils to the point where they are no longer suitable for growing plants. Moreover, water, as with other resources, is being depleted, especially in those areas that have little rainfall. For example, the Ogallala aquifer, discussed in Chapter 2, is being drawn down at a much faster rate than it can be recharged by natural processes.

The drought of 1988 in the Midwest demonstrated a cascade of consequences when normal rainfall no longer occurs: fossil energy powered irrigation was hopelessly inadequate; rivers, lakes, and ponds dried up quickly, foreclosing water transportation; wildlife and fish either died or failed to reproduce; a virtual epidemic of forest fires broke out; and of course, crop losses were huge.

People typically have established settlements where the land was fairly level and good farm land was nearby. As cities have expanded in size, prime farm land has been taken over for building lots, parking lots, and roads. "Right now in the United States, around 2,000 acres of prime farm land are gobbled up by urban sprawl each day." (Jackson, 1980, p. 24) This amounts to more than one million acres per year. Urban sprawl in the United States has removed from cultivation an area the size of Ohio. In a world hungry for food, can you think of a less foresighted way to prepare for our need for good farmland? Urban sprawl into good farmland is a worldwide phenomenon. Population pressure in China makes every hectare of farmland highly treasured, but it also forces cities to expand into farming areas. We can only expect it to get worse as human population continues to grow.

This review of practices and problems has shown how lack of foresight, or lack of an understanding of the interconnectedness of natural things, or just plain greed, has been making our food prospects poorer rather than better. It is obvious that we cannot continue in our present ways if we wish to have a sustainable food supply for all humans. Population limitation must become a crucial aspect of any future policy. What else can we do?

Strategies For A Sustainable Food Supply

The idea of a sustainable agriculture was discussed thoroughly at several recent conferences; the following definitions came out of those deliberations: "Sustainable agriculture does not deplete soil or people... Sustainable agriculture protects soil and water and promotes the health of people and rural communities... Sustainable agriculture needs to be ecologically sound, economically viable, socially just, and humane." (Quoted in *The Land Report,* Spring 1986, p. 8)

These definitions are stated more as goals than as programs, but they can guide programmatic choices. It is clear from our previous discussion that limiting human population and easing away from our dependence on fossil energy will be essential. What else can we do to maintain an adequate supply of food in our sustainable society? Actually, much can be done but it will require hard work.

Lessons From The Past

We could study cultivation that has been persistently successful for thousands of years. Wendell Berry did just that when he visited Peruvian peasants high in the Andes Mountains. (1981, pp. 3-46) They had practiced agriculture in that region for 4,000 years. He summed up the themes of Andean agriculture this way: frugality, care, security in diversity, ecological sensitivity, and correctness of scale.

China combines the vastness of America with traditional agricultural practices that have been worked out by trial and error for thousands of years—as in the Andes. Much of China is mountainous or desert: the cultivable land constitutes only 11 percent to 14 percent of the total, according to various best estimates. One-half of that land is cultivable only because of irrigation. In the United States, by contrast, 21 percent is now cropland and, if we took "China-like" measures to use land, that figure could rise to 30 percent. The ratio of total population to cultivable land is *seven times as dense* in China as it is in the United States. Furthermore, the Chinese use very little fossil energy (mainly for chemical fertilizer and some pesticides). How do they manage to feed their 1,000,000,000 people?

Mainly, they follow the same themes as the Andes farmer: plots are small, they are frugally and carefully tended, they rotate crops, they recycle human and animal wastes back to the fields, they do not waste anything, and they tune their usage to climate, topography, and soil. No land is planted to lawns. No habitat is saved for wildlife in farming areas (they do maintain wildlife habitat in preserves in the mountains). They grow two or three crops per year, by rotation, on each plot that is suitable. The revolutionary communist government tried to improve production by importing methods from the Soviet Union: large fields with big machinery, organized into communes with an emphasis on grain. The reforms failed, all have been abandoned; the Chinese have returned to traditional methods, including allowing peasants to sell their produce in free markets. Agricultural productivity has risen and is now sufficient to feed their people.

In 1985, the Chinese Academy of Sciences invited a "Pollution Control Delegation," of which I was a member, to visit China. We visited a pig farm on the outskirts of Guangzhou where the pigs are fed with food wastes from the city. Some of the pig feces go into underground digesters to produce methane that is used for cooking and to run a small generator for electric lights. Some of the feces go into a fish pond to nourish fish that are periodically harvested for food. The remainder of the feces are spread on the fields where vegetables are grown. China leads the world in developing technology for biogas generation and in methods for maximizing yields from aquaculture. Food supply worldwide could be greatly increased if intensive cultivation of the land were practiced as in China. We should keep in mind, however, that this strategy takes over the entire ecosystem to serve humans; the needs of other species are given little weight.

Food Grown In Water

What are the prospects and limitations of using fishing, aquaculture, and mariculture to produce food? The deep sea fishing industry has now developed such technical proficiency and capital equipment that it overfishes, reduces reproductive capability of fish populations, and forces fish stocks into decline. This industry illustrates the "tragedy of the commons." It was the victim of linear thinking aimed only at increasing immediate yield and ignored the maxim: "We can never do merely one thing." Fishing yields, worldwide, are down so drastically that we can expect less rather than more food from that sector.

In some places, the trend toward fish farming is growing. Salmon farmers in Norway raise young salmon in cohorts—each age group within a separate large round net shaped like a bowl. They are tended with plenty of food just as cattle are fattened in feedlots in the United States. Because the food for the fish comes from plants grown on land, it is not clear that there is any gain in total food availability from fish farming. Eating salmon is perceived as superior to eating grain, just as eating beef or pork is perceived as superior to eating corn; therefore, the market is good and the enterprises are profitable. Their whole investment was threatened by red-tide algae in May 1988. They towed their nets deep into the fjord and managed to escape the poisonous red-tide. Now the fjords are becoming so eutrophicated that sport fishing and other water recreation is losing its attraction.

Growing fish for food rather than sport in freshwater ponds and lakes is not practiced much outside Asia. Additional food could be obtained from aquaculture in many areas that are not well-suited for farming by impounding streams on steep slopes or digging ponds in marshes. Much additional research is needed to guard against ecosystem degradation from these human intrusions into nature.

Food from algae can also be grown in ponds. Normally, we think of algae as a smelly unpleasant nuisance. Actually, many marine animals live off these tiny plants that use sunlight and nutrients as they grow in water; their energy is passed up to predators at the top of the food chain, like ourselves. Certain kinds of algae, Spirulina for example, have a protein content as high as 66 percent and can be eaten directly as food by humans, as is done by some African tribes. Entrepreneurs are beginning to grow algae commercially to be used as a food supplement for cattle, hogs, and chickens; they have not yet made it sufficiently palatable that people in the West are willing to consume it directly. Spirulina algae reproduce quickly with high yields in very hot weather. An almost continuous input stream of nutrients and almost continuous harvesting must be programmed into the process to make it work effectively. At this early stage of development, algae growing apparently can be as productive, and possibly more productive, as growing plants in soil.

Becoming Vegetarians

Feeding plants to animals and then eating the animals is one of the oldest ways of acquiring food that humans have used. It also is a much less efficient use of plant nutrients than eating the plants or seeds directly. A person receiving nutrients from eating meat requires six to eight times as much plant nutrients as would be required if the plants were consumed directly. Most world hunger could be alleviated if food grains were diverted from feeding animals and distributed instead to needy people in other lands. This simple fact has been understood for a long time; yet, this strategy has never been implemented. The reason is economics. The poor people in far away lands have little money to buy grain. The much richer people in developed countries are willing to pay a premium to be able to eat meat instead of grain. Some nutritionists argue that meat and fish are needed for a well-balanced human diet; others claim that a perfectly adequate diet can be derived from eating only plants. Certainly, the vegetarian option could someday be implemented but people would have to learn and value new dietary patterns.

Grasses and leaves cannot be digested by humans; therefore, it would make sense to continue to graze cattle, sheep, and goats on land that cannot be cultivated for grains—herds of chickens, ducks, or geese also can be grazed. We could, of course, leave such lands to wild animals if we could disabuse ourselves of the notion that such lands are unproductively "wasted."

Weaning Ourselves From Fossil Energy

As fossil energy becomes scarcer and scarcer, costs are likely to rise so high that energy intensive agriculture will no longer be feasible. How do we shift away from that style of agriculture? As we contemplate the restructuring of agriculture, we should keep in mind there is only one imperative value: the life and health of the world's ecosystems. For them to be kept viable, healthy topsoil must be regenerated and maintained.

A second reason for turning away from energy intensive agriculture is that chemical inputs become less and less effective for their intended purpose but more destructive to the ecosystem. As noted, heavy inputs of nitrogen fertilizer kill energy-fixing bacteria; soils then require increasing amounts of fertilizer to maintain productivity. Insects and weeds develop immunity to pesticides and larger doses or different chemicals are required to kill them. These chemicals sink into the soil and form a barrier that roots cannot penetrate; when dry spells come along their shallow roots cause the plants to wilt. Experiments show that plants on nearby plots that have not been treated with chemicals grow deeper roots enabling them to withstand dry spells better. More and more farmers are discovering that chemicals are destroying the ecological vitality of their farms.

Some farmers are turning to regenerative agriculture; between 30,000 and 40,000 organic farmers are found throughout the United States. Time and thoughtful husbandry are needed to restore ecological vitality to a farm that has been heavily treated with chemicals. A number of organizations and research institutes are developing the appropriate knowledge to help guide farmers through this transition.

Institutions Promoting Regenerative Agriculture

In the United States alone, dozens of research and training centers are devoted to regenerative and sustainable agriculture. The Rodale

Press (Emmaus, Pennsylvania 18049) has spawned a cluster of institutions devoted to regenerative agriculture. It publishes the magazines *The New Farm, Prevention,* and *Organic Gardening,* for example. It has established the Rodale Research Center at Kutztown, Pennsylvania. Its new Regeneration Project is an ambitious social change initiative for solving the problems identified in this chapter; it also sponsors a newsletter *Regeneration* and Rodale's Regenerative Agricultural Association.

The Land Institute lies on prairie land near Salina, Kansas (2440 East Water Well Road, 67401). Founded and led by Wes and Dana Jackson, it is a nonprofit organization, devoted to the study of sustainable alternatives in agriculture, energy, waste-management, and shelter. It gives special attention to developing high-yielding, seed-producing perennials. The Jacksons work with between eight and ten student interns each semester. The institute also publishes *The Land Report* three times each year.

The Meadowcreek Project was started in the mid-1970's by brothers David and Wilson Orr. They selected a 1,500 acre site in the Ozarks near Fox, Arkansas (72051) to establish "a center for education and research in applied ecology, agriculture, renewable energy systems, forestry, wildlife, as well as the ethical, social, economic, and political aspects of sustainability." The project emphasizes learning by doing and now conducts a very active educational program, hosts conferences, and can support scholars in residence. The farm is oriented toward establishing a permaculture.

Other institutions, with which I am less familiar, are: The Malachite Small Farm School (A.S.R. 21, Gardner, Colorado, 81040), which emphasizes organic methods; Tillers Small Farm Program (Kalamazoo Nature Center, 7000 North Westnedge Avenue, Kalamazoo, Michigan 49007), which emphasizes agricultural techniques for Third World countries; The New Alchemy Institute (237 Hatchville Road, East Falmouth, Massachusetts 02536), which emphasizes greenhouses, aquaculture, gardens, solar energy shelters; The American Farm Foundation (P. O. Box 288, Vienna, Missouri 65582); The Farallone Institute (15290 Coleman Valley Road, Occidental, California 95465); The Kerr Center for Sustainable Agriculture (Poteau, Oklahoma 74953); The Center for Rural Affairs (P. O. Box 405, Walthill, Nebraska 68067); The American Farm Foundation (1712 Riback Road, Columbia, Missouri 65201); The Committee for Sustainable Agriculture, which publishes *Organic Food Matters,* (P. O. Box 1300, Colfax, California, 95713); and Terra Nova (2919 Oakland, Number 2, Ames, Iowa 50010).

This minimovement also has spawned a new journal: *American Journal of Alternative Agriculture* (9200 Edmonston Road, Greenbelt, Maryland 20770). A new guide written by Susan Sansone has recently been published: *Healthy Harvest: A Directory of Sustainable Agriculture Organizations* (Potomac Valley Press, 1424 16th Street NW, Washington, D.C. 20036), which lists more than 300 pro-organic groups across the United States; more than 100 are described extensively (some with illustrations).

Wendell Berry favors a return to traditional agriculture with smaller farms, smaller machinery, fewer chemical inputs, diverse crops, and an emphasis on regenerating the topsoil. He studied and wrote about Amish rural culture (1981, pp. 249-63). The Amish have deliberately separated themselves from the cultural influences of modern North America. They have flourished by maintaining traditional farming methods with few fossil energy inputs. While the rest of American agriculture was transforming to larger and larger corporate-style farms, and more and more farmers were going broke and leaving the land, the Amish have doubled their population in twenty years. They have taken over hundreds of run-down farms and transformed them into viable homesteads, mainly by dint of hard work and sensible husbandry. Six to eight Amish families can support themselves on the same amount of land (about 500 acres) that an energy-intensive farm requires to support one family. Berry's description of Amish farming was similar to the style of farming that I grew up with as a boy on a 150-acre farm in Minnesota more than fifty years ago. Much of the relevant knowledge is still available; reverting to it would not be that difficult.

A major reason Berry likes smaller, less energy-intensive farms is that they nourish rural community life.

To farmers who give priority to the maintenance of their community, the economy of scale (that is, the economy of *large* scale, of "growth") can make no sense, for it requires the ruination and displacement of neighbors. A farm cannot be increased except by the decrease of a neighborhood. What the interest of the community proposes is invariably an economy of *proper* scale. (W. Berry, 1981, p. 261)

The culture that sustains agriculture and that it sustains must form its consciousness and its aspiration upon the correct metaphor or the Wheel of Life. The appropriate agricultural technology would therefore be diverse; it would aspire to diversity; it would enable the diversification of economies, methods, and species to conform to the diverse kinds of land. It would always use plants and animals together. It would be as attentive to decay as to growth, to maintenance as to production. It would return all wastes to the soil, control erosion, and conserve water. To enable care and devotion and to safeguard the local communities and cultures of agriculture, it would use the land in small holdings. It would aspire to make each farm so far as possible the source of its own operating energy, by the use of human energy, work animals, methane, wind or water or solar power. The mechanical aspect of technology would serve to harness or enhance the energy available on the farm. It would not be permitted to replace such energies with imported fuels, to replace people, or to replace or reduce human skills. (W. Berry, 1977, pp. 89-90)

The February 1986 issue of *Mother Earth News* headlined a story about a farmer in southern Minnesota who had lost his farm to the bank because he could not meet payments on his loan; he was left with only five acres. He decided to farm those five acres intensively, but organically, concentrating on producing a variety of fresh vegetables. He used only small and appropriate technology to lower capital and operating costs. He developed his markets and emphasized high-quality fresh produce. In 1985, he derived $27,000 net income from those five organic acres. His model has proved so successful that Minnesota has invested several hundred thousand dollars to help him carry his model to another 100 displaced farmers, hoping that they too can become productive once more on small acreage. This "small is beautiful" success story illustrates how we could wean ourselves from our dependence on fossil energy while still maintaining an agriculture that would be highly productive of nutritious food.

Land Use Planning And Controls

Land, and the uses people put it to, is so fundamental to society that every society has found it necessary to control the use of land. The more crowded our planetary surface becomes, the greater the necessity for landuse planning and controls. When the Europeans colonized North America, they found such a plenitude of land that planning seemed unnecessary and controls were kept to a minimum. Many Americans have now come to believe that people should be free to do anything they wish with their land. That assumption really is not true, because no society can allow people to do **anything** they choose with their land, but the myth stands in the way of sensible planning and controls.

If we are to preserve farmland from being "developed" into building lots, roads, parking lots, and so on, we must plan the use of land and employ legal strategies to ensure that good land keeps producing food. Some countries utilize very tight controls. In Norway, for example, the national Ministry of Agriculture must grant permission to take good farm land out of production. In China, title to all land–and therefore ultimate control–is held by the government. Such tight controls would be unpalatable in North America, but other incentives are being used to keep farms producing food. For example, New York and several other states have passed Agricultural District Laws enabling the farms on the periphery of a metropolitan area to form a district and be recognized as a legal entity. The farmers in the district pledge not to sell their land for development for a certain period of time (for example, seven years); in return, the local and county governments assess and tax their land for its agricultural value, not its development value. Such weak controls may not be adequate to the need. When the need to protect prime land becomes powerfully manifest, we can be sure that government will have no other choice than to plan and control rigorously.

A Biotechnical Solution?

Because they live and die within a single season, most cultivated annual crops must be more enterprising than perennials and produce myriad seeds to ensure survival. Humans, desiring lots of food from seeds, propagate the more productive annuals, but in doing so they must use till agriculture with its attendant problems of erosion and loss of topsoil. Wes Jackson (1980) of the Land Institute wants to find and propagate perennials that will produce as many seeds as annuals. Because they would not need to be planted each year, we could dispense with till agriculture and rebuild the topsoil.

Jackson's ideal is to propagate a group of high-yielding, seed-producing, herbacious perennials in a polyculture. His dream system would reorient agriculture toward the ways of nature; it would emphasize diversity; it would rebuild and maintain the topsoil; it would use natural controls for weeds and pests; and it would diminish fossil energy inputs. Alas, nature has not visibly and abundantly produced the kinds of plants he seeks. He hopes to assist nature by finding special mutant varieties that can be crossbred with currently used varieties to develop new ones coming closer to the high-yielding perennial ideal (a perennial corn has been found, for example). Even though he is cautious about gene splicing, he sees that new technology as possibly the only way to develop species with the qualities he seeks. If perennials could successfully be propagated to produce high yields of seeds, it would stimulate a revolution in agriculture—promising a sustainable food supply not only for humans but many other species as well in a diversified ecosystem that draws from and nourishes a healthy topsoil.

Bioengineering

Bioengineering is a special case of biotechnology that is so important it has become a new concept in our lexicon; it goes by such other phrases as *recombinant DNA, genetic engineering, and gene splicing.* Scientists discovered the DNA structure in the 1960's. They have since learned how to "splice genes," in effect, remanufacturing biological structures to emphasize characteristics more to the liking of humans; this procedure is called *recombinant DNA.* Genetic structures are extremely complicated. A tiny microorganism contains 3,000 to 5,000 genes; more complex organisms have approximately 50,000 genes. Humans are estimated to have about 500,000 genes. Scientists are slowly linking specific genes to specific traits in organisms. By adding, subtracting, or substituting genes they can engineer organisms to have desired traits or perform desired functions.

This new line of inquiry has spurred a vastly expanded biological research effort. Many firms see the possibility to make a lot of money by inventing, patenting, and selling new chemicals, cures, plants and animals. They have hired scientists and established laboratories specifically for this purpose. They have sent gene prospectors to the far corners of the globe looking for genes with special characteristics that can be used in their recombinant research. Less developed countries have begun to worry about their genetic treasure being stolen to make more wealth in developed countries and are beginning talks to establish worldwide controls on gene trafficking.

As is usual with powerful new knowledge, some people make strong claims for the benefits, while others make dire predictions of calamity if something goes wrong; both may be correct. The potential for changing things is enormous, hence, the list of potential benefits is lengthy, with no end of new ideas in sight. Examples of immediate prospects that would expand food production are: A bovine growth hormone that could increase milk production by 30 percent per cow; an ice-minus bacteria that could reduce frost damage; plants engineered to resist herbicides that would kill neighboring weeds.

Genetic engineering signals the most radical change in our relationship with the natural world since the dawn of the age of pyrotechnology. Up until the past decade, we were still wholly dependent on the gift of fire bestowed on us by Prometheus.... With recombinant DNA technology it is now possible to snip, insert, stitch, edit, program and produce new combinations of living things just as our ancestors were able to heat, burn, melt and solder together various inert materials creating new shapes, combinations and forms. The transition from the Age of Pyrotechnology to the age of Biotechnology is the most important and disturbing technological change in recorded history. (Rifkin, 1985, p. 41)

While few benefits have yet been produced, genetic engineers hope to create: microbes that eat pollution, that leach useful minerals out of ore, that accelerate the production of cheap energy through biomass conversion.

They hope to develop plants that are resistant to certain diseases or perennial plants that could produce grain yields as high as annual plants. They hope to design new animals that produce leaner meat as well as animals that convert their food to meat (or milk or eggs) more efficiently. Recently, a newspaper carried a story about a bioengineer who is trying to produce a cow that will give only low fat milk.

These are not hybrids but animals that reproduce just as natural animals; they are new beings. The U. S. Patent Office recently decided that new creatures are patentable. Bioengineers even hope to use cloning to reproduce millions of identical copies of desirable biostructures. Genetic surgery is also contemplated for humans. Current efforts focus on correcting gene abnormalities for victims of crippling hereditary diseases such as sickle cell anemia. If this line of research is successful, however, we are likely to see efforts to correct the gene structures of people who tend to gain weight too easily, who are too short or too tall, and so on. Is there no end? Should there be?

We are in the early dawn hours of a new epoch in history, one in which we become the sovereigns over our own biological destiny. Though reluctant to predict a timetable for the conversion of our species from alchemist to algenist, those involved in the biological sciences are confident that they have at last opened the door onto a new horizon, one in which the biology of the planet will be remodeled, this time in our own image. Our generation, they say, stands at the crossroads of this new journey, one whose final consequences won't be fully grasped for centuries to come. . . .

The question of whether we should embark on a long journey in which we become the architects of life is, along with the nuclear issue, the most important ever to face the human family. The ecological and environmental questions raised by such a prospect are mind boggling: the economic, political and ethical questions that accompany this new technology are without parallel. (Rifken, 1985, p. 43-45)

This line of research is very powerful, with the potential to change the life of every living creature; it puts humans very close to the position of playing God. Those who urge us to go slow, or desist altogether, point out that we cannot possibly know the ramifications of what we are doing. Some have suggested that a virus or microbe created in a laboratory could accidentally find its way into the environment. Because it would have no natural enemies, and other organisms would not have developed natural defenses, it could create an epidemic. Obviously, this is a powerful agent of social change. For example, when European colonists came to New England, they displaced Native Americans not so much by force of arms as by infecting them with smallpox, to which most of the colonists were already immune. In some villages, 90 percent of the native inhabitants succumbed to smallpox; their culture was devastated and their power to resist the Europeans was almost nonexistent (Cronon, 1983). National defense establishments, or terrorists, probably are trying to create new critters to wield more effective biological warfare against their enemies.

History tells us what happens to an ecosystem when new species are introduced from another continent. They usually fit in well with those already present, but sometimes the consequences have been nearly catastrophic. North Americans wish the following species had never been introduced: Gypsy moth, Kudzu vine, Dutch elm disease, chestnut blight, starlings, Mediterranean fruit fly. Natural evolution in Australia had never produced mammals; they were introduced by humans from other continents. When rabbits were introduced, they had no natural enemies and reproduced at epidemic rates, nearly stripping the land of vegetation. The epidemic was slowed only when humans introduced a virus that specifically targeted rabbits. Now the rabbits have developed immunity to the virus and are once more reproducing in epidemic numbers. Foreseeing deleterious consequences is always difficult; but doubly so for species with which we have had little experience.

Some of the longer-range effects of genetic engineering are more subtle and have the potential to profoundly change society. Genetic researchers intend to bring more and more of the biosphere under human control; it is the ultimate expression, to date, of human domination of nature. Increasing numbers of wild animals and plants will be crowded from their niches, their life support appropriated for human purposes. Unless humans learn to restrain their growth in numbers and use of resources, we will turn the biosphere into an overwhelmingly human-dominated entity. Our special sense of the wonder and magic of nature will be diminished; our view of it will be more akin to the way we relate to machines.

Genetic engineering clearly has the potential to produce our food more efficiently. Keep in mind we are expanding biological efficiency but not life support systems; limits of water, nutrients, and energy will still limit productivity. The socioeconomic-political consequences are difficult to foresee but they will be great. For example, a 30 percent increase in milk productivity from new cows raised with a bovine growth hormone could produce such a glut of milk as to drive thousands of dairy farmers, already on the brink of insolvency, to leave their farms. Moving into this new era could destroy our rural way of life. Predictably, we will not confront that choice directly. Our linear mode of thinking will accept each increment of technological change for its visible benefit and will ignore the hidden cost until it becomes painfully manifest many years later.

This new technology cannot be stopped; the genie cannot be stuffed back in the bottle; our only hope is to find effective controls. Scientists and public officials in the United States have recognized the potential dangers of genetic engineering and have instituted an elaborate set of reviews before a new organism can be cleared for release. Most bioengineers proclaim confidence in the safety of these procedures, but here are some cautions concerning them: 1) The proposed new entity should be considered "guilty" until proven innocent; exhaustive testing must precede development—even if it adds delays and increases development costs; 2) we must carefully license who can do this kind of research; it is not appropriate for high school biology classes; and 3) we must move quickly to worldwide controls; it will do little good to have effective controls in one's own country if lax controls in another allow an epidemic to start.

We should be as foresighted as possible about the long-range consequences of these new technological developments. We should consider them just as carefully as we consider new legislation, for they will shape our lives just as permanently as legislation—perhaps more so. It is easier to repeal a law than to repeal a technology or a new creature. We should require assessments of their environmental, social, and value impacts before allowing deployment. We urgently need to learn how to do that.

Reconceptualizing Agriculture

Agriculture traditionally has been considered an economic activity to which economic analysis applies. A new wave of thinking views agriculture in a much broader perspective. Dahlberg (1986) reports on a conference that considered new directions for agriculture; it pointed to four major but often neglected elements: 1) inadequate attention to externalities—social, environmental, and health effects; 2) inadequate attention to societal goals, ethics, and values; 3) local and regional concerns and developments that do not take into account the larger national setting; and 4) the way all of these considerations are integral parts of the larger global setting.

In a similar vein, Lowrance, Hendrix, and Odum (1986) identify four levels of sustainability that must be considered if we are to have a sustainable agriculture: 1) *agronomic sustainability*: the ability of a tract of land to maintain acceptable levels of production over a long period of time; 2) *microeconomic sustainability*: the ability of a farm, as the basic economic unit, to stay in business; 3) *ecological sustainability*: maintenance of life-support systems of larger-scale landscape units (forests, river basins) far into the future; and, 4) *macroeconomic sustainability*: monetary and fiscal policy at the national and international level. Policies could be aimed at any level but need to consider implications at other levels if they are to be successful.

Agroecology is the name given to a new scientific subdiscipline concerned with sustainable agriculture. Altieri (1987) defines it as being concerned with the optimization and stability of the agroecosystem as a whole. Altieri believes the transition to a sustainable agriculture can be accelerated by 1) focusing research attention on long-range problems; 2) integrating agricultural planning with an ecological perspective for all land use; 3) encouraging local producer-consumer cooperatives to set production goals; 4) organizing small farmers into a strong political constituency; 5) subordinating agricultural resource interests to broader political and economic interests; and, 6) consumers becoming more effective in challenging agriculture research that ignores nutrition and environmental issues.

These are but three examples of the kind of rethinking that is needed, and now being carried out, if we are to develop a sustainable agriculture.

Conclusion

A sustainable society can flourish only if it is founded on a sustainable agriculture. An agriculture that is trying to support more people than its carrying capacity is not sustainable. An agriculture that wastes and loses its topsoil is not sustainable. An agriculture dependent on heavy inputs of fossil fuels is not sustainable. An agriculture that poisons soils, waters, and the air is not sustainable. Modern agriculture does all those things and must be turned around if it is ever to become sustainable.

We need a regenerative agriculture that works with nature rather than against it. Regenerative agriculture will help the soil to renew its vitality. It will propagate plant communities that are diverse and will serve other animals in addition to humans. Such an agriculture will need many hours of human labor in loving nurturance of natural elements and systems. We can expect a surplus of labor in the future, therefore, many people could turn to doing their own work nurturing plants and animals on the land. These husbandmen of regenerative agriculture will have to understand ecological principles and pursue strategies that enhance the vitality and sustainability of ecosystems. We will have to re-vision man as a member of that system, not its master. Our very lives and those of our children will depend on how well we understand what we are doing and on how faithful our stewardship is as we nurture the land and its creatures.

Work That Is Fulfilling In A Sustainable Society

Chapter 10

How will we organize our work in a sustainable society? First, I examine the dilemma of modern society with respect to work and unemployment. Then, I propose redefining and restructuring work to better provide fulfilling life activities for everyone. Finally, I discuss some innovative ideas for restructuring economic relationships.

Jobs: The Imperative For Growth

Virtually every society in the contemporary world is caught up in a pressing, persistent demand for jobs. This is as true for the affluent Western societies as it is for the teeming societies in the Third World. No matter what national leaders do, there never are enough jobs. Many other societal ills, such as crime, poverty, psychological disorders, and poor health, are rooted in the lack of jobs.

Providing fulfilling work is constrained by two indisputable facts: 1) the human population continues to grow, and 2) the biosphere, in which all of these people must live and work, has finite resources and a finite capacity for absorbing our wastes. The way that a society tries to solve the dilemma of jobs within these constraints depends on its beliefs about the way the world works—on its DSP.

The paradigm that is currently dominant in most societies, developed and developing alike, has an overly simple answer to the dilemma: spur economic growth. Let us examine the ramifications of this proposed solution.

In the mid-1980s, the International Labor Organization estimated that one billion jobs must be created by the turn of the century to accommodate the burgeoning population that is moving into the work force worldwide. Meanwhile, economies around the world are losing rather than winning the battle to create enough jobs. The Organization for Economic Cooperation and Development reported in 1983 that the average unemployment rate in developed countries had risen from 3 percent in 1970 to 8.7 percent in 1983. The International Labor Organization reported in 1982:

However, the deteriorating employment situation in the Northern hemisphere, where unemployment has now reached from 6 to 10 percent of the active population, is relatively insignificant compared to the situation outside Europe. In Africa, Asia and Latin America, no less than 40 to 45 percent of the total labor force, almost 330 million adult men and women, are either out of work or underemployed, especially in rural areas. (ILO, 1982, p. 3)

Why is it so difficult to absorb new workers? Rural areas already have a surplus of workers, thus few unemployed can work on the land. In most Third World countries, rural people are flocking into cities to find work. Chinese peasants must receive permission before they can move to a city because the government would have great difficulty accommodating them with either jobs or housing. All Third World cities are losing the struggle to provide employment for immigrants from rural areas.

Many people look to technological development to provide new jobs. It can result in new products, new processes and new jobs; but it also produces new machines that do more swiftly, more reliably, more continuously, and more cheaply, the same work that humans used to do. Everywhere in the world, factory workers are being displaced by automated robots on assembly lines. Manufacturers feel compelled to automate as much as possible in order to remain competitive in the world market. Fierce international competition has also created excess capacity. A Harvard Business School study of the auto industry (Dyer, et al., 1987) estimated that the U. S. auto industry had 4 million units of excess capacity in 1987. They project that international competition will create 7 million units of unneeded capacity by the early 1990s. There is overcapacity in nearly all industries; every major city in the world has experienced a decline in manufacturing employment over the past two decades.

Can we not look to high tech production, such as computers, to provide jobs? Even those jobs have proven to be ephemeral; productivity in computer manufacturing and servicing is so high that unemployment is now being experienced in that sector of the economy. There is no escaping the conclusion that the percentage of people employable in industry will continue to decline for several decades.

Although manufacturing now accounts for less than 20 percent of U. S. employment, *nearly half of all workers displaced from 1979 to 1984 worked in manufacturing industries, especially those hard hit by international competition. . . . Manufacturing jobs—especially production jobs—probably will continue to decline. Within manufacturing, the most vulnerable jobs are those of unskilled and semiskilled production workers.* These jobs are not only the easiest to automate, they are also the easiest to move overseas. (U. S. Office of Technology Assessment Report, 1986, emphasis in the original)

Firms in developed countries that make products requiring hand labor feel compelled by competition to move that component of their operations to developing countries where hourly wages are much lower. It is even possible to continue the fiction that a manufacturing operation is "domestic" by locating dual manufacturing plants along national borders. For example, many U. S. manufacturers have plants on the Mexican/United States border. The labor-intensive components of the manufacturing process are carried out across the border in Mexico; the semifinished product is shipped to a plant on the United States side where the product is assembled and shipped.

A country could refuse to admit products made with cheap foreign labor. It also could forbid automation, requiring more use of hand labor. Most countries have been reluctant to do those things because it would raise the cost of products and limit their quality and variety. Furthermore, trade restrictions invite retaliation. Many products could not even be manufactured without materials from abroad. Small national economies would have difficulty surviving without relatively free trade. The world would probably experience a general economic decline, and unemployment would get worse.

Even if agriculture and industry need fewer workers, could we not achieve full employment by expansion of the service sector? That sector is growing but not fast enough to provide full employment. Much service is provided by government (teaching, research, counseling, regulating, and so forth) and there is a limit on the willingness of citizens to pay taxes. Professionals in private practice (consultants, therapists, lawyers, doctors, dentists, and so on) face problems of oversupply; the level of prosperity also affects the demand for professional help. Many service jobs in private business are low paying and probably not very fulfilling (such as food service, cleaning service, and custodial service jobs). The Office of Technology Assessment (OTA), a staff arm of the U. S. Congress, recently estimated that by the turn of the century, personal computers would be as common on office desks as telephones are now. They predicted growing productivity from these machines but also a decline in the demand for office workers (OTA, 1985).

The only answer that seems available to those believing in the DSP is to press ever harder for economic growth, hoping it will bring full employment. Whether it succeeds, a thrust for jobs and growth creates regrettable anomalies: A large automobile firm, such as Chrysler Corporation, must not be allowed to go bankrupt because we need the jobs, although we do not need the cars. Advertising tries to make us dissatisfied with the products we have, even if they still do the job, and it promises us happiness if we buy new ones. Fashions are changed regularly so that perfectly serviceable clothing must be replaced. Firms and governments invest large sums in research hoping to develop new products we did not know we needed. Then advertising convinces us we cannot be happy without them.

Some people believe that maintaining a high level of military preparedness and armaments is an "answer" to the employment problem; some even claim that we cannot "afford" to disarm for fear of unemployment. I leave aside the question of the need for military preparedness to give us security and address only the question of its utility in providing work. No doubt, many jobs are found in the military/industrial complex, but supporting that sector merely for jobs is no answer to our need for fulfilling work. Nearly all of that effort is supported by taxes and constitutes a net loss for human welfare. If military material is not used, it is waste. If it is used, it is destruction. The money spent in this sector could have been spent to make life better for humans and to preserve the integrity of the biosphere. Society can provide fulfilling work in many more constructive ways.

Our persistent thrust for growth constantly increases throughput. At an ever more frenetic pace, we dig up materials, process them, consume the products, and throw the wastes into the biosphere. At each stage, we consume more fossil energy, encourage the greenhouse effect, further damage ecosystems, and drive other species from their niches.

Based on the analysis in Chapter 2, this scenario cannot be sustained and cannot provide full employment. Energy, minerals, and other natural resources are insufficient to support a constant increase in throughput. Accelerating extraction, processing, consumption, and waste disposal would so damage ecosystems as to weaken their carrying capacity. Distortion of biospheric systems can devastate our economies and our health. Trying to solve the unemployment problem by economic growth constitutes a dangerous chase after a goal that probably is unattainable by that strategy.

Lessons From Other Societies

A few societies in the world have achieved something close to full employment. What can we learn from them? Some Eastern European countries and China have a policy of providing jobs for everyone. In practice, this means that a sizeable proportion of workers (10 percent to 20 percent) are placed in "make-work jobs" that have little economic value and provide few opportunities for learning and development. Scholars of employment do not consider this to be full employment (Ginsburg, 1983).

The Scandinavian countries come closer to "real" full employment. In Scandinavia, private firms and labor unions are prominent economic factors whereas both those sectors are weak in Communist countries. In Sweden full employment is a highly prized national goal that is actively pursued by all sectors of the economy. The main mechanism is a proactive National Labor Market Board on which labor, management, and governmental officials serve; it is supported by the national government. Bargaining over wages and working conditions takes place between a national coalition of labor unions and a similar coalition of employers, with government watching over their shoulders. The board has the power to set labor market policy and utilizes such instruments to implement policy as investment relief funds, government grants, subsidies, job training and a vigorous job placement service that mandates listing of all positions (see Ginsburg, 1983, for details).

It may be that Scandinavia has lower unemployment due to the special conditions seldom found elsewhere: 1) the countries are small and homogeneous with high agreement on national goals; 2) a high level of trust is found among major economic actors; 3) their population grows very slowly; 4) they have been careful to minimize abuse of their environment; 5) they have a good information system to monitor closely what is happening in their economy; and, 6) they expect government to guide economic activity (government owns and operates some enterprises).

The Scandinavian experience provides lessons for us, but we should not expect most other countries to be able to follow their example. Realistically, we must expect continuing high unemployment in most parts of the world and probably it will go even higher. It is unlikely that people will be content with high unemployment; rather they will start to make changes. What kind of restructuring might we imagine?

Restructuring Work

The restructuring must start with the way we think about work. The DSP of modern society teaches all of us to think of work as holding a paying job. Actually, employment was uncommon prior to the Industrial Age. Before then, people worked in family groups or tribes; no one expected a wage for contributing to the welfare of the group. Medieval serfs or servants usually received no wage for their labor. Craftsmen made and sold a product, but they were not paid a wage. History shows us, then, that there is no necessary connection between employment and work; therefore, we will separate the two concepts. If we define our goals as paid jobs, we are confusing a means with an end.

Instead, consider quality of life and human activity. Only a few thousand years ago (only a few hundred years ago in North America), hunter and gatherer tribes could supply their life necessities by working, on the average, three or four hours a day (Cronon, 1983; Schmookler, 1984). The rest of the day could be spent on socializing, games, cooking, crafts, etc. Even though they had few material goods, it is not clear that they desired more. Their routines were well worked out and tied to nature's rhythms. We cannot know if they experienced high levels of stress but the very simplicity of their lives suggests that it was not very stressful. Could we claim that they had lower quality of life than we who have more material goods, more comfort, and longer lives, but also more work and high stress?

I am not recommending that we return to a hunter-gatherer way of life; we are far too many people for that to be possible. But, reflecting on earlier times tells us something ironic about our own age. We have built a civilization that, presumably, should make for a better life. We turned to cultivation to ensure our food supply; yet, we now labor much longer and expend much more energy to obtain food. Some of us have plentiful food while millions of others are hungry. We invented machines to save us from drudgery, but now we serve the machines as much as they serve us. We believed we could conquer nature sufficiently to control our destiny but now our destiny is controlled by faceless powerful forces that none of us can control.

Work that produces unnecessary consumer junk or weapons of war is wrong and wasteful. Work that is built upon false needs or unbecoming appetites is wrong and wasteful. Work that wounds the environment or makes the world ugly is wrong and wasteful. There is no way to redeem such work by enriching it or restructuring it, by socializing it or nationalizing it, by making it "small" or decentralized or democratic. (Roszak, 1978, p. 220; quoted in Capra, 1982, p.231)

I *am* recommending that we make quality of life our goal, not simply jobs. Living a good life means engaging in activity that is pleasurable and fulfilling. In a job-oriented society, many of us do things we dislike (typically, work) in order to do other things we like (typically, leisure). Our society imposes that structure on us; it is designed to maximize power and wealth; it is not the natural or the best structure. Rather than maximizing quality of life, it probably diminishes it. Let us rethink society's activity structure—keeping our eye on quality of life, not on jobs or wealth.

James Robertson, who has done some of the clearest thinking about future work (1985), sees three possible future scenarios for the way we approach work in the future. The first he calls "business as usual" in which we would continue on our present path, emphasizing economic growth in the hope of creating more jobs. For reasons already discussed, he sees little hope that it will achieve full employment. He (and I) believe most leaders will continue that policy, however.

He sees the possibility of moving toward two other kinds of postindustrial society: 1) a HYPER-EXPANSIONIST (HE) program, which might be described as superindustrial, embodying a vision of a future based on big science, big technology, and expert know-how—it is industrial society accentuated and writ large; and, 2) we also might pursue a Sane, Humane, Ecological (SHE) vision of life that would require a change in direction, a new breakthrough, that will be primarily psychological and social, not technological and economic. It seeks fundamental social and personal change oriented to amplifying our capacity to develop ourselves as human beings, together with the communities and societies in which we live. The outstanding features of these two visions are contrasted in the following table:

Table 10.1
Contrasting Features of Alternative Visions of the Future

HE	SHE
quantitative values and goals	qualitative values and goals
economic growth	human development
organizational values and goals	personal and interpersonal values and goals
money values	real needs and aspirations
contractual relationships	mutual exchange relationships
intellectual, rational, detached	intuitive, experiential, empathetic
masculine priorities	feminine priorities
specialization/helplessness	all-round competence
technocracy/dependency	self-reliance
centralizing	local
urban	countrywide
European	planetary
anthropocentric	ecological

Source: Robertson, 1985, p. 5.

Readers should note the similarity of the ideas in this table to those in the DSP/NEP contrast set forth in Table 6.1.

If the HE vision were fully realized, most work would be done by machines. Only a fraction of the work force (30 percent or 40 percent) would have paid jobs. Robertson (1985) foresees such a society as divided into two groups: 1) a minority would be highly skilled, highly responsible, highly regarded, and highly paid—a technocratic elite, and, 2) the majority would have no useful work to do and would live lives of leisure. Even if such a society were possible to achieve (Robertson and I doubt that the physical infrastructure needed to support it could be assembled), it would probably not be the paradise it might appear to be at first glance. There would be inevitable status differences between the working elite and the idle masses leading to grumbling about unjust differences in obligations and opportunities; neither group would be satisfied with the social arrangement.

The growth of the exploiter's revolution on this continent has been accompanied by the growth of the idea that work is beneath human dignity, particularly any form of hand work. We have made it our overriding ambition to escape work and as a consequence have debased the products of work and have been, in turn, debased

by them. Out of this contempt for work arose the idea of a nigger: at first some person, and later some thing, to be used to relieve us of the burden of work. If we began making niggers of people, we have ended by making a nigger of the world. . . .
    But is work something we have a right to escape? And can we escape it with impunity? We are probably the first entire people ever to think so. All the ancient wisdom that has come down to us counsels otherwise. It tells us that work is necessary to us, as much a part of our condition as mortality; that good work is our salvation and our joy; that shoddy or dishonest or self-serving work is our curse and our doom. (W. Berry, quoted in Schleuning, 1986, p. 4)

Probably, some people will move in a HE direction while others move in a SHE direction. Employers will try to minimize their dependence on employment by finding other ways to get work done—the HE vision. Simultaneously, because job opportunities will become scarcer, people will restructure their lives to do many things for themselves; Robertson calls it "ownwork." He extensively discusses how ownwork could become the dominant way people will work in the future. It is a work pattern and lifestyle that fits into a sustainable society. Robertson summarizes why he believes people will become disenchanted with employment and turn to ownwork:

    . . . two key features of the path of development that has characterized all industrialized and industrializing societies have been: the increasing dominance of employment over other forms of work; and the increasing subordination of local work and local economies to outside control. The principle in both cases is the same. Just as individual people in their own households have lost the freedom, and the capacity, and the habit to meet their own needs directly by their own work, and have become more and more dependent on paid employment outside their own control to provide them with money to buy goods and services, so local communities have lost the freedom, and the capacity, and the habit, to meet their needs directly by their own work. They too have become more and more dependent on externally organized work to provide an inflow of money from outside, which is then spent on buying in from outside the goods and services required to meet local needs.
    In short, as the industrial-age way of organizing work as employment has become ever more deeply engrained in the structures of industrialized and industrializing societies, it has turned work into a form of activity which is dependent, remotely controlled, and instrumental. Work in the post-industrial age will have to develop in a positively different direction. It will have to become more autonomous, more self-controlled, and more directly related to the needs and purposes of those who are doing the work. (Robertson, 1985, p. 38)

Robertson recognizes that a transformation of the magnitude and depth that he envisages will take a long time. We must realistically expect all three of his visions of the future (business as usual, HE vision, and SHE vision) to be pursued simultaneously, by different people. The ownwork revolution may be far from full realization, but it has already begun.

Robertson cites the following as evidence of the new trend: 1) a rise in self-employment, 2) a rise in part-time jobs, 3) more people adopting self-sufficient lifestyles, 4) new community business ventures, 5) cooperatives, 6) job sharing, 7) early retirement, and 8) paternity leave. He believes that people will turn to ownwork as a positive way of approaching life. If that turns out to be the case, it will help to avoid some of the malaise and other social ills that plague people who chastise themselves for being unemployed in a society that glorifies employment.

Elgin (1981) advocates a lifestyle of voluntary simplicity and believes it would transform our approach to work in ways similar to those foreseen by Robertson: 1)our relationship with our work is simplified when our livelihood makes a genuine contribution both to ourselves and to the human family; 2) the quality of simplicity—realized by greater clarity and directness of involvement—is enhanced by more human-sized, less complex, places of work; we might revitalize our institutions by redesigning them to be more comprehensible; and, 3) the quality of simplicity is also manifest in more direct and meaningful participation in decisions about our work. (Elgin, 1981, pp. 172-74) "Men, by their nature, seemingly cannot be happy unless engaged in enterprises that make them feel useful. They must, therefore, be returned to participation in such enterprises. . . . I propose that men and women be returned to work as controllers of machines, and that the control of people by machines be curtailed." (Kurt Vonnegut, quoted in Schleuning, 1986)

Living Life To The Fullest

What do we want to achieve as we struggle to transform work? What brings fulfillment in work? This topic can be examined from the perspective of the working individual as well as from the perspective of the overall society.

Work That Is Fulfilling For The Individual

One of the best definitions I have encountered of the quality of working life, from the perspective of the individual, was put forth by Arne Mastekaasa (1984). It assumes employment, not simply working, but is of interest for its clear delineation of several valuable facets of work activity:

1. The challenge facet
   a. possibility to develop one's special abilities
   b. interesting work (variety and stimulation)
   c. freedom to decide how the work should be done

1. d. work tasks sufficiently hard to solve
   e. enough authority to get the work done
   f. possibility to do the work one knows best
   g. seeing results of one's work

2. The convenience facet (ease and comfort facet)
   a. easy transportation to and from work
   b. not being asked to do excessive amounts of work
   c. pleasant physical surroundings
   d. chances to forget abut personal problems
   e. enough time to do the work
   f. freedom from conflicting demands
   g. good working hours

3. The financial rewards facet
   a. good pay
   b. good job security
   c. good fringe benefits

4. The relations with coworkers facet
   a. chances to make good friends
   b. friendly and helpful coworkers

5. The resources facet
   a. enough help and equipment
   b. enough information
   c. competent superior
   d. clearly defined responsibilities

Of course, these categories could be elaborated further and other categories added, but these are sufficient to clarify what I mean by satisfying work.

Quality in ownwork also is many-faceted but, being oriented to widely varying personal and local needs, it is more difficult to define. Robertson (1985, p. 189) defines ownwork as "forms of work, paid or unpaid, which people organize and control for themselves; in order to achieve purposes which they perceive as their own; as individuals, in groups, and in the localities in which they live." There is no clear distinction between *ownwork* and *leisure* as there is between *employment* and *leisure*; it would make more sense to distinguish enjoyable from not enjoyable tasks. If a task is inherently not enjoyable, a person who performs it to meet her own needs will more likely find it satisfying than if she is paid a wage to perform it to meet someone else's needs. Reading a book is fulfilling ownwork in the same sense as cooking dinner or bringing in the firewood. The pattern of activities that a person finds fulfilling comprises that person's lifestyle preference. A person who gets involved in politics, for example, is doing ownwork.

Referring to Mastekaasa's categories, a person can select many challenging ownwork tasks; the freedom to do so is probably greater than when employed. Ownworkers also can do things at their convenience. Financial rewards would typically be weaker when doing ownwork. Coworkers would be just as important in ownwork as in employment. Having sufficient resources could be a problem in ownwork, but it frequently is a problem with employment, too. Persons may seek employment (perhaps only part-time) in order to get better access to resources.

A society that accepts easy movement between employment and ownwork would be structurally superior for finding personal fulfillment in work. What else can we say about the structure of work in a sustainable society?

Structurally Redefining Work In A Sustainable Society

As noted, it is doubtful that modern societies can provide full employment; but, all societies *can* provide meaningful opportunities for people to do their ownwork. Ownwork-oriented societies would not encourage people to become wealthy. Paying someone for services means subjugating one person's desires to that of another. A society that lives mainly by ownwork would provide more equal access to goods and services. I do not deny that people would need a specialist occasionally, but society need not admire or encourage wealthy lazy drones who hire others to pick up after them. People who become accustomed to fulfilling many of their needs by their ownwork probably will place less emphasis on buying consumables. They will more easily be comfortable with the idea of limits to growth.

Hawken (1983, pp. 76-79) sees the next economy as an informative economy that will be much less consuming. It will substitute information for material things by raising the demand for information. Whereas mechanical tools extend our muscle power, microelectronics extend our mind into matter. In order to put more information into products, companies will have to put more information into their employees through education, communication, and honesty. We will produce better designs with greater utility and durability. Better-quality products will reduce maintenance, operating expenses, and the amount of material in the product. Because we will make matter more intelligent, we will require less to maintain a high standard of living.

Ownwork covers a wider scope of life than employment and will vary greatly from individual to individual, from group to group, and from community to community. Society should be open to this variety and encourage learning as well as readiness to change. People will gradually learn to perceive ownwork as a creative opportunity and not a second best alternative to employment.

An ownwork society will give less deference to specialized expertise. Specialization encourages tunnel vision and adherence to work definitions that protect the positions of specialists. Specialization serves corporate purposes well in an employment society; but it subordinates the needs of the specialists to those of the corporation. The sustainable ownwork society needs many more integrative thinkers than specialists. While we undoubtedly will continue to train specialists, we must simultaneously teach people how to think integratively. A person can easily learn to use both modes of thinking.

A sustainable ownwork society would encourage people to develop several talents and to rotate the use of their talents. Doing ownwork is likely to be more fluid than being employed where one is required to utilize the specific talents that the employer needs. People should be encouraged to develop any talent they may possess, irrespective of their gender. A system of social facilitators (respect, examples, lessons, relief from burdens) should be developed to encourage the flowering of a multitude of talents. Recall our earlier discussion of *paideia*, the idealized Greek society that emphasized lifelong transformation of the *person* as an art form. An ownwork society should aspire to the *paideia* ideal.

Obviously, very few individuals will be able to do everything for themselves. Some tasks require teamwork and a multitude of talents. Cooperative relationships could substitute for many monetary transactions. A cooperative work pattern was quite prominent in rural America as recently as fifty years ago and is still prominent in many parts of the world today. In an ownwork society, people would place less emphasis on winning and more on treating people justly and doing one's share of common tasks. Material wealth would cease being the gauge by which we measure success.

Could science and technology advance without an emphasis on specialization? Of course it could; we would continue to have plenty of specialists in an ownwork society. We would probably see science and technology in a more utilitarian, less worshipful, perspective, however. In contemporary society we glorify science and technology. We forget that they are means and not ends; they can be used for evil or foolish ends as well as for good ends. They should not be worshiped and they should not become our masters. A sustainable society will continue to support science because it is so important to know how our world works, especially the intricate systemic processes in the biosphere. However, society should distinguish science from technology; it is always important to learn but not necessarily to do something.

Technological development should be oriented more toward "appropriate technology" that helps people to be more self-sufficient. Robertson suggests that a proposed technology should be appraised to see if it fits with the ownwork agenda.

A useful criterion ... is to ask whether the new material, or equipment, or process, or system is likely to enlarge the range of competence, control and initiative of the people who will be affected by it; or whether it is more likely to subordinate them to more powerful people and organizations, and make them dependent on bureaucracies and machines which they cannot themselves control. (Robertson, 1985, p. 180)

With appropriate technology people will be able to do many more things for themselves. People who know how to use solar energy, for example, will be more self-sufficient, and do less damage to the biosphere, than people dependent on large utility plants to supply them with heat and power. The development of the personal computer has freed many persons from their dependence on mainframe computers and from having to use secretaries to type letters and manuscripts. In writing this book, I have learned to do my "ownwork" on a personal computer.

A Guaranteed Basic Income?

Some writers who recognize that future society will not be able to provide employment for everyone have recommended a guaranteed basic income (GBI) (Handy, 1984; Robertson, 1985; Roberts, 1982; and Miller, 1984). They remind us that modern industrial societies already do this through their welfare programs; we do not let people starve or freeze. The GBI, however, would go to each individual as a matter of right; a person would not have to suffer the indignity of proving destitution in order to receive it. Robertson suggests that the GBI, which would be sufficient only for subsistence, should not be taxable; any income beyond it would be taxable. Poor people would be motivated to earn something extra because the basic income would not be threatened. Much of the current welfare structure, including Social Security, could be dismantled because the GBI would routinely go to everyone. Persons who choose to do much of their ownwork could use the GBI to purchase certain necessities in the market that their ownwork could not provide.

Keep in mind that many people will turn to ownwork only because they cannot find satisfactory employment; others will prefer ownwork and may never hold a job. A sizeable proportion of the population will prefer to work for wages (if they can find a job), and will never enter the ownwork pattern. Thus, the ownwork society will be only partially realized. The present economic structure would not be dismantled, but it would be significantly amended.

The present economic accounting system, which mainly tracks money flows and is expressed in the gross national product (GNP), captures only a portion of the economic activity. It misses nearly all housework, volunteer work, child care, community activities, politics, much education and other forms of self-improvement, pastimes that do not involve purchases, and much more. Henderson (1981) estimates that the total value of these activities is larger than that captured by the income accounts.

A GBI would, in effect, support people as they do the basic things of life. Robertson summarizes the pros and cons of the GBI as follows:

Arguments For GBI

1. Citizens will no longer be divided into two classes (those with jobs and those on welfare).

2. The poverty trap will be abolished—poor people will be able to work without losing their basic benefits.

3. Much of the activity now in the "black economy" will be legitimized (it will not be necessary to conceal various income sources for fear of losing benefits).

4. Greater equality will be created between the sexes (think of the GBI as wages for housework for all persons).

5. It will be easier for people to spend time on community activities.

6. It will have an economic liberating effect. For example, employers would pay whatever wage the market set since a minimum wage would no longer be required. Also, people would be disinclined to take unpleasant menial work, forcing wages to rise for those jobs.

Arguments Against The GBI

1. The rates of tax needed to finance it would be too high. Robertson discusses the income tax, a value added tax (VAT) (similar to a sales tax), and a land tax. He concludes that financing is manageable.

2. What is required for subsistence would vary considerably with circumstances (type of community, type of dwelling, the stage in one's personal lifecycle) so why pay the same GBI to everyone?

3. Lazy people will live off the hard work of others. In general, people will slack off and work less hard. (He does not believe this will happen due to arguments 2 and 3 in favor of GBI.) (Robertson, 1985, pp. 167-70)

A GBI would have so many ramifications throughout the social system that it is difficult to anticipate them all. It should be put into effect cautiously and experimentally. A GBI would require quite a productive economic structure with efficient machines to do much of the work. A society probably could not support a large armament and military program in addition to a GBI. Some writers suggest we could find ways to pay people for ownwork without going to a full-fledged GBI for everyone (Harman, n.d.). The main impediment to adopting a GBI would be the present pattern of beliefs and values that emphasizes wage employment as the key to being considered a contributing member of society. Failure of modern society to provide adequate employment would be a powerful stimulus to rethink our basic beliefs and values.*

Bioregionalism

Bioregionalism is an interesting new concept for restructuring society that is especially relevant for the Gaia Hypothesis and for the ownwork approach to structuring work. Barely a decade old, it has already become something of a movement that claims sixty affiliated groups, a dozen regional congresses, two North American congresses, a nascent continental organization, approximately fourteen publications, and several books.

The kernel of the idea is that economic, social, and political life should be organized by regions that are defined by natural phenomena. "Bioregion is a life-territory—a region governed by nature, not legislature" (Sale, 1985). Some would determine the boundary of a region by 'biotic shift'; if between 15 and 25 percent of the species where I live are different from where you live, we occupy different regions. Watersheds could also be a criteria; the people in a river valley would comprise a region. Unfortunately, our forefathers typically used rivers to draw political boundaries; the people on one side of the river end up in a different state (or country) from the people on the other side. I live in Buffalo, New York on the Niagara River; the people in this part of New York State feel greater kinship with their Canadian neighbors across the river than with people in the New York City area. Regional thinking would have put the people along the Niagara River in the same community. We actually do feel as though we are the same community, but the national boundary is a significant barrier. It is ironically tragic for ecosystems that the 110 major, internationally shared drainage ecosystems in the world are broken up by political boundaries.

Regions also could be differentiated roughly by landforms; a frequent distinction made in Illinois or New York is *upstate* versus *downstate*. Some people urge that region be defined in a cultural/phenomenological sense; you are where you perceive you are. Others suggest that the force you feel

---

*Space does not permit a complete discussion of the GBI because it would encompass the whole book. Interested readers should consult Robertson (1985) and the writers he cites.

within you defines your region; they speak of "spirit places" or "psyche-tuning." Another common distinction is between hill people and flatlanders. Although most bioregionalists probably would not subscribe to this notion, I believe region is best defined by the reach of a television signal. A sense of region must be a shared perception, and television is our main medium for sharing perceptions. Obviously, if different criteria are used to define the region, one could belong to several "regions" simultaneously.

The specifics of defining the boundaries of a region are less important than the central idea that people in a region should do most things for themselves. So far as possible they should grow their own food, make their own products, protect their own biosphere, provide their own services, collect their own taxes, control their own institutions, and run their own government. An ideal region would have a central city with agricultural and other natural areas surrounding it. The city would nourish the countryside and vice versa. Intraregional trade would be high but, interregional trade would be deemphasized. If region in this full sense could be realized, it would be an ideal setting to nourish ownwork. Self-reliance would be more practical at the regional level than at the individual level. Residents would identify with their region; identification with nation would decline in impor-tance. Obviously, the transition to bioregionalism would require considera-ble shift in perceptions and values.

The most difficult transition to make is from a homo-centric to a bio-centric norm of progress. The entire community must progress. Any progress of the human at the expense of the larger community of the living must ultimately lead to the diminishment of human life itself. A degraded habitat will produce degraded humans; an enhanced habitat will assist toward an elevated mode of the human. This is evident not only in the economic order but throughout the entire range of human affairs. The splendor of the earth in the variety of its land and its seas, its life forms and its atmospheric phenomena; these constitute in color and sound and move-ment that great symphonic context that has inspired our sense of the divine, given us our emotional and imaginative powers and has evoked from us those entrancing insights that have governed our most sublime moments. . . .

The solution is simply for us as humans to join the earth community as partici-pating members, that we foster the progress and prosperity of the bioregional com-munities to which we belong. Here we note that a bioregion is an identifiable geographical area of interacting life systems that is relatively self-sustaining in the ever-renewing processes of nature. The full diversity of life functions is carried out, not as individuals or as species, or even as organic beings but as a community that includes the physical as well as the organic components of the region. *Such a bioregional community is self-propagating, self-nourishing, self-educating, self-gov-erning, self-healing, and self-fulfilling.* Each of the component life systems must integrate their own functioning within this community functioning to survive in any effective manner. Within this perspective the human can be identified as that being in whom the community reflects on and celebrates itself in conscious self-awareness. Yet this distinctive role of the human requires our appreciation of the full range of the community life processes. (T. Berry, 1984, pp. 3-4, emphasis in the original)

Advocates of bioregions expect them to diminish such problems as: food being shipped thousands of miles into the region; factories being closed by faceless corporations trying to compete in a world market; factories threatening to close rather than clean up their pollution; young people having to leave the region because they cannot get a good education or meaningful work; most people working in a single industry (for example, a steelwork town), resulting in a distorted and vulnerable economic structure; soil being depleted to produce a single crop for export; inability to get investment capital because bankers do not believe a business can maintain itself in world competition. The more self-sufficient the bioregional economy is, the less vulnerable it would be to external business cycles. Differences in resource reserves, however, could lead to striking differences in standard of living from region to region. A region also would be quite vulnerable to climate change, but that danger would hold for any locality at the present time as well.

Thomas Berry (1984) identifies six functions that bioregions should perform: 1) *Self propagation*. Niches would be preserved for all species so they can propagate in balance with others in the community. 2) *Self-nourishing*. Members of the community would nourish each other in established patterns of the natural world. 3) *Self-education*. The entire evolutionary process is a remarkable feat of self-education via billions of experiments. Humans can educate themselves for survival through the instruction available in the natural world. 4) *Self-governing*. Governing is an intrinsic bonding of the community, not an extrinsic imposition. In human deliberations, each species of the community should be represented. 5) *Self-healing*. The community carries within itself the special powers of regeneration. 6) *Self-fulfilling*. The community is fulfilled in each of its components. The human has a special role, for in the human, the community celebrates itself in reflexive self-awareness. "This is something beyond environmentalism which remains homocentric while trying to limit the deleterious effects of human presence on the environment." (T. Berry, 1984, p. 7)

Bioregionalists urge local residents to walk the territory to get to know the land and what lives there. They should develop a resource inventory for the region. They should learn the details of trade and resource-dependency between city and country. In the city they should explore possibilities for rooftop gardens, solar energy, urban forestry, recycling, and so forth. They should learn the historical lore of both the human and nonhuman inhabitants of the region. These actions should develop a sense of rootedness, help people to feel they are members of a community, and enjoy the values of cooperation, participation, and reciprocity.

Bioregionalism tugs at many of our tradional values: nature, rootedness, cooperation, compassion, self-reliance, participation, sustainability. Most of us would like to see bioregionalism work, but is it likely to come to pass? Walter Truett Anderson (1986) says flatly that it will not. He reminds us that natural elements do not neatly define regions; instead, natural systems overlap and interpenetrate. Second, we have learned that we are part of the world's biosphere; The Gaia Hypothesis tells us the whole planet is a system. Third, bioregionalists underestimate the extent to which we already are inextricably linked into national and international networks. We are grounded not only in our region but also in much broader cultural and religious traditions, and in worldwide communities of like-minded people (scientists, environmentalists, and so on). Fourth, economies interpenetrate so inexorably that it would require the most draconian measures to force economic actors to think and act regionally. (My wife just purchased a bottle of California sherry because it was cheaper than New York State sherry; should she have been forced to buy the more expensive local product?) Fifth, if people really found regions to be meaningful units for organizing their lives and to which they would like to give allegiance, they already would have done so.

Bioregional Literature and Activities

The most thorough statement of bioregionalism is by Kirkpatrick Sale, *Dwellers in the Land: the Bioregional Vision* (San Francisco, Sierra Club Books, 1985). The *Whole Earth Review* is a quarterly journal for the movement (27 Gate Five Rd. Sausolito, California 94965); regional magazines are *Rain* (3116 North Williams, Portland, Oregon 97227); *Tilth* (4649 Sunnyside North, Seattle, Washington 98103); *Raise the Stakes* (c/o Planet Drum Foundation, Box 31251, San Francisco, California 94131); *Konza* (Box 133, Whiting, Kansas 66552); *Annals of Earth Stewardship* (c/o Ocean Arks International, 10 Shanks Pond Road, Falmouth, Massachusetts 02540). The Planet Drum Foundation in San Francisco publishes information on the relationship between human culture and the natural processes of the planet, devoting considerable attention to bioregionalism.

Some of the stronger bioregional organizations are: EarthBank, which financially and morally supports cooperatives and like-minded businesses (Box 87, Clinton, Washington 98236); The Ozark Area Community Congress (Box 3, Brixey, Missouri 65618) is a federation of 200 groups across the Ozarks emphasizing political ecology; it has organized two North American Bioregional Congresses; The E. F. Schumacher Society (Box 76A, R.D. 3, Great Barrington, Massachusetts 01230) develops model programs for land tenure and appropriate financing; and it sponsors a Self-Help Association for a Regional Economy (SHARE) in which members make deposits in a local participating bank to be used as collateral for loans to member farmers or businesses.

(Much of the information on bioregionalism was taken from the *Utne Reader* No. 14, February/March, 1986)

Even though we are likely to see few existing political boundaries give way to bioregionally defined boundaries, we can expect that several elements of bioregionalism will be adopted as we move toward a sustainable society. For example, the people in the Great Lakes basin have begun to see themselves as members of a regional community even though the region includes eight states, two Canadian provinces, and covering portions of two countries. They have developed regional coordinating entities such as The International Joint Commission, to give expression to their joint concern. Governments have signed treaties and other agreements obligating themselves to care for the ecosystem of the basin. Most importantly, the people in the basin have formed binational citizen organizations such as Great Lakes United and Great Lakes Tommorrow to give concrete and sustained attention to their regional concerns. The governments in the Connecticut River Valley, urged on by their citizens, have signed an agreement to regenerate and protect Atlantic salmon spawning grounds.

When resources are scarce, people may be more attracted to regional organizations and solutions. Bioregional thinking will likely be strongest in rural areas. Regional self-reliance is probably easier to achieve than local self-reliance. In the short term, I am doubtful that most economic activity will become regionally oriented. Some economic activity will be local, some regional, some national, and some international. Regional activity and identity will coexist with national and international activity and identity. In certain areas, such as the Great Lakes Basin or the Mediterranean Basin, regional thinking and regional efforts at management and governance are likely to override and erode some purist thinking about national soveriegnty because coordination across political boundaries is necessary to preserve the vital integrity of crucial ecosystems. We have much social learning ahead of us; the thinking that is going into bioregionalism will become increasingly valuable as we struggle to find ways to live in harmony with our ecosystems.

Summary

These are the main points produced by the analysis and discussion in this chapter:

1. Full employment will be a much sought after but unattainable goal for many decades because three conditions will prevail: a) growth in human population, b) increase in the productivity of machines, and c) resource limits that will stunt economic growth.

2. We should distinguish *work* from *employment*; our concern should be to make work personally fulfilling.

3. We should cease trying to achieve the unattainable goal of full employment and focus instead on trying to make our everyday activities as fulfilling as possible; our goal is quality of life and not simply jobs.

4. Many will find fulfilling daily activities by learning how to do our ownwork. We can restructure our socioeconomic patterns to foster ownwork, thereby becoming less dependent on paid jobs.

5. Work that is fulfilling has five important facets: challenge, convenience, rewards (money or goods), satisfying relationships with coworkers, and sufficiency of resources.

6. A society structured to nourish ownwork and be sustainable will:
   a. not seek wealth, or growth, or high material consumption;
   b. be open and encourage change, learning, and variety;
   c. emphasize integrative thinking over specialization;
   d. encourage people to develop multiple talents;
   e. blur sex roles—let everyone become all they are capable of being without confinement by traditional sex roles;
   f. shift emphasis from competition to cooperation;
   g. find ways to share paid employment opportunities more widely: shorter work week, job sharing, part-time jobs, or early retirement; and
   h. employ technology in a more subdued fashion tuned to be appropriate to the needs of people doing their ownwork.

7. A GBI for every person may possibly become part of an ownwork-oriented society; its implementation would require a strong basic economy and a large shift in public beliefs and values.

8. Over time, a society striving to become sustainable may restructure itself into bioregional units. Whether bioregions could be self-sufficient, however, is unclear. It is probable that elements of bioregionalism will be adopted and coexist with activities oriented to a national and world level.

Enjoying Life Without Material Indulgence

Chapter 11

"Simply to see to a distant horizon through a clear air—the fine outline of a distant hill or a blue mountain-top through some new vista—this is wealth enough for one afternoon. (Henry David Thoreau, from his journals)

The Emotional And Ecological Cost Of Materialism

"That happiness is to be attained through limitless material acquisition is denied by every religion and philosophy known to man but is preached incessantly by every American television set" (Bellah, p. 135). The magazine advertisement for the ocean cruise promises feasting, pampering, and adventuring. Schlitz beer used to proclaim, "You only go around once in life so grab all the gusto you can." Politicians constantly equate prosperity with quality of life. *The Wall Street Journal* calls itself, "The Daily Diary of the American Dream." People are relieved of any lingering guilt for indulging themselves by the continual admonition to keep buying to keep the economy going. Such messages encourage a widespread belief in the goodness of perpetual growth. Unfortunately, few recognize that this thinking creates a society that cannot be sustained.

Chapter 2 demonstrated how economies possess an inherent dynamic that leads to continual expansion. As resources become scarce, an economy will contract because the expense of extracting resources will approach and then surpass their use value. Even if substitute resources can be found, continual expansion will generate geometrically increasing quantities of wastes. The expansive economy will either starve from a dearth of resources or disrupt the natural cycles of the biosphere with its pollution.

Consumer indulgence in material goods, by a population steadily increasing in size, simply cannot continue for very long. Affluent people will have to change the way they live. In a previous chapter, we distinguished material wealth from quality of life and declared our true goal as quality of life, not wealth. Exposing the fallacies of the supposed linkage between consuming and quality of life will help us find a better way.

The Poverty Of Affluence

Paul Wachtel's 1983 book title, *The Poverty of Affluence*, expresses well the ironic fallacy of our current preoccupation with consuming. Paraphrasing Mishan (1980), we can summarize well-known arguments refuting a presumed connection between economic growth and a better life for people:

1. A market economy cannot ensure an excess of goods over bads. Growth leads to resource depletion, ecosystem damage, congestion, permissiveness, and breakdown of social order.

2. To ensure absorption of the products of modern industry, considerable resources must be devoted to the creation of dissatisfaction with existing possessions. The fashion industry is a prime example. The advertising industry promises happiness, but mainly delivers dissatisfaction.

3. Sales become primary; no matter what people are trained for, many are forced into sales. Salespersons even blatantly intrude on the privacy of the home to literally thrust a presumed good on someone who has not expressed an interest in acquiring it.

4. A feeling that one is well off is coming to mean not simply absolute command over private goods, but command over private goods relative to those of others. Hirsch (1976) shows that there are social limits to growth. If one stands on tiptoe in a crowd, others soon do so as well and the advantage is lost. A deserted beach loses its value if people begin to flock to it. Andre Gorz (1980, p. 69) distinguishes ordinary use goods (vacuum cleaners, radios, bicycles) which retain their use value if everyone possesses one, from luxury goods which lose their use value if everyone tries to possess one. No one would sensibly propose that every family be given beachfront property on the coast; cottages would be jammed so close together that their use value would be nil. Gorz sees cars as analogous to beach houses. The proliferation of cars in Paris has sharply reduced their use value and severely damaged the environment.

5. "A mass consumption economy, one of rapid obsolescence and replacement, cannot but breed a throwaway attitude toward man-made goods irrespective of quality. There is no time to grow fond of anything, no matter how well it serves. And in any case, it will soon be superseded by a new model. In consequence, everything bought comes to be regarded as 'potential garbage' and treated as such." (Mishan, 1980, p. 271)

6. "The ideal public for the modern economy is one that is uprooted from all conventional constraints; one that is free-floating, volatile,

plastic; one that can be molded and segmented and pulled hither and thither by the carefully planned campaigns of modern advertising agencies. This ideal public is, of course, that found in the permissive society. For a society in which 'anything goes' is *ipso facto* a society in which anything sells." (Mishan, 1980, p.278) Mishan sees permissiveness as a threat to the social order. If the moral order is scrapped in the name of emancipation, people will turn to the state for security and accept its repression as preferable to the anarchy and anxiety of the permissive society.

Mishan then asks,

How did we come to this sad impasse? The simple answer is man's *hubris*. In his search for mastery over nature, and addressing his intelligence to specific and immediate ends, man has been all too successful. And his appetite has fed on his success. Today there are no bounds to his ambition, and no limits to his capacity. He has begun to wreck the social order as surely as he has begun to wreck the natural order. . . . The unavoidable frustration resulting from this act of social vandalism has, alas, only aggravated man's lust for power and sparked his hopes with technological fantasies that can only remove him further from human fulfillment. (Mishan, 1980, p. 279)

Large differences in property and income between families has an inherently undemocratic influence. The ability of the wealthy to hire other people's time, or to proffer rewards, conveys disproportionate political power. High incomes also encourage wasteful consumption that has little relevance for maintaining life or well-being. The previous chapter considered the possibility of a minimum income for everyone. Should a good society, which recognizes limits to growth, also place a maximum on income or property? Daly (1977, pp. 54-55) cites passages from John Stuart Mill and John Locke to show that they believed it should. Daly would limit income by a highly progressive income tax. He recognizes that it would have a negative incentive for those in the top brackets but would open up opportunities and provide incentive for those at lower levels (Daly, 1977, p. 56). He suggests that no American family should have more than five-times the average income, but adds that it is less important to decide where we draw the line than to establish the principle that such lines must be drawn. (Daly, 1977, p. 73)

The drive for affluence in an expansionist economy encourages a preoccupation with money and property. One cable television channel is dedicated to investment information. Money is supposed to provide freedom but its diligent pursuit can be enslaving. Property must be managed, protected, insured, repaired, and replaced. Instead of helping us to feel free, caring for property can make us feel that we are on a treadmill.

Our preoccupation with money and wealth has yet another insidious consequence: we inappropriately try to convert many human values and aspirations into money and prices. Economists have developed "shadow prices" as a way of monetizing values that currently are not monetized. Lester Thurow (1980, p. 108) asked his readers to imagine that someone could sell them an invisible , completely comfortable, facemask that would guarantee clean air. Adding up the prices each of us would be willing to pay for such a mask would give us the shadow price of clean air. Wachtel thought Thurow's example better illustrates the irrational overvaluation of money that pervades the lives of almost all of us. He suggests that the proper question should be:

How much money would I be willing to take in return for letting someone blow polluted air into my child's face every day? . . . There is no amount of money I would take in free exchange under such circumstances. . . . What we want is far more complex than what is revealed just in acts of buying or borrowing. . . . Life holds infinite variety, but if everything is reduced to one number, if how we are doing can always somehow be added up, then the only real value can be "more," and growth in some quantity the only acceptable sign of progress or of doing well. . . . And so we experience an imperative to grow. The word "growth" becomes for us synonymous with the good and is trotted out in an enormous range of contexts. . . . With "growth" such an omnipresent symbol of the good, it is very difficult for us to accept any idea of a limit to growth as implying anything other than stagnation. (Wachtel, 1983, p. 91)

Even though these tendencies are pervasive, they do not dominate the lives of everyone. Duane Elgin (1981) estimates that as many as 10 million Americans have voluntarily redirected their lives toward simplicity. National survey studies in the United States have shown that more than one-half of the adult population has adopted values that emphasize concern for others, and the community, rather than mere concern for self and one's affluence (Yankelovich, 1981; Milbrath, 1984). Nevertheless, the dominant consumer-oriented system continues to create conditions that threaten to undermine simplicity-oriented changes or make their further development nearly impossible. Will people change their values and lifestyles before conditions force them to? Probably not! If not, will it then be too late for the transformation to be anything less than deeply wrenching—perhaps even futile?

New But Old Ways To Enjoy Life With Fewer Material Goods

The toy industry is now a very big business. The inventiveness of designers using advanced technology has produced some fantastic creations. Seemingly, there is no limit to what they can create nor to the prices they charge. Children with a closet full of such toys can have stimulating and happy days (although a poor child could envy a rich one). But what did children do to enjoy life before they had the largess of affluent parents and the modern toy industry? While travelling recently in a developing country, I watched some boys rolling an old auto tire, guiding it with a stick; they seemed every bit as happy with their "toy" as their modern American counterparts with closets full of expensive toys.

The same question applies to adults. Thousands of generations of people enjoyed life with only a small fraction of our material goods. Were they less happy than we? We all have inner resources for meditation, conversation, loving, communion with nature, reading, writing, playing music, dancing, and engaging in sports. These talents may need to be developed further because our present society lures us to buy and consume, buy and be entertained, buy and be pampered. Those who have given in to those inducements have become more bystanders than participants in life's unfolding drama.

While writing these chapters in winter, I tried to take an hour each day that the weather permitted for cross-country ski trips. I took about twenty-five such trips in 1986. (In the winters of 1987, 1988, and 1989,the snow fall was only enough for five or six trips. Is global warming due to the greenhouse effect going to deprive me of this simple pleasure?) My route takes me up and down a few gentle hills, follows beside a swift-flowing stream, and through some lovely woods. The sun sparkling on the snow and the brisk air refresh me, body and soul; the exercise helps keep me fit. The rewards are rich but the lifestyle is simple. My ski outfit, amortized over its useful life, costs me about $5 per year; my cash outlay in 1986 was $5 for ski wax which I am still using.

About twenty miles away there is a downhill ski resort that has ten lifts and makes snow every night. Many people go there because they get a greater thrill from downhill skiing. The difference in fossil energy consumption between the two experiences is enormous. The downhill skier wears out an automobile and consumes gasoline to get to the slopes. All of the skiers' equipment, in addition to that of the resort, required energy and material in manufacture. The snow-making machine and chairlifts burn energy in operation. Each trip to the slopes to get a greater thrill costs the downhill skier between $35 and $40 for the use of that energy versus my trip cost of 40¢.

When we someday must learn to enjoy life while consuming less, will it be such a deprivation to convert from downhill skiing to cross-country skiing? The enjoyment can be just as great, our health will be better, nature will be much less abused—quality of life may go up rather than down. Just a footnote to this story; when I first started skiing on my cross-country trail twenty years ago. I frequently was the only user. Now it is not uncommon for 100 people to be out there on a nice day. There seems to be a growing readiness to turn away from the high consumption, big thrill lifestyle.

Goods That Are Not Zero Sum

Economists characterize most goods exchanged in the market as "zero sum." Because I have it, you cannot have it—that is zero sum. Our conditioning toward material consumption inclines us to think of all enjoyment as zero sum. Actually, many of the most satisfying and fulfilling things in life are enhanced when shared. Let us take a few minutes to reflect on some of these nonzero-sum goods.

Recently, I viewed a televised lecture by Leo Buscaglia on "The Politics of Love," which was a rebroadcast of a lecture delivered in Chicago. Buscaglia is a well-known professor, author, and lecturer who colorfully and effectively conveys the message that love is good for people and society. That message is welcomed by many people; 10,400 attendees paid admission to his lecture in Chicago. Countless more viewed the televised broadcasts of his lectures, and his books are best sellers.

You would think that everyone would know that love is good for people, that it is easy to give and to share, that fulfillment from loving is enhanced, not diminished, by sharing. Buscaglia's message is so popular because many people sense that our modern affluent society has somehow lost its understanding of the meaning of love. Buscaglia reports numerous instances where persons reject his claims for the virtue of love. If someday our society turns away from trying to find fulfillment in material goods, we may, indeed, find much greater fulfillment in love. We should be actively learning from each other how best to love.

About ten years ago I conducted a study of quality of life on the Niagara Frontier (Milbrath, 1976) and discovered that the ways people sought fulfillment in life clustered into lifestyle patterns. As might be expected, some persons emphasize a consumer lifestyle; their greatest enjoyment came from buying and consuming. They were a minority, however.

Another lifestyle, favored by many, emphasized fulfillment in interpersonal relations. These people loved to socialize with friends and relatives. Rewarding companionship with friends is not difficult to find and most of these people felt quite fulfilled. Most important, this lifestyle is not zero-sum, is not highly consuming of goods, does not waste scarce resources, and does not injure the environment. If we slowed down our frantic production pace, demanded less and consumed less, we would have more time for enjoying companionship; chances are, we also would enhance our quality of life.

Enjoyment of nature emerged as another lifestyle in our study; it is not consumed in the same way as restaurant meals, automobiles, or tickets to seats in a football stadium and, thus, is not zero-sum. Normally, my enjoyment of nature does not detract from your enjoyment (see boxed insert), but, nature can be overrun and destroyed by too many people. Having to contend with a crowded beach, or bumper-to-bumper traffic heading for a national park, or elbow-to-elbow fishing in a trout pool is not a fulfilling experience. Many U. S. National Parks have had to ration nature experiences by advance reservations, quotas, and admission tickets. They are so crowded in China that they have had to assign people to take holidays on different days. The obvious demand for nature experiences makes it all the more important that nature be protected and, where necessary, restored to beauty. Nature protection and beautification is a fulfilling activity that many people can join in, derive satisfaction from, and strengthen rather than diminish by their sharing. Urgent joint action also is needed to obtain and maintain such vital natural elements as clean air, water, and soils. Cutting back on consumption would help a lot, but collective political action to assure environmental protection also is imperative.

The Outdoor Way of Life in Norway

In Norway, the outdoor way of life is a deep and abiding part of the culture. Norway's mountains and forests are vast and beautiful. Until a century ago, most of these areas were inaccessible and unknown to many people. In 1868, the Norwegian Mountain Touring Association was founded. To encourage hiking enthusiasts to get to know and use the great outdoors, volunteers built and marked trails, built footbridges, built and stocked cabins, prepared detailed maps, and made rowboats available for crossing mountain lakes. Many of these areas are not accessible by vehicles, so most of the tools and some materials had to be carried in by hand. At its fiftieth anniversary in 1918, it could declare most of the mountains open to the public. The work of establishing trails and cabins continued. Today, the association maintains 260 mountain cabins offering very inexpensive accommodations.

The mountain wanderer arriving at a cabin will find wood neatly stacked in the fireplace. The cabin will be stocked with solid and good, if not luxurious, food. Beds are made ready for sleeping. Only a few of the cabins have electricity. Upon leaving, the hiker is expected to make things ready for the next visitor, and payment must be left for what is consumed. There is a price list on the wall and a box in which to deposit the money. The arrangement is voluntary and based on mutual trust. Most of these cabins are open and maintained year round. A few of the more popular stopping places have larger chalets, with a staff, and the visitor can get bed, breakfast, dinner, and a packaged lunch. The association provides members with keys for cabins with no staff.

The association also runs an extensive training program for glacier walking, mountain climbing, dog sled driving, and winter survival. It publishes books, maps and magazines, and runs its own travel agency. The association works politically to protect the environment; with 90,000 members, it has considerable political clout. It has contributed significantly to establishing several national parks.

Another Norwegian organization, the 65,000-member Association for the Promotion of Skiing, maintains more than 1,000 miles of ski trails in the 50,000 acres of nature preserves that ring Oslo. On a winter Sunday, 100,000 people from the city ski on these countless trails; some of them are lighted for night skiing. In the summer, they are used for hiking and berry picking. The land in this natural preserve is held in trust for the people and is not open for development; no vehicles or dwellings are allowed. The association sponsors ski schools and arranges cross-country tour races. (Source: *News of Norway*, April 27, 1984; and personal experience.)

Learning is another pleasurable and fulfilling activity that is developed rather than diminished by sharing. Philosophical understanding, especially, is deepened by interpersonal discourse. Cultivation of the mind has been emphasized in many cultural traditions and surely would be an important activity to emphasize in a sustainable society. Deepening one's understanding requires time and periods of quiet contemplation; ironically, these are scarce goods that many frantically busy people today fervently wish they could have. If we slowed down, produced less, and consumed less, perhaps we could find more quiet times for learning and for deepening our understanding.

Enjoyment in creating and appreciating literature, music, and art similarly are not diminished if shared and should be emphasized in a sustainable society. Instead of life being bleak and cold when we are forced to slow down, it could be a flourishing period of creativity and learning.

If we can understand how our possessions have failed us, we can more readily decrease our thralldom. Turning instead to a focus on the quality of our relations with others; on the clarity and intensity of our experiences; on intimacy, sensuality, aesthetic sensibliity, and emotional freedom, we can see how a more ecologically sound society can be a more exciting and enjoyable one as well. (Wachtel, 1983, p. 143)

Play is another pleasurable and fulfilling activity that typically consumes few resources and need not damage nature. I do not speak of energy consuming and nature destroying thrill contests such as off-road vehicle racing. Nor do I speak of sporting events with large crowds of spectators; they should be seen as a branch of the entertainment industry. Rather, the sustainable society should emphasize widespread participation by nearly everyone in games that bring pleasure and are not wasteful or destructive; there certainly is sufficient variety to serve almost any taste. Games requiring vigorous activity not only pass the time pleasurably but also nurture good health.

Self-governance also is nonzero-sum in the sense that everyone benefits when better laws are passed or when better community programs are undertaken. (Many elections in this and other countries are zero-sum when the winner takes all.) Self-governance does require interest, concern, and time from people. Those caught up in the rat race for money often claim they are too busy to participate. However, if life were structured to emphasize ownwork, or were organized into bioregions (as discussed in the previous chapter), people could more likely see the relevance of their participation for a better life; furthermore, schedules would be more flexible, allowing people to take the time for political affairs—it could become a natural and expected aspect of everyday life.

Leisure?

So far I have not given specific attention to leisure, although I have strongly urged people to take time for personally fulfilling activities. Entrepreneurs in modern affluent society try to sell expensive goods and services to help people use their leisure "to the fullest"; this approach to leisure appropriately could be called an industry; it fits with our delusion that happiness must be bought. Most of the activities discussed above that people do to fulfill themselves might also be thought of as leisure but they do not comprise an industry. People engage in such activities to enjoy their leisure but they consume few leisure goods. If life is organized to give greater emphasis to ownwork, the distinction between leisure and work becomes blurred. The sustainable society would have little need for a "leisure industry."

Voluntary Simplicity

Duane Elgin's (1981) book, *Voluntary Simplicity: Toward a Way of Life that is Outwardly Simple, Inwardly Rich*, is a much deeper examination of philosophy, lifestyles, social forces, and revolutionary change than one might expect from the title. His central thesis is that people voluntarily choose a life of simplicity because it is richer than modern consuming lifestyles. The discussion in this section is summarized from the book. I recommend reading the book to savor the full development of Elgin's thought.

The phrase *voluntary simplicity* came from a 1936 essay by Richard Gregg, a former student of Gandhi; it contains two complementary concepts: voluntary and simplicity. To live *voluntarily* means to consciously live more deliberately, intentionally, purposefully. "We cannot be deliberate when we are distracted from our critical life circumstances. We cannot be intentional when we are not paying attention. We cannot be purposeful when we are not being present. Therefore, crucial to acting in a voluntary manner is being aware of ourselves as we move through life." (Elgin, 1981, p.32)

He distinguishes "embedded consciousness" from "self-reflective consciousness." Embedded consciousness is our normal or waking consciousness so embedded within a stream of inner-fantasy dialogue that little attention can be paid to the moment-to-moment experiencing of ourselves. Self-reflective consciousness is a more advanced level of awareness in which we are continously and consciously "tasting" our experience of ourselves. It is "marked by the progressive and balanced development of the ability to be simultaneously concentrated (with a precise and delicate attention to the details of life) and mindful (with a panoramic appreciation of the totality of life)." (Elgin, 1981, p. 151)

Living more consciously has several enabling qualities: (1) Being more consciously attentive to our moment-to-moment experiences enhances our capacity to see things as they really are; thus, life will go more smoothly. (2) Living more consciously enables us to respond more quickly to subtle feedback that something is amiss, so that we can move with greater speed toward corrective action. (3) When we are conscious of our habitual patterns of thought and behavior, we are less bound by them and can have greater choice in how we will respond. (4) Living more consciously promotes an ecological orientation toward all of life; we sense the subtle though profound connectedness of all life more directly. These four enabling qualities are not trivial enhancements of human capacity; they are essential to our further evolution and to our survival.

Our civilizational crisis has emerged in no small part from the gross disparity that exists between our relatively underdeveloped "inner faculties" and the extremely powerful external technologies now at our disposal. . . . Unless we expand our interior learning to match our technological learning, we are destined, I think, to act to the detriment of both ourselves and the rest of life on this planet. (Elgin, 1981, p. 158) A greater degree of conscious simplicity is of crucial relevance for revitalizing our disintegrating civilizations. (Elgin, 1981, p. 125)

Self-reflective consciousness can open the door to a much larger journey in which our "self" is gradually but profoundly transformed. The inner and outer person gradually merge into one continuous flow of experience.

The capacity to ultimately experience the totality of existence as an unbounded and unbroken whole is not confined to any particular culture, race, or religion. This experience of ineffable unity is sometimes referred to as the "perennial philosophy" because it appears throughout recorded history in the writings of every major spiritual tradition in the world. . . . Each tradition records that if we gently though persistently look into our own experience, we will ultimately discover that who "we" are is not different or separate from that which we call God, Cosmic Consciousness, Unbounded Wholeness, the Tao, Nirvana, and countless other names. (Elgin, 1981, p. 153)

Simone de Beauvoir said, "Life is occupied in both perpetuating itself and in surpassing itself; if all it does is maintain itself, then living is only not dying." To live with *simplicity* is not an ascetic but rather an aesthetic simplicity because it is consciously chosen; in doing so we unburden our lives to live more lightly, cleanly, and aerodynamically. Each person chooses a pattern or level of consumption to fit with grace and integrity into the practical art of daily living on this planet. We must learn the difference between those material circumstances that support our lives and those that constrict our lives. Conscious simplicity is not self-denying but life-affirming.

Simplicity, then, should not be equated with poverty. Poverty is involuntary whereas simplicity is consciously chosen. Poverty is repressive; simplicity is liberating. Poverty generates a sense of helplessness, passivity, and despair; simplicity fosters personal empowerment, creativity, and a sense of ever present opportunity. Poverty is mean and degrading to the human spirit; simplicity has both beauty and functional integrity that elevate our lives. Poverty is debilitating; simplicity is enabling. (Elgin, 1981, p.34)

Simplicity is not turning away from progress; it is crucial to progress. It should not be equated with isolation and withdrawal from the world; most who choose this way of life build a personal network of people who share a similar intention. It also should not be equated with living in a rural setting; it is a "make the most of wherever we are" movement.

Voluntary simplicity is an integrated path for living. . . . To live more consciously is to live in a "life-sensing" manner. It is to directly "taste" our experience of life as we move through the world. It is consciously open—as fully, patiently, and lovingly as we are able—to the unceasing miracle of our "ordinary" existence. To live more simply is to live in harmony with the vast ecology of all life. It is to live in balance— taking no more than we require and at the same time giving fully of ourselves. . . . A self-reinforcing spiral of growth begins to unfold for those who choose to partici- pate in the world in a life-sensing and life-serving manner. As we live more voluntar- ily—more consciously—we feel less identified with our material possessions and thereby are enabled to live more simply. As we live more simply, our lives become less filled with unnecessary distractions, we find it easier to bring our undivided attention into our passage through life, and thereby we are enabled to live more consciously. Each aspect—living voluntarily and living simply—builds upon the other and promotes the progressive refinement of each. . . . Life sensing and life-serving action become one integrated flow of experience. (Elgin, 1981, pp. 174-176)

Elgin includes an epilogue titled, "East-West Synthesis: the Source of an Emerging Common Sense." He briefly examines the central theses of East- ern and Western thought. Even though they are highly contrastive, they are complementary; the practice of voluntary simplicity provides an opportu- nity for a creative and workable synthesis of the two streams. The Western stream emphasizes rationality, materialism, progress, the individual unique and alone—separated from God and the ecosystem. The Eastern stream sees life as more spiritual and holistic; knowing comes as much from intui- tion and meditation as it does from reason; the individual is organically identified with the social order, the universe and the infinite.

The West has pursued material and social growth without a balanced regard for the development of interior human potentials. The result has been the emergence of a life-denying and self-serving social order that has exhausted both its vitality and its sense of direction. The East has pursued spiritual attainment without a balanced regard for the development of the exterior potentials of social and mate- rial growth. The result has been to render the development of human conscious- ness a spiritual luxury for the few while making the lives of the many a stultifying struggle for subsistence. (Elgin, 1981, p. 233)

If the views of East and West were to merge, a more powerful flow of evolutionary potential would be released that would move beyond the worldly passivity of the East and the material obsession of the West. Broadening the material capability of Eastern cultures would support a widespread realization of higher human potentials. Interior growth in Western cultures would increase our capacity for compassionate self- regulation which lessens the need to rely upon paternalistic bureaucracies to manage the affairs of everyday life. This, in turn, encourages freedom and creativity which we need for the continuing refinement of the material side of life. Voluntary simplicity would evolve both the material and the conscious aspects of life in balance with each other—allowing each aspect to infuse and inform the other synergistically.

A progressive refinement of the social and material aspects of life—learning to touch the Earth ever more lightly with our material demands; learning to touch others ever more gently and responsively with our social institutions; learning to live our daily lives with ever less complexity and clutter; and so on.

A progressive refinement of the spiritual or consciousness aspects of life— learning the skills of touching life ever more lightly by releasing habitual patterns of thinking and behaving that make our passage through life weighty and cloudy rather than light and spacious; learning how to "touch and go"—not to hold on—but to allow each moment to arise with newness and freshness; learning to be in the world with a quiet mind and an open heart; and so on. . . .

If the human family is able to consciously bring these two, long-separated streams of human learning together into a synergistic whole, we can then embark on an evolutionary journey that would not have been possible, and could not have been imagined, by either East or West working in isolation. . . . Who we are as an entire human family is much greater than who we are as the sum of isolated cultures. (Elgin, 1981, pp. 234-236)

Elgin frequently suggests that we will gradually make this transition through social learning, as I have myself often suggested.

Wachtel proposes (pp. 163-165) that the "human potential movement"— recast in a less individualistic form—could help lure people away from the consumer orientation to life. A variety of therapies have been promoted as part of this countercultural movement but they have certain themes in common. One is an emphasis on the body; change cannot occur just in the head. The stategies range from direct manipulation and massage to interpretations of and constant calling attention to posture, gestures, facial expression and so forth in order to break up chronic, locked-in patterns. Developments in biofeedback technology point us toward new methods of overcoming constricting and deadening modes of thought and experience.

"Marathon sessions" are a second common feature; they range from several hours to several days. Continuous challenges to customary patterns of relating eventually erode a social facade and new perceptions and experience can break through. People often return from marathon weekends with a "high" and see the world in a fresh exciting way. It is difficult to maintain this fresh outlook, however. The context for daily life has not changed; the behavior of other people toward the "renewed" person will follow old patterns. Thousands of encounters tend to pull the person back into old patterns and the effect of the marathon may dissolve. His analysis suggests that psychological change techniques should address whole networks of people who can mutually support the changes until they become stabilized. This interdependent approach to change might be seen as "psychoecological."

Conclusion

Life in a sustainable society could be a realization of the Greek concept of *paideia*—the lifelong transformation of our own person as an art form—that we discussed earlier; it will have as good or better quality than the life we know today. It is ridiculous to characterize life with fewer material goods as "freezing in the dark," as some environmental critics have painted it. It would be a *very different* way of life: more contemplative, less frantic; more serene, less thrilling; valuing cooperation and love more, valuing competition and winning less; with more personal involvement, less being a spectator; more tuned to nature, less tuned to machines. Changes this sweeping may take several generations to come about. Many people have already begun the journey and their learning can help others find the way. Necessity may well hasten our relearning.

Science And Technology In A Sustainable Society

Chapter 12

We in Asia, I feel, want to have an equilibrium between the spiritual and material life. I noticed that you have tried to separate religion from the technological side of life. Is that not exactly the mistake in the West in developing technology, without ethics, without religion? If that is the case, and we have a chance to develop a new direction, should we not advise the group on technology to pursue a new kind of technology which has as its base not only the rationality, but also the spiritual aspect? Is this a dream or is this something we cannot avoid? (Speaker from the floor at hearing in Jakarta, Indonesia, March 26, 1985, held by the World Commission on Environment and Development, p. 111)

Science And Technology Cannot Be Separated From Values

Science and technology are so interdependent that we often use them as interchangeable words; we forget that humans were using technology long before the scientific mode of inquiry was well-developed. The swift growth of science has spurred an even swifter growth of technology. Even though the two phenomena are distinguishable, they are so intertwined that I will combine their discussion in the same chapter.

Technology is usually developed with some purpose in mind; therefore, technology cannot be value free. *Science,* however, claims as a matter of doctrine, that it is value free. I argued in Chapter 4 that science cannot be value free. This is such an important point, and runs so counter to our conventional wisdom, that it will be useful to repeat the essence of the argument here.

Before the last two centuries, a large proportion of inquiry was cast in a religious framework that encouraged some kinds of questions and discouraged others. Furthermore, religious authorities enforced "doctrinarily correct" interpretations of natural phenomena. Scientists fought to be freed from these shackles to pursue their inquiries no matter where they might lead. They were able to demonstrate that knowledge grew much more swiftly if inquiry could be conducted freely. The slowly acquired freedom of inquiry grew into the maxim that science should be free of value control, that science should be value free. Scientists soon began to declare that all science is, in fact, value free.

We have now had two centuries of experience with science. What has actually happened with science? Who controls it? Is it value free? Can humans, who are beings who hold values, conduct inquiry which is value free?

Certainly science has many fewer controls from religion or government than it did two centuries ago. But that does not make it value free. Every expenditure of energy, time, and money in scientific inquiry is an expression of one or more values. A scientist choosing a line of inquiry is expressing a belief in that line of inquiry as being more valuable than others. A scientist may recognize his choice of inquiry to be a personal value choice but may still believe that the conduct of science is value free.

Institutions employing scientists clearly use values in choosing lines of inquiry. Private firms or government agencies seek new knowledge to solve problems or to make money. Universities expect their research to attract students or grants. The beliefs of governmental officials, business executives, granting agencies, and students as to what is valuable will affect the choice of line of inquiry by scientists, as well as the amount of support they will receive. President Eisenhower understood this and warned us of it in his "Farewell Address":

Today, the solitary inventor, tinkering in his shop, has been overshadowed by task forces of scientists in laboratories and testing fields. In the same fashion, the free university, historically the fountainhead of free ideas and scientific discovery, has experienced a revolution in the conduct of research. Partly because of the huge costs involved, a government contract becomes virtually a substitute for intellectual curiosity. For every blackboard there are now hundreds of new electronic computers.

The prospect of domination of the nation's scholars by federal employment, project allocations, and power of money is ever present—and is gravely to be regarded.

Yet, in holding scientific research and discovery in respect, as we should, we must be alert to the equal and opposite danger that public policy could itself become the captive of a scientific-technological elite.

Societies also express their value in the pattern of their support of science. Both the United States and the Soviet Union have chosen to maximize their power by supporting science that enhances power. Inquiry that does not have much potential for increasing power is deemphasized or neglected altogether. Some people will say, "We do not seek power for power's sake but rather for the ability of power to serve other values: survival, freedom, wealth, comfort, etc." That statement emphasizes my point: *the pursuit of science is always directed by values.*

The popular image of science is that it is a coldly rational deductive process in which meaning flows clearly from observations. That image does

not recognize the true nature of science. It is much more accurate to perceive science as an evolutionary process in which accepted "truth" is that
which survives vigorous conflict about the perceptions of meaning derived
from observations of phenomena. Ideas must compete for acceptance. Interpretations that do not withstand repeated truth testing are eventually
discarded. Human observations, then, probably cannot be completely free
of value bias, but it is worthwhile to keep inquiry as free of value biases as
possible. Mutual criticism among scientists helps to correct for biases. "The
critical attitude may be described as the conscious attempt to make our
theories, our conjectures, suffer in our stead in the struggle for the survival
of the fittest. It gives us a chance to survive the elimination of an inadequate hypothesis—when a more dogmatic attitude would eliminate it by
eliminating us." (Popper, 1962)

Despite the correctives of criticism, scientific inquiry frequently is used
in a biased way to serve special values. This bias is dramatically illustrated
when each side in a controversy hires scientists to present "facts" in support of its position. An interested organization not liking the "facts" currently presented by scientists on an issue may hire its own scientist(s), or
finance its own study, to ensure that scientific "facts" favorable to its point
of view are generated. Having witnessed a plethora of such confrontations,
I am forced to conclude that the scientific enterprise cannot avoid being
embedded in values.

Lest the reader misunderstand, I am not antiscience. Scientific knowledge has been extremely useful to me as I try to understand my world and
the way it works. I conduct scientific research myself, and I support many
lines of scientific inquiry. Distinguishing facts from values and keeping our
inquiries from being biased by values is very useful. But we must not delude
ourselves; we should be quite conscious of the way science is shaped by
values. Science is not a sacred cow; it does make mistakes; it can lead us
into predicaments we would do better to avoid. We had better learn how to
control it or it will lead us in directions that have the effect of controlling us.

The belief in the myth of a value-free science has many ramifications.
Most seriously, it gives scientists the luxury of abdicating responsibility for
choosing their line of inquiry or for the consequences flowing from their
discoveries. "Science for science's sake" is a phrase we have all heard. That
assertion is equivalent to saying that science is a core value. I reject that
assertion because it confuses means and ends. Science, rather, is a means
to realize the core value of a better quality of life. If scientific inquiry leads
to consequences that diminish rather than enhance quality of life, science,
per se, is not to be valued.

For scientists to reason deeply about values is difficult because scientific specialization encourages tunnel vision. When scientists decide whether

to proceed with a line of inquiry, they perceive only a small part of the relevant reality. Many scientists, in fact, report *feeling crippled* when called on to make value judgments. Their training and work has not given them sufficient experience in thinking about ethics to be able to foresee the value implications of their choices. The typical result is that crucial value questions are ignored.

I recognize that these generalizations do not apply to all scientists. Such organizations as The Union of Concerned Scientists, Physicians for Social Responsibility, The Educational Foundation for Nuclear Science, and the Atomic Scientists Forum are willing to address value questions and act upon them. The scientists who developed the atomic bomb established a Committee on Social and Political Implications that submitted a report to the U. S. War Department one month before the first atomic bomb was exploded in the New Mexico desert in 1945. Their report was kept secret for many years. This subverted their attempt to take some responsibility for their discovery. A quotation from the report is relevant here:

In the past scientists could disclaim direct responsibility for the use to which mankind had put their disinterested discoveries. We now feel compelled to take a more active stand because the success we have achieved in the development of nuclear power is fraught with infinitely greater dangers than were all the inventions of the past. All of us, familiar with the present state of nucleonics, live with the vision before our eyes of the sudden destruction visited in our own country, of a Pearl Harbor disaster repeated a thousand-fold magnification in every one of our major cities. (quoted in Cousins, 1987, p. 45)

Engineers and other technologists typically are no better than scientists in understanding values and how they influence their work. They, too, tend to believe that their work is value free. They are poorly trained in value analysis or in the ability to forecast the consequences of their projects. They prefer to abdicate responsibility for the consequences of their actions. The CBS television show *60 Minutes* recently ran an episode in which a private arms dealer declared in an interview that his guns were neutral and that he had no responsibility for the motives or values of those who used them. Rifkin has a sharp rejoinder to that way of thinking:

There has never been a neutral technology. All technologies are power. . . . The purpose of every technology is somehow to enhance our well-being. . . . Technologies change the equation of nature by giving human beings a distinct advantage over each other and the other species. . . . The tools we create are saturated with power because their whole reason for being is to provide us with "an advantage." (Rifkin, 1985, p. 92)

The myth of a value-free science frequently leads scientists and engineers to confuse facts with values. In effect, they are so accustomed to thinking their work is value free that they cannot recognize values and the way they creep into their inquiry. To illustrate: one of the principal speakers at a conference on energy held a few years ago was the leader of a team from a famous "think tank" in Austria; the team had studied energy needs for Europe up to the year 2040. During conference discussion, the major conclusion of their report was questioned (as it also was in several later reports). He replied, "It is a *fact* that Europe must have [X amount of energy] by the year 2040 and the only way they can get that is to build the breeder [reactor]." (emphasis in the original) Had he been trained in value analysis, he would have understood that a perceived need for energy must be a value and not a fact. He was blind to the way his analysis was biased by his values.

The notion of a value-free science can also be used as a weapon in political battle, as illustrated by a current controversy in my community. The lawn-care industry has grown swiftly in recent years and increasingly uses chemical pesticides to control weeds and insect pests. Some people are made ill by the ubiquitous use of these chemical poisons. Environmentalists also are worried about injurious long-run consequences of this practice for the biosphere. The county legislature responded to these concerns and called a "fact-finding" session at which each side turned up to clobber the other with their science. The public officials apparently believed the issue was a matter of fact rather than value. The lawn-care industry spokespersons claimed that it was a purely scientific problem and should not be discussed in a political arena. If they could have this argument accepted, the controversy would be resolved by scientific authorities (read, their authorities). They cannot or will not recognize that the controversy really is about what we should value in our society. Is easy (but expensive) lawn and garden care more important than the health of a few people who have adverse reactions to chemicals? Are jobs and profits more important than the undiminished health of ecosystems?

What Are The Long-Run Consequences Of Our Love Affair With Science?

The posture of people in modern societies toward science is akin to religious worship. Science is mysterious, exciting, powerful, magical; people become intoxicated with its power and potential. Little wonder they have a religious-like faith that science can know everything. This adulation of science is even more pronounced in Third-World countries. It is dangerous for people to put unwarranted faith in science. A problem of value may mistakenly be converted into a problem of fact, or a problem for technology, thereby delaying or frustrating more meaningful attempts to solve the problem. Instead of looking to their own resources, their own values, their own common sense for solutions, many people look to science for miraculous solutions.

Scientific successes have been so impressive that people expect more and better in the future. Could science and technology be so successful as to work to our detriment? Most of us do not ask that question, although some of us may feel uneasy about both the question and the possible answer. Perhaps it would be helpful to review some instances where the success of science and technology may work to the detriment of our well-being.

Would humanity and the biosphere have been better off if science had failed to unlock the atom? The full story is not in yet, but forty years of experience provide some basis for an answer. The atomic bombs dropped on Hiroshima and Nagasaki produced such horror that people have been fearful of nuclear war ever since. Some people say that is good; the fear of mutual destruction has stopped war. Actually, we cannot say that fear of nuclear weapons has prevented war; nor can we say that it has drastically changed the way war is conducted. We still believe we must maintain exceedingly expensive conventional forces.

What about the peaceful uses of atomic energy? They were prominently highlighted in the early 1950s. Electric power, generated by nuclear energy, was projected to be so cheap that it would not be necessary to meter it. Radioactive tracers would help medical scientists to understand better how bodies work; and so forth. Who could tell what additional marvelous uses might be found?

The only significant peaceful use turns out to be generating electric power; and it is not cheap. Building nuclear power plants with sufficient safeguards has cost so much energy that the energy derived is barely net positive. A group of university professors and engineers in Lyon, France, the Diogenes Group, studied the French electronuclear program which generates approximately 60 percent of its electricity. They calculated the construction and operating costs of the plants, the reprocessing plants, decommissioning costs, and the heavy costs of the distribution network, the new highway network, the fuel, the enrichment plant, research and teaching institutes. They concluded that until the end of the twentieth century, the program will consume more energy than it will produce. Overall, seven plants under construction consume annually as much energy as can be produced by four plants in full operation. (Gorz, 1980, p. 112)

Because of its massive power and capability for damage, nuclear power organizations must be failure-averse and strive for nearly perfect operation. Electronuclear plants in the United States have been plagued by cost overruns and extended periods of "downtime." The nuclear emergency at Three Mile Island in Pennsylvania, which resulted in a meltdown of the core of the fuel rods, occurred as a result of a string of six errors; if any one of those errors had been averted, the meltdown would not have occurred. The probability of six errors occurring in succession is extremely remote; yet it happened. The plant explosion at Chernobyl also has been traced to operational error.

In the early 1960s, engineers developed probabilistic nuclear risk assessment. They concluded that nuclear core-damaging accidents would occur once every 10,000 reactor years. (A reactor year is one reactor operating for one year; the world's 1986 complement of 366 power reactors accrued 366 reactor years in 1986). However, after the reactor melt-down at Three Mile Island, a new study raised the risk to once every 4,000 reactor years. Yet, core damaging accidents are occurring at a faster rate—Chernobyl came only 1,900 reactor years after Three Mile Island. If we assume core-damaging accidents every 1,900 reactor years, we can expect three more by the year 2,000. Swedish and West German scientists estimate that there is a 70 percent chance of another accident in the next 5.4 years. (Flavin, 1988). With nuclear power, the costs of errors are greater than the value of the lessons learned; we cannot look to trial and error learning as a mode of policy improvement.

The uncertainties of electronuclear plants in the United States are so great that construction has not begun on any new ones since the late 1970s. Some partially completed plants were never finished and some completed plants have not been allowed to go into operation. Taxpayers have borne the main costs of nuclear power. Government-funded research and development, promotion, and guarantees of limited liability got the nuclear power program going. In all likelihood, taxpayer money will fund the decommissioning of these radioactive plants and the storage of nuclear wastes, which must be safeguarded for hundreds of thousands of years.

The uncontrolled radioactive release from the Chernobyl plant in April 1986 not only killed more than thirty people, but will injure the biota, and humans, in Eastern and Northern Europe for many decades. The area surrounding the plant still has not been reinhabited as of this writing (more than 3 years). Ironically, the Soviet authorities had announced a few months before the explosion that they planned to greatly expand their nuclear program; they had no doubts about its safety because there was only one chance in 10,000 of an accident. Nuclear plants in the cluster at Chernobyl have been reopened, and they are proceeding with their nuclear buildout. The Italians, who experienced frightening fallout, seem to have learned, however; they approved referenda in November 1987 that effectively preclude further nuclear construction. Swedes and Austrians had come to a similar decision even before Chernobyl.

The Chernobyl explosion illustrates how development of a technology can have unanticipated but devastating consequences. The radioactive clouds spread widely, but high radiation gradually declined to levels the authorities declared "safe" in most areas. At a conference on the effects of Chernobyl, held in the summer of 1987, Professor Batjer of the University of Bremen in West Germany, estimated that only 10 percent of the released radiation that would be absorbed had been absorbed by a year after the accident (Harding, 1987). Thus, whether those exposed to radiation will later develop health problems remains to be seen.

Human health problems were in everybody's mind, but no one anticipated that the accident would totally disrupt the Sami culture. The Sami (often called *Lapps*) live in mid to northern Scandinavia and have built their culture around herding and utilizing reindeer. The deer feed mainly on lichens which are remarkable "radioactive sponges" because they draw most of their nutrients from the air and thus incorporate airborne contaminents much more than other vegetation. Radioactive cesium 137, airborne from Chernobyl, built up quickly in the lichens and bioaccumulated in the reindeer. The deer became unfit for human consumption; some were fed to animals being raised for fur but most were simply buried in a "radioactive dump." Since cesium 137 has a half life of thirty years, it will be a very long time before reindeer can safely be grazed on the contaminated land.

The Scandinavian countries moved quickly to help their unfortunate Sami citizens, but they could not save their culture. The Sami's own plaintive words tell the story.

Our men care for the deer and know them. When deer are slaughtered, it is done with respect. We women know how to care for the meat, to use every bit, the blood, the head, even the feet in soup. We know how to make thread from sinew and how to prepare the skin and furs for clothing and shoes. The work of our hands puts food on the tables and clothing on our backs. We give our food to our guests and send dried meat to our children when they are away in school. Even if there comes a time when we can eat the deer again, it may be too late to pass the knowledge of how to take and use *niestti* on to our children.

This is not just a matter of economics, but of who we are, how we live, how we are connected to each other. Now we must buy everything. Thread, material, food, shoes are now all different things, when they used to be part of one thing.

It seems sometimes that things have become strange and make-believe. You see with your eyes the same mountains and lakes, the same herds, but you know there is something dangerous, something invisible that can harm your children, that you can't see or touch or smell. Your hands keep doing the work, but your head worries about the future. (quoted in Stephens, 1987, p. 68)

The destruction of their culture energized the Sami, especially the women, to take political action. Informal networks arose to discuss the effects of

Chernobyl not only on the Sami's reindeer economy and their political situation, but also on men's morale, the roles of women, and the health of their children. Some even sought and won political office. Note once again the tendency for women to take the lead in protesting ecological destruction which in turn destroys a way of life.

Early in the nuclear era people did not worry about safe disposal of nuclear waste. Now they worry, but a completely operational method for isolating and safeguarding nuclear waste (for hundreds of thousands of years) has yet to be put in place. (Scientists and engineers believe they are close to having a workable system.) Where should they put the nuclear waste repository? The U. S. Department of Energy identified some suitable sites and held hearings near those areas to gauge citizen reaction. Frightened and angry citizens do not want such a facility near them. When Department of Energy officials were closely questioned about how long they could guarantee security of such a facility, they said they could not promise security beyond 100 years. The waste will be dangerous for more than 100,000 years. Could any human institution guarantee that it would be in place and in control for even 100 years?

A new problem now looms on the horizon: What should be done with nuclear plants that have passed their useful lives? (The expected life of such a plant is between thirty and forty years.) They must be dismantled and safeguarded for centuries. The materials cannot be used elsewhere and the sites may be too contaminated for any other use. If we continue to use nuclear energy, these costs can only escalate. Will the energy we derive be greater than the energy we will put into this monstrously complicated system?

When all of these factors are considered, my judgment is that we would have been better off if the power in the atom had never been discovered. Many people would agree with that today; but then, none of us were asked. There is no way in any modern society for such a question to be asked and no way that the citizenry could give a meaningful answer had they been asked.

Nanotechnologies

In Chapter 9 dealing with food supply, I discussed the development of recombinant DNA. This will have the same level of impact on the ability to transform society and the biosphere as did the discovery of fire and later of nuclear power. The relationship of mankind to nature was altered forever with each of these discoveries. The problems of managing the consequences so that they do not destroy us become increasingly larger with each discovery. This management problem applies even more impressively to the next breakthrough that scientists envisage as just over the horizon; it is called *nanotechnology.*

As of this writing, few people have even heard of the concept. I drew most of my understanding from an exciting and disturbing 1986 book by K. Eric Drexler, *Engines of Creation*. Inquiry into the potentiality and practicability of nanotechnology is proceeding at several universities and think tanks; The Massachusetts Institute of Technology, for example, has a nanotechnology study group. Governments and private firms are putting sizeable research funds into developmental work. Whoever wins the race to the breakthrough will win unimaginable gains in power and wealth.

Thinkers in the nano mode distinguish *bulk technology* from *molecular technology.* Bulk technology handles molecules in bulk. To shape things the bulk way we heat, hammer, saw, fasten and so forth. Learning how to use fire was the breakthrough that enabled us to make tools and do these things. Nanotechnology differs in that it constructs or disassembles things atom by atom. This is the way nature builds its structures.

Most of us have heard of microelectronic circuits that can be placed on small microchips; a micrometer is one-millionth of a meter. Molecules are measured in nanometer units—one thousand-times smaller. A current estimate is that 100,000 more nanoelectronic circuits could be put on a computer chip than is now possible. Nanotechnology and molecular technology are interchangeable terms.

Imagining how small nanomachines really would be is difficult. An atom is 1/10,000 the size of bacteria and bacteria are about 1/10,000 the size of mosquitoes. Yet, there is a lot of space inside an atom—the atomic nucleus is 1/100,000 of the atom itself. An intelligent nanomachine the size of a virus could travel inside our smallest capillary; compared to the machine, the capillary would seem like a 150-line highway. New materials, constructed the nanoway, would be lighter, stronger, and more durable than anything we know today. Because most of what we think of as nature, and more, could become a human creation, some of the mystery of nature will vanish. Drexler thinks we will still be in awe of the complexity of the whole.

Modern day microcomputers are still bulk technology. Researchers are now working to develop molecular computers. Their circuits would be so small and so precise that they would be one million-times faster than present day computers. With components a few atoms wide, they would be many billions of times more compact than todays microelectronics. Even with a billion bytes of storage, a nanocomputer could fit in a box the size of a bacterium. The plan is to use protein machines to build nanomachines of tougher stuff than proteins.

A flexible, programmable protein machine will grasp a large molecule (the workpiece) while bringing a small molecule up against it in just the right place. Like an enzyme, it will then bond the molecules together. By bonding molecule after molecule to the workpiece, the machine will assemble a larger and larger structure while keeping complete control of how its atoms are arranged. This is the key element that chemists have lacked. . . .

These second generation machines—built of more than just proteins—will do all that proteins can do, and more. In particular, some will serve as improved devices for assembling molecular structures. Able to tolerate acid or vacuum, freezing or baking, depending on design, enzyme-like second generation machines will be able to use as "tools" almost any of the reactive molecules used by chemists—but they will wield them with the precision of programmed machines. They will be able to bond atoms together in virtually any stable pattern, adding a few at a time to the surface of a workpiece until a complex structure is complete. Think of such nanomachines as *assemblers.*

Because assemblers will let us place atoms in almost any reasonable arrangement, they will let us build almost anything that the laws of nature allow to exist. In particular, they will let us build almost anything we can design—including more assemblers. . . . Assemblers will open a world of new technologies. . . . With assemblers we will be able to remake our world or destroy it. (Drexler, 1986, pp. 13-14)

These molecular assemblers will be controlled by molecular computers. But where will they get the detailed instructions needed for assembly? Nanocomputers with molecular memory devices will store data generated by a process that is the opposite of assembly—disassembly. Nanomachines directed by nanocomputers will disassemble an object, record its atomic or molecular structure and then direct the assembly of a perfect copy. Working at lightning speed, this ability to copy and multiply carries the potential to change everything. Fifty years with this technology could bring more change than humankind has seen since the dawn of civilization.

Such an assembler, working at one million atoms per second, could copy itself in about fifteen minutes—about the time a bacterium takes to replicate under good conditions. It could make thirty-six copies in ten hours. In one week, it could make enough copies to fill the volume of a human cell. In a century, it stacks up enough to make a respectable speck. This does not sound very powerful or threatening; what is there to worry about?

But this is linear growth, what happens with exponential growth when each copy builds yet more copies? Two build two more, four build four, eight build eight, and so on. At the end of ten hours, more than 68 billion would exist (and they would weigh a ton). In fewer than two days, they would outweigh the earth; in another four hours they would exceed the mass of the sun. Obviously, there are limits to growth—this replicator process would have stopped long before, due to lack of raw material. Bacteria can replicate just as fast; they would go into overshoot and dieback. Just as we must be concerned about exponential growth of bacteria or humans, we must be concerned about controlling rapid new nanoreplicators.

Drexler (1986, pp. 60-62) spins a scenario for making a rocket engine. Assemblers controlled by molecular computers would "grow" the engine inside a vat as chemicals are fed to it via pipes. Obeying instructions from a seed computer, a sort of assembler crystal grows from the chaos of the liquid. The rocket engine is created in less than one day with almost no human attention. It is a seamless thing, very strong–able to withstand stresses that no metal engine built by bulk technology could withstand– yet weighing less than 10 percent of the mass of a metal engine.

In short, replicating assemblers will copy themselves by the ton, and then make other products such as computers, rocket engines, chairs, and so forth. They will make disassemblers able to break down rock to supply raw material. They will make solar collectors to supply energy. Though tiny, they will build big. Teams of nanomachines in nature build whales, and seeds replicate machinery and organize atoms into vast structures of cellulose, building redwood trees. There is nothing too startling about growing a rocket engine in a specially prepared vat. Indeed, foresters given suitable assembler "seeds" could grow spaceships from soil, air and sunlight.

Assemblers will be able to make virtually anything from common materials without labor, replacing smoking factories with systems as clean as forests. They will transform technology and the economy at their roots, opening a new world of possibilities. They will indeed be engines of abundance. (Drexler, 1986, p. 63)

Obviously, this creative process operates with vast knowledge that extends to the minutest detail. How could mere humans amass that knowledge? Using nanocomputers we will develop artificial intelligence (AI) that will do most of the research for us. AI will be lightning fast, have a vast and precise memory, and may even teach itself to be creative. It will be the ultimate tool because it will help us build all possible tools. Advanced AI systems could maneuver people out of existence, or they could help us build a new and better world. Aggressors could use them for conquest, or thoughtful politicians could use them to stabilize peace. They could even help us control AI itself.

Drexler's imagination may sound like science fiction, but his projections are based on solid scientific knowledge. He deliberately exposed his ideas to experts to see if they could survive critical review. I conclude that we must take this possible development very seriously, it is probable and, if true, unstoppable.

The power of this technology *for good and evil* dwarfs anything humans have ever imagined. It could be a realization of the "genie' imagined in the *Tales of the Arabian Nights.* Whatever we might ask of a "genie machine," it could produce. This technology would enable people to maintain youthful vigor and health for many more decades, even centuries. Nanomachines could have detailed knowledge of the proper structure of each of our cells and could dispatch rescue machines to make direct repairs by rebuilding cells on the spot. This is not magic, these repair processes already operate in nature. "Physicians aim to make tissues healthy, but with drugs and surgery they can only encourage tissues to repair themselves. Molecular machines will allow more direct repairs, bringing a new era in medicine." (Drexler, 1986, p. 104)

Drexler projects that people will routinely live several centuries. He seriously suggests that people in the 1980s should arrange to have their bodies put in biostasis at so-called clinical death in order to preserve their structure. At some future time, when new knowledge has developed cell repair machines, intelligent nanomachines could analyze the cell structure of a body, dispatch cell repair machines, and restore the body active. Drexler would reserve the word *death* to the state we now know as permanent death. Present-day medicine concentrates on maintaining function, as that is essential for healing. Cell repair machines would change the requirement to preserving structure. Biostasis involves preserving neural structure while deliberately blocking function, otherwise all memory would be lost. He even suggests that scraps of tissue may contain enough DNA information to be able to restore extinct species.

Drexler forecasts equally powerful "planet mending" machines. The dangerous molecules in toxic wastes are made up of innocuous atoms. Intelligent cleaning machines could seek out toxics and render them harmless by rearranging their atoms. As a matter of fact, present-day technology uses selected bacteria to do something very similar. Radioactive isotopes could be isolated from living things by building them into stable rock. Radioactive materials also could be sealed in self-repairing, self-sealing containers the size of hills and powered by desert sunlight.

Replicating assemblers will make solar power inexpensive enough to eliminate the need for the fossil fuels that presently pollute the biosphere so drastically. Replicating assemblers could even remove the excess carbon dioxide building up in the earth's atmosphere; as with trees, solar-powered nanomachines will be able to extract carbon dioxide from the air and split off the oxygen. Unlike trees, they could restore the carbon to seams in the earth. Planet mending machines could mend torn landscapes and restore damaged ecosystems. After the cleanup, we will recycle most of the mending machines, keeping only enough to protect the environment. Environmental protection will be easier because we will live in a cleaner civilization based on molecular technology.

If life could be significantly prolonged, what will we do with all the people? Drexler projects that we will use replicating assemblers, directed by advanced AI machines, to build new worlds in space using power from the sun and raw materials from asteroids. These new worlds may be smaller, or possibly larger, than Earth and would have gravity, water, atmosphere, plant, and animal life similar to that on Earth. He estimates that it will take quite a long time to fill up the empty space in our solar system before humans find it necessary to expand into other solar systems.

If assembler-based replicators are able to do all that life can, and more, this technology poses an obvious threat to the rich fabric of the biosphere and to humans themselves. As Schmookler portrayed so eloquently in the *The Parable of the Tribes* (1984), powerful technology forces all of us into a race for power and control. Drexler also foresees the problem:

Knowledge can bring power, and power can bring knowledge. Depending on their natures and their goals, advanced AI systems might accumulate enough knowledge and power to displace us, if we don't prepare properly. And as with replicators, mere evolutionary "superiority" need not make the victors better than the vanquished by any standard but brute competitive ability.

This threat makes one thing perfectly clear: we need to find ways to live with thinking machines, to make them law-abiding citizens. (Drexler, 1986, p. 173)

Life as we know it could be destroyed in a new kind of germ warfare utilizing programmable computer-controlled "germs." Replicating assemblers could improve on contemporary weapons and replicate them so swiftly that they could become abundant in a few days. A single state, or even a terrorist group, possessing such power could enslave or exterminate the rest of humanity at will. States with advanced technology could simply discard people, because they would no longer be needed as workers, soldiers, doctors, administrators, leaders, even scientists. Because of this threat, we cannot afford to allow an oppressive state to take the lead in racing toward the breakthrough. A mistake or an accident with such powerful technology also could have devastating consequences not only for humans but for all forms of life.

We are, then, caught on the horns of a terrible dilemma. Nanotechnology, or some other unforeseen but equally powerful technology, could bring unheard of abundance as well as solutions to so many of today's problems. Yet, that power simultaneously carries the threat of destroying everything we value. Buckminster Fuller made the point some years ago that utopia and oblivion may coincide; I leave the reader to ponder his meaning.

Could Mankind Decide Not To Develop A Technology?

As we have seen, there are certain technologies that we wish had never been developed and others that we are unsure should be developed. Could we prevent a technology from being developed? Could we halt further development of a technology just getting underway such as life-prolonging or genetic engineering techologies? Trying to answer these questions carries us into deeper questions: Could society be designed to handle such questions? Would it be willing to do so? Could it be that foresighted and creatively governed? Although this will be a very difficult line of analysis, and the policy proposals deriving from it are likely to be equivocal, just thinking about the problem will help us to face the future better.

Focusing on nanotechnology is useful because the technology has not yet been developed, and because it has the potential for such great benefits as well as such great threats. If left to me, I would choose not to develop the technology. I am not persuaded that a future life of abundance, power, and control would be better than the life of people today. I am not sure a long life would be better, even if we could remain healthy. Facing the possibility that life, all life, could be snuffed out in a few hours or days detracts a great deal from the enjoyment of a long healthy life. I am not sure that the "nature" humans would devise would be better than the nature we now enjoy.

Whether nanotechnologies are developed will not be my choice or that of my country. The competition for knowledge, power, and wealth will continue relentlessly. Scientific inquiry is not centralized; it is carried on in millions of localities and by millions of individuals and teams. Of course, a government could issue an order to halt a given line of research, but would it be obeyed? An order given and obeyed in one nation would not halt research in other nations. Because scientific research confers power, and given the competition for power between nations, it is extremely unlikely that any nation would try to halt research that is needed to keep abreast in the power race.

Suppose we were able to form a world family of nations, a world government. Could it order a halt to a given line of research? How could it enforce the order? Halting nanotechnology would be possible only if the world government had the capability to reach into every locality to search for violators and the police power to enforce compliance with the order. In effect, only a worldwide police state could halt a line of scientific research. Is it better to live in a planetary police-state with the capability to control the pace and direction of scientific development or is it better to allow science to develop freely, even if dangerously, in a free society? Most of us would choose the free society.

In December 1987, then-President Reagan and Soviet Union General Secretary Gorbachev signed a treaty to ban short- and intermediate-range nuclear missiles. That treaty assumed that nuclear weapons could be banned and their destruction verified. Drexler (1986 p. 193) makes the point that nuclear reactors and weapon systems are large, thus they are definable and visible; and in principle can be banned.

However, nanotechnology will be different; dangerous replicators will be microscopic and AI software will be intangible. Modern biochemistry leads in small steps to nanotechnology and modern computer technology leads in small steps to AI. No line defines a natural stopping point. Furthermore, each advance brings benefits that humans would not wish to forego. How could anyone be sure that some laboratory somewhere is not on the verge of a breakthrough? Ordinary verification measures will not work, thus negotiation and enforcement of a worldwide ban would be almost impossible.

The promise of technology lures us onward, and the pressure of competition makes stopping virtually impossible. As the technology race quickens, new developments sweep toward us faster, and a fatal mistake grows more likely. We need to strike a better balance between our foresight and our rate of advance. We cannot do much to slow the growth of technology, but we can speed the growth of foresight. And with better foresight, we will have a better chance to steer the technology race in safe directions. (Drexler, 1986 p. 203)

Because conventional notions of disarmament will not work for nanotechnology, Drexler (1986, pp. 187-188, 196-199) proposes the development of "active shields." These shields would be purely defensive, in contrast to conventional defenses which also carry the potential for offensive use. They would be active in the same sense that our white blood cells rush to defend against invaders in our bodies. The defenders would be carefully designed and programmed to be fast, selective, and strategically effective. He has not worked out the details of active shields, but he believes that a contingent of intelligent humans using fast nanocomputers and AI could manage to build an active shield system.

If we recognize the reality that some controls must be placed on science and technology, how can we make those controls most compatible with our own nature? How can we preserve the greatest amount of freedom? My answer:"Emphasize social learning." Somehow, we must learn to learn faster. Controls that come from within are more acceptable than those from without. Controls that make sense in terms of our own recognition that they are essential are more acceptable than those that seem to make no sense. Controls that restrict some freedom are better than anarchy or uncontrolled power (for example, terrorism) that leave us with neither freedom nor security. As we think about these matters, we must constantly keep in mind that *there is great danger in the unrestrained development of science and technology which confers great power.* If science and technology are given free rein, it means that ordinary people will inadvertently slide under the control of those who control science and technology.

Directing Science And Technology To Serve A Sustainable Society

What kind of learning do we need to emphasize in order to develop moral controls that work from within? I have repeatedly made the point that the highest value for any society hoping to survive is to maintain the integrity of its ecosystem, its whole biotic community. This deep valuation on all of nature's creatures can and must be learned on both an intellectual and emotional level. Surely, almost all of us recognize that a developmental thrust resulting in humans crowding out of existence thousands, even millions, of other species would be morally wrong. Our unique capacity to understand the workings of our ecosystem confers on us the responsibility to preserve its integrity. We can learn how the developmental thrust of science will have the consequence of putting humans in an overwhelmingly dominant position in nature, threatening other species with extinction. It is especially important that scientists themselves learn of this connection.

As a first step, scientists must recognize the value content of science; that science cannot be value-free. If we want to keep science as free of external controls as possible, we must get scientists to study values and impose their own internal controls. They should take the time to learn how to do value analysis. They should study the value implications of their inquiries. Scientists tend to accord greatest respect to those whom they perceive as truly understanding science—usually other scientists. The high regard scientists hold for each other needs to be accompanied by a recognition that scientific work is embedded in values. In other words, scientists should discontinue disparaging the work of other scientists who have thought through the role of values in their work—they should stop believing that science can be value free.

Scientific research should be planned for and supported in a context that takes a careful look at the long-range consequences of a proposed line of development. This thorough examination should include an estimate of its impact on our way of life, on our values, and on the feasibility of controls. We should be persistent in asking Hardin's ecolacy question: "And then what?" A thorough analysis of long-range impacts should be conducted by an interdisciplinary team of scientists, scholars, and policy analysts. They will bring multiple perspectives and can point out aspects that might be overlooked if only a single perspective were to be taken. Another way to ensure multiple perspectives would be to have two or more review teams examine each major developmental question. As these teams write their scenarios, they may well envision several possible modes of scientific/technological development. They should vigorously criticize each others' projections so that the final product of the inquiry represents the best possible thinking on the subject.

Scientists often dispute the meaning of findings, leaving the public and policymakers in a quandary about what to believe. The critical procedures of science eventually resolve most such disputes; but public policy usually must be made long before the disputes are resolved. Dr. Arthur Kantrowitz, a member of the U. S. National Academy of Sciences, advocated (1975) a "Scientific Adversary Procedure," which is based on principles of legal due process. He hoped that this proceeding would establish facts, separate them from values, and accelerate social learning. Various sides in a dispute would present the facts as they see them before a well-informed referee who would seek statements of agreed-upon facts. Where disagreement persists, a technical panel would seek further evidence and write an opinion outlining what seems to be known and what still seems to be uncertain. Every aspect of the proceeding would be public. This proceeding need not be conducted by the government.

Kantrowitz's idea for adversarial proceedings was quickly picked up in both scientific and science policy circles, because a vigorous debate was just getting underway about the possible dangers of recombinant DNA research. The proposed proceeding soon became known popularly as a "Science Court." President Ford appointed an Advisory Group on Anticipated Advances in Science and Technology, chaired by Kantrowitz, which delivered a report in the summer of 1976. This report was followed in September 1976 by a conference called by scientific and government groups and attended by more than 250 scientists, lawyers, business leaders, and government officials. A consensus was reached that the idea should be tried; a proposed procedure was laid out; and specific steps were planned for getting underway. The experiment was endorsed by several scientific associations; Chief Justice Burger appointed a judicial task force to explore its feasibility; yet, no experimental court was ever convened. By the end of the decade, the idea had virtually disappeared.

Cole (1987) suggests that interest waned because institutional changes in government in the 1970s had more adequately come to grips with the dangers of nuclear and recombinant DNA technologies; hence, there was less pressing need for another body. Additionally, as the science court idea was scrutinized more closely, doubts were created about several of its features. Most troublesome was the doubtful premise that facts could be separated from values in assessing the wisdom of science policies. (Kantrowitz does not assert that the separation can be clean but he maintains that it would be useful to try to separate values from facts when evaluating facts—and that wiser policies will result from the increased certainty and clarity of the facts.) There was doubt that truly disinterested judges could be found. People were concerned that Science Court judges would issue verdicts that would carry the aura of scientific disinterestedness, and political officials, not being able to make a similar claim, would be fearful of overriding them. Finally, the court would only hear currently controversial issues. It would have no capability for anticipating problems or for uncovering existing ones.

Both Kantrowitz and Drexler believe adversarial proceedings can be adapted for use in a nongovernmental context: they avoid using the term *science court*. Kantrowotiz has conducted adversarial proceedings at various universities in order to show the feasibility of the idea. I believe their attempt to separate what is widely agreed to be known from what is still in dispute is useful for social learning. I agree with Drexler that we must carefully distinguish *examining facts* from *making policy*. Carefully determining facts can help us choose policies that will serve our values. However, we must take care not to allow scientists to declare a policy perspective to be factual when it really incorporates values, and use it as a club to defeat those who oppose the policy.

Others have come forward with related ideas. Krimsky (1978) proposed a "Citizen Court" based on the experience with a citizen review panel established in 1976 to examine a proposal made in the City Council of Cambridge, Massachusetts, to discontinue recombinant DNA research within the city (which includes Harvard University and the Massachusetts Institute of Technology). A Cambridge Experimental Review Board, appointed by the City Manager and comprised of only citizens in order to avoid scientific elitism, studied the problem at great length. Six months later they recommended that the research should be allowed to continue, but only under certain safeguards; this policy was adopted by the City Council (Dutton, 1984).

Krimsky proposed that citizen courts, comprising between eight to fifteen lay members, be appointed by local governments to consider specific scientific controversies that would come before local communities. The court would try to separate facts from values but also would make recommendations to steer policy. This instrument is weak in making science policy for these reasons: 1) facts are not readily separable from values in policymaking; 2) lay people might not understand complex scientific issues; 3) there might be little connection between understanding an issue and the ability to persuade others to one's point of view; 4) having different communities determine policy on the same scientific issue could lead to confusion and dangerous inconsistencies in policy.

Cole (1983, 1987) has proposed a National Science Hearings Panel that would have three or four scientists and an equal number of nonscientists appointed by the president for five-year terms. If concerns about the safety or wisdom of a science or technology were expressed to it in a convincing way, it would convene a forum of experts and nonscientists to be participants and observers. The forum would employ adversarial proceedings but it would offer no verdict or decision. The information it generated would be disseminated through professional journals and the mass media. Its purpose would be to expose the issue and facilitate social learning. Regular public officials would still make authoritative science policy decisions. This institutional structure would not assume that facts could be separated from values. There would be no claim that science policy issues could be resolved "above" politics. The panel could solicit information and expose anticipated problems (not simply recommend with respect to current controversies). Finally, it would constitute a continuing procedural mechanism; therefore, it could evolve practices that proved workable and that would become systematic and predictable.

Cole's proposal for a science hearings panel has several similarities to a proposal I make in Chapter 14 for a Council for Long-Range Societal Guidance that would assess the long-range impacts of all major public policy initiatives. We need similar assessments of the long-range impacts of proposed scientific developments and new technologies. A thorough review of the future consequences of a scientific or technological development will help all parties, especially scientists, to become aware of the full context, including value components, of a possible line of inquiry. It is important to ensure that the findings of these teams of reviewers will be widely circulated and discussed, especially in scientific communities. The results of such studies should develop a growing awareness of appropriate controls for scientific or technological initiatives that would guide public as well as private funding for research.

Drexler also worries that our thrust to learn faster would be stymied by an overload of studies and information. Needed information may be avilable but the time and effort required to find and retrieve it frustrates effective utilization. Drexler proposes (1986 pp. 220-230) swift development of a new technology proposed by Nelson (1981) called "hypertext publishing." When fully developed, this technology not only would retrieve a given text but would show linkages to all other writing that had commented on the text. Using the linkage network embedded in the technology, the text of the commentators also would be instantly retrievable. This technology could be put in place with current "bulk" computers. Using hypertext would be greatly eased by nanotechnologies, however. Nanocomputers would be so fast and so small with so much storage that one could retrieve and digest the most up-to-date information on a subject in the space of a few hours without having to leave one's office or home. Later readers would be able to benefit from the insights and judgments of earlier readers without wasting intellectual time on search and retrieval.

Morone and Woodhouse (1986) propose a set of decision rules that they call "sophisticated trial-and-error" for cautiously dealing with hazardous technologies.

1. Take initial precautions to protect against the worst consequences of errors.

2. Err on the side of caution.

3. Actively prepare to learn from error by establishing monitoring mechanisms to report and interpret negative feedback.

4. Use analysis in support of this strategy, as by advance testing to speed up negative feedback, and by setting priorities so that key uncertainties get most attention.   •

5. Adjust the initial precautions as uncertainties are clarified: reduce them if the threat is less serious than anticipated, or enhance the precautions where warranted.

These rules recognize the reality that the greatest latitude of choice exists prior to the very first time a particular technique, instrument, or system is introduced. Once economic investment, material equipment, and social habit are in place, alternatives already have been selected and the original flexibility vanishes. We must come to see that deployment of a technology is not merely an economic or technological action; it also is a social and political action. We need to examine technologies for their social and political elements during the earliest trials of the proposed initiative.

We discussed above the difficulty of trying to control science by passing laws that would be enforceable in only a single nation. Scientists, more than almost any other social grouping, easily cross national boundaries as they communicate and think with others in their scientific community. If the scientists within a scientific community, drawn from many nations, come to agreement about a policy for development of their science, that policy is much more likely to be implemented in many nations than if only political authorities were making the decisions. Collaboration among scientists of many nations on these questions also will help lay the foundation for international cooperation in protection of the biosphere. If the integrity of the biosphere is to be protected, deep and broad international cooperation must eventually take place. A recent proposal to launch an International Geosphere-Biosphere Program: A Study of Global Change was quickly approved by several nations after being strongly recommended by the International Council of Scientific Unions.

Although I have placed primary responsibility for the thoughtful control of the development of science on scientists themselves, there also is a role for ordinary citizens. Their stake in the outcome is just as crucial as that of the scientists. We should be wary of the notion that scientists know better than the people themselves what is good for the people. Science policymaking should be in an open enough forum, and be given sufficient time, so that citizens can meaningfully play a role in the development of policy.

Conclusion

Evaluating science, with a view to controlling it, inevitably leads to some very difficult choices. Science has been so successful that it has delivered great power to those who control its development. Unfortunately, most scientists labor under the delusion that scientific activity is, and can continue to be, value free; this delusion disables the scientific community for the task of policing itself. Political authorities also have difficulty effectively controlling science. Thus, the control of science, and the power it creates, will go to those societal entities that control funding for its development. In a market society, most control will flow to large corporations.

Allowing science to develop without constraint would likely lead to a future society (50 to 100 years from now) that would be densely populated, nearly all its resources would be devoted to human purposes, wilderness and wildlife would be in sharp decline, and the biosphere would be in great danger.

Scientists themselves provide the best possibility for preventing that scenario from coming true. They need to realize the consequences of their failure to see where their own scientific development is carrying them. They could learn how to project possible future scenarios of the development of their science; they could learn how to analyze values; and they could devise systems of control for the development of their work. They should work with citizens in the development of this policy. In essence, we all need to do a lot of thinking and learning. If we fail to devise adequate controls, science, our greatest achievement, may turn out to be the greatest threat to the biosphere and to the continuation of the human species. "We may fail. Replicating assemblers and AI will bring problems of unprecedented complexity, and they threaten to arrive with unprecedented abruptness. We cannot wait for a fatal error and *then* decide what to do about it; we must use these new technologies to build active shields before the threats are loosed." (Drexler 1986 p. 237)

Science pervades our lives; we cannot turn back. As we strive to direct the power of science to protect the biosphere and to develop a society that can live harmoniously with nature, our energy should be focused on enhancing and speeding up social learning. A well-developed understanding of our values and our place in the biosphere will make it easier for us to work together to control the power we are swiftly unleashing. We should expect debate concerning controls on science and technology to become the dominant issue of the twenty-first century. We should be developing our knowledge and understanding in anticipation of the debate and be prepared to participate vigorously.

Recognizing And Avoiding The Detrimental Effects Of Enticing Technologies

Chapter 13

Technological Imperialism Of The Developed Countries

Many technologies have an enticing allure that spreads them quickly to all societies in the world. Is there something imperialistic about technologies? Societies that develop powerful technologies can use their greater power to conquer or dominate less technologically developed societies. If a society does not wish to be dominated or exterminated, it must race to develop technology that is as powerful as that of its potential enemies. The technology development race is a driving force for social change as well as the central component of power politics.

Invaders used change-inducing technologies to subject indigenous people to their rule; they deliberately used technologies to displace many elements of their traditional way of life. Even when a traditional society had not been invaded, the contact with advanced technology proved to be such a powerful stimulus that indigenous people quickly saw its potential for enjoyment, power, and control and moved to acquire it. In recent decades, commercial agricultural methods and massive road and dam building projects have spread through the Third World, aided by international development agencies—most prominently the World Bank.

In the past three decades, television has proven to be the most attractive of all technologies, having been adopted widely in the Third World. Less developed countries (LDCs) that adopt television also import a large proportion of their programming from the most developed countries (MDCs). This practice carries images of wealth and power to the far corners of the globe and creates a desire for the benefits of technology in nearly all people. Television thus becomes a "seed" technology stimulating greater desire for technological development.

In order to attract and hold our attention, and sell advertised products, television programming in the United States conveys a super-dramatic picture of life. A study by the National Institute of Mental Health (NIMH) showed that murder is 200 times more prevalent and crime is twelve times more violent on television than in actuality. The average young person witnesses 18,000 murders on television by age sixteen. Those who frequently watch television perceive the world to be more dangerous, violent, and untrustworthy than those who watch it infrequently. Apparently, people gradually come to believe the distortion is the reality.

Television shapes public perceptions of appropriate sex roles—that typically portray men as dominant. Racial stereotypes also are reinforced by television. Although these generalizations are valid for the most part, social learning by the public feeds back into the system. For example, I recently saw an advertisement on television that portrayed a tough-looking lineman wearing a helmet, high on a pole, making repairs to a power line. The lineman climbs down the pole, takes off his helmet for lunch, and a cascade of long hair tumbles out—he's a young woman. I do not even remember what product the advertisement wanted me to purchase, but I do remember the young woman playing a "man's role." This advertisement clearly carried two messages—"buy our product" and "look at what women can do if we let them"—while also attracting attention. I also have noticed a significant increase in blacks playing sympathetic roles in advertising.

Advertising obviously is a powerful social force; it shapes people's perceptions of the way the world works or should work. If social movements such as the feminist, civil rights, and environmental movements can increase awareness among television programmers and advertisers, or can get persons of their persuasion into programming situations, they have an effective ally for reshaping public awareness. This type of programming will only work as long as an advertisement is likely to increase sales. No advertiser will push an image change that would have the effect of reducing consumption. The subtle, background, insidious objective of advertising is to constantly nourish the belief that quality of life can be realized via consumer indulgence. The more successfully advertisers inculcate that belief in the people, the greater the damage to the ecosystem.

Technologies, then, are powerful, even overwhelming, social forces that we adopt, hoping they will be good for us—but they are not always good for people or for the multitude of other life forms on this planet. They have enticed us into lifestyles and patterns of production and consumption that are wasteful. They have been coopted by greedy people seeking power, conquest, and wealth. They have deterministically shaped our social structure and behavior patterns just as effectively as if those structures and behaviors had been legislated; in this sense, technologies can be seen as legislation. All in all, they have hardened our compassionate human qualities and turned us into less caring persons than most of us aspire to be.

The Imperialism Of Large Technology-Based Systems

The more attractive a technology is, the greater the change it will bring to our social patterns and social structure. Even though we are often frustrated and annoyed by having to conform to technological demands, we find that we cannot live without them. They lock us into some scenarios and out of others. We have become so dependent on our automobiles that we cannot slow down on making and servicing them. Our addiction to the television and telephone is eliminating the development of writing skills in most people. The development of computers and automation is eliminating many jobs, profoundly shaping business practices, and changing settlement patterns and home life. Long (1985, p. 49-51) interviewed Joseph Weizenbaum, one of the pioneers of the computer revolution, who is perceived as a turncoat by his colleagues because of his probing challenges:

> The computer–logical, linear, rule-governed–encourages a certain kind of thought process. Call it scientific rationality. The introduction of the computer has driven man to an ever more highly rationalist view of his society, and an ever more mechanistic view of himself. We are now, as a society, close to the point of trusting only modern science to give reliable knowledge of the world. I think this is terribly dangerous. The dependence on computers is merely the most recent–and most extreme–example of how man relies on technology in order to escape the burden of acting as an independent agent; it helps him avoid the task of giving meaning to his life, of deciding and pursuing what is truly valuable. . . .
>
> Now what we have to fear is that inherently human problems–for example, social and political problems–will increasingly be turned over to computers for solutions. And because computers cannot, in principle, ask value laden questions. The most important questions will never be asked. . . . I think we are absolutely intoxicated with science and technology. Massively distorted perceptions of reality are everywhere. We euphorically embrace every technological fix as a solution to every human problem, which we have, of course, first converted to a technological problem. . . . You used the phrase "the computer causes" . . . a computer doesn't cause anything. You see, the habit of speech that makes instruments responsible for events leads directly to speaking and *thinking* of science and technology as autonomous forces and to the idea of technological inevitability. And it can lead, finally, to the belief that man is simply too powerless to struggle with these powerful impersonal agencies, even though they are of his own making.

Another aspect of the imperialism of large technological systems is the way large corporations, many of them multinational, crowd smaller companies out of existence. My city of Buffalo, New York, for example, used to have fourteen local breweries; now it has none. Probably as much beer is consumed as before, but now the beer is trucked in from regional breweries that are part of larger nationwide corporations. Larger corporations are able to invest in research and development and thus obtain a technical edge over smaller companies. They also are usually better able to attract the necessary financial backing to produce and market new products or to use new technologies of production. The trend from recent history is clear:

we are inexorably developing toward larger and larger organizational units. The ability of individuals to influence or control these units correspondingly weakens.

Technologies, by their very presence, force us into searching for solutions to problems that become defined as technological, even though they many not have been technological or a problem to begin with. Weizenbaum is talking about computer technology, but he could as well have been talking about nuclear technology:

These discussions are carried out in a mode of thought that has become traditional among computer technologists: it begins with a great many solutions and then looks for problems. And the inevitable consequence of this way of thinking is that it obscures real problems. The aimlessness of everyday life experienced by millions in modern society has deep roots in the individual's alienation from nature, from work, and from other human beings. . . . What's needed more than anything else, I believe, is an energetic program of technological detoxification. We must admit that we are intoxicated with our science and technology and that we are deeply committed to a Faustian bargain that is rapidly killing us spiritually and may eventually kill us all physically. . . . I think the world is overly mechanical and that it invites the substitution of machines for human beings in fairly intimate areas of life. I think that maybe we should look upon genuine warm human relationships, together with all the tensions that they bring, as endangered phenomena and should carefully try to enhance these relationships while we have the opportunity. . . . To be completely honest with you, my own view is that we cannot recover without the help of a miracle. . . . I think our job now is to work with all our energy to prepare the ground for such a miracle. (Long, 1985, pp. 76-78)

Structural Wastefulness

Most western societies have market economies that emphasize production, growth, wealth, and power. Very often these factors structurally force people to be wasteful, as is shown by the following examples. The railroad was invented in the mid-nineteenth century and widely adopted in many lands. Railroad transport is still one of the most energy-conserving ways to move people and goods, requiring about one-sixth the energy used by automobiles and trucks. Yet, railroads are structurally being forced out of existence, especially in the United States. Automobiles and trucks are more versatile and, most of the time, more convenient. Automobile and truck use also has been encouraged by our growing per capita wealth that allows us to pay high energy costs for the convenience of an automobile. Furthermore, the richer we are, the bigger and more wasteful our automobiles can be. Now that the greenhouse effect and disastrous climate change loom over our heads, people are beginning to think about bringing back the rail-

roads in order to save fossil energy and reduce carbon dioxide buildup–
that is painful learning.

As railroads were phased out, we became increasingly dependent on our
automobiles, not only for transport but also to supply employment for factory workers and service people. A few years ago the U. S. government
decided not to allow Chrysler Corporation to go bankrupt because it felt
the nation could not afford to lose so many jobs. Because we cannot stop
making automobiles, our junkyards for old cars get larger and larger–more
materials and energy are wasted and the biosphere becomes more devastated. We are locked into a pattern that enslaves us as much as it liberates
us. On many trips, the automobile wastes more time than it saves and creates
more distance than it overcomes. People can live in distant suburbs and
drive fifty miles to work, taking the better part of two hours to do so. A
good part of each day's work goes to pay for the travel necessary to get to
work. Because public transportation is seldom used, everyone must have
an automobile and struggle to pay for it. The ambiance which once made
the city the only place worth living is now so degraded by automobile
congestion that many consider the city to be a "hell." "Since cars have
killed the city, we need faster cars to escape on superhighways to suburbs
that are even farther away. What an impeccable circular argument: give us
more cars so that we can escape the destruction caused by cars." (Gorz,
1980, p. 74)

As another example, consider the variety and volume of packaging that
clutters and dominates our retail outlets, especially in the United States.
Packages are carefully chosen to attract attention to a product so as to
outsell competitors, and also to be convenient for the retailer. The wasteful
use of packaging materials and the energy required to make and dispose of
packages seems not to be considered in designing packages. Waste disposal costs Americans $4 billion per year. Typically, the package also pre-
vents the consumer from carefully examining the product, thereby
interfering with rather than facilitating wise consumer choice. Packaging
constitutes 40 percent of American household refuse. Plastics production,
much of which goes into packaging, has grown by 2,000 percent in the past
fifteen years–and plastics are made of oil which is nonrenewable. Furthermore, plastics require about 450 years to biodegrade.

Another unfortunate pattern of modern society is to encourage the use
of cheap but nonrenewable resources in the place of renewable resources.
The most obvious example is the substitution of plastics for wood and paper.
The use of plastic carrying bags to tote goods away from retail outlets is
ubiquitous in Western Europe. They were aggressively introduced in my local
supermarket about three years ago (paper bags were simultaneously made
unavailable); however, consumer resistance was so strong that consumers

can now choose between plastic or paper bags. Not only do plastic bags utilize a nonrenewable resource that is better used for other purposes, but people cast them into nature, marring the landscape for many years.

For many decades, beverages were offered for sale in containers that could be returned and reused. The systems for recycling containers worked well. In the 1950s "throwaway" containers were introduced and aggressively marketed by distributors. These containers, naturally, were soon thrown away everywhere, cluttering up roadsides and nature trails. They piled up in solid waste dumps that were filling up far too swiftly. Throwaway containers consume about ten times the energy of containers that can be recycled fifteen or twenty times. By the time the manifold costs of this throwaway system became clear to many people, the new system was so well-entrenched that it strongly resisted changing back to a recycling mode. In state after state, the container and beverage industries have bitterly fought proposed legislation that would mandate deposits on containers to encourage their recycling. Only very slowly has the pressure of public opinion been able to force a return to recyclable containers.

Recycling an article is always more troublesome than throwing it in the trash heap. If people are to be willing to recycle, the opportunity to do so must be made readily available. Recycling is not simply an individual act, it requires a recycling system. Typically, that means governmental sponsorship of the appropriate infrastructure. A vital part of that infrastructure is to legally deny pick-up of garbage unless the materials have been separated into recycling streams: paper, glass, metal, and food wastes—this is called *source separation.* Our current society in the United States systematically discourages people from recycling. Virgin materials may be given regulatory or tax preferences over used materials. Market incentives may not be sufficient to draw used materials back into production. For example, paper made from recycled fibers sells at a higher price than paper from virgin fibers. Recycling depots are troublesome to maintain. The one in my community recently refused to accept metal cans and glass because, "they were too much trouble." Recycling will not be successful if it is totally dependent on a market.

Modern consumer-oriented societies also have a variety of incentive structures to encourage wasteful rather than conserving consumption patterns. Persons who buy a lot of something typically get a better price than people who buy only a little. One of the most wasteful practices is charging a lower rate for electricity to those who consume a lot. Heavy users with cheap rates have no incentive to reduce consumption at times of peak demand, and they thus force the power company to build standby generating capacity to cover the peak demand. If we want people to conserve, it would make much more sense to charge heavy users more than those who make few demands; or else require them to make their demands at nonpeak hours.

An Infrastructure Designed For Sustainability

Our beliefs about what makes for a good life and a good society find expression in socioeconomic-political structures. Designing a sustainable society requires us to rethink these structures. A society that emphasizes voluntary cooperation, that strengthens the ability of communities to live by their own efforts and resources, and that encourages them to determine their own future will foster the development of technolgies and methods of production that:

1. can be used and controlled at the level of the neighborhood or community;

2. can be jointly controlled by producers and consumers;

3. encourages the development of more autonomous local production units;

4. avoids biospheric injury; and,

5. uses resources efficiently, encourages their recycling, and conserves them for future generations.

Rifkin (1985, pp. 93-95) distinguishes empathetic from controlling technologies:

It is time to develop a new set of tools that are empathetic to the environment they are used in: tools that cajole rather than grab; tools that select rather than pillage; tools that operate at a speed commensurate with the rhythms, beats and tempo of the environments they are engaged in. Appropriate technologies are technologies that are congenial with their surroundings, that create the least amount of disturbance, and that are used sparingly enough to insure that the environment can be allowed to replenish itself.... With controlling technologies, the emphasis is on maximizing present opportunities. With empathetic technologies, the emphasis is on maximizing future possibilities. With controlling technologies, a high premium is placed on optimizing efficiency for the present generation. With empathetic technologies, a high premium is placed on maintaining an endowment for future generations....

An empathetic approach to technology starts with the assumption that everything is interrelated and dependent on everything else for its survival, and that technological intervention should be minimized in order to do the least damage to the myriad relationships that exist in the natural world. There is an acknowledgement that some form of expropriation is always necessary. All things desire to live, and it is a law of nature that for something to live, something else must die. But it is also true that too much expropriation can result in destroying the very life support systems we rely on for our future survival. There is a distinction to be made, then, between mutual give and take, which exists in every set of relationships, and the kind of blind expropriation that strains and eventually severs the relationship altogether.

Examples Of Structures And Technologies That Enhance Sustainability

In this section, I discuss goals and principles that should guide the ways we design, build, use, and discard things so that our way of life will last far into the future. It would not be possible to cover every possible strategy, because we will learn more and more about how to achieve sustainability. I would much rather stimulate your own creative energies—this is a task that needs many creative minds.

A production structure that emphasizes sustainability will avoid waste of resources and be more likely to produce long-lasting and beautiful things that people can love and enjoy for many decades. Such a structure would avoid fads and fashions that encourage obsolescence. It would use renewable materials, like wood and paper, instead of nonrenewable materials, like plastics. It would emphasize simple and efficient production facilities as well as simple products that are both beautiful and functional.

A planning structure that supports the NEP will avoid large projects. For one reason, their sheer size and financing forecloses other options. Another reason is to keep our man-made structures to human scale. Gigantic buildings or projects intimidate people and leave them with a sense that they cannot control their destiny. Huge installations typically impact the environment much more severely than smaller installations. What should we do with huge structures that are no longer useful, such as obsolescent nuclear power plants? We cannot just abandon them; abandoned structures are always a problem for someone. Planners designing large projects should also plan for their decommissioning.

Buffalo, New York, was once the dominant flour milling city in the world. Many firms built huge grain elevators on the Buffalo waterfront. When the St. Lawrence Seaway made flour milling in Buffalo less attractive, some firms simply stopped paying their taxes and abandoned their elevators. The property reverted to the city, which now has many abandoned elevators on land that cannot be used for anything else without incurring a high cost to remove them.

Soft energy is produced in smaller units, that are widely dispersed, and readily controlled by ordinary people. In contrast, hard energy structures are huge, centralized, and difficult for ordinary people to control. (Lovins, 1977, originated the distinction and is the leading proponent of the soft energy path.) Hard energy is epitomized by nuclear power which presupposes and imposes a centralized, hierarchical and police-dominated stucture. The technology is too sophisticated and the process too dangerous to allow many dispersed units to become providers. Solar energy is softer in that it utilizes comparatively simple technologies that can be understood and constructed by many people; the same is true of such solar derivatives as wind power, water power, and biomass energy. Soft technologies can be dispersed widely throughout an environment and thus can be locally constructed, maintained, and controlled. Their impact on the environment is modest and seldom highly injurious. Because there are many producing units, perhaps using a variety of technologies, the total energy of a community is not vulnerable to the breakdown of a single centralized system. Because the producing units are small and dispersed, the technology is more adaptable to social change. Large, centralized structures, in contrast, lock a society into their long-term use and maintenance.

Cogeneration uses fossil fuel to produce heat and electrical power at the same time. This method is economical only when large and relatively constant amounts of heat are required. Manufacturing plants requiring both heat and power are obvious sites for cogeneration installations. As better small-scale systems are developed (they are called *modular units*), cogeneration is likely to be installed in schools, hospitals, shopping malls, fast-food restaurants—wherever a substantial amount of heat is needed. Flavin (1985, p. 38) reports a study estimating that 40,000 small-scale, modular cogeneration systems will be operating in the United States by the year 2000.

Because electricity captures only 35 to 40 percent of the energy value in the fuel, most electrical generating plants produce a large quantity of waste heat that is dissipated in nearby water bodies or via huge cooling towers. It has been estimated that the waste heat from electrical power plants would be sufficient to heat all the homes in the United States; the same estimate has been made for Great Britain. It is not easy to use this waste heat because swift cooling is necessary to maximize efficiency in electrical generation. With proper planning, waste heat from electrical plants could provide space heating to nearby homes, factories, and offices. Few people wish to live near a huge power plant, however. Another variation on this theme is exemplified by a chemical manufacturing plant in Niagara Falls, New York, that burns solid waste from nearby communities to provide heat and power for its manufacturing processes. We need to be cautious about some of these activities, however; even though they conserve resources, they could damage the ecosystem in other respects. In the case of waste to energy plants, many environmentalists have expressed concern that dioxins will form in the burning process and be spread into nearby neighborhoods. Furthermore, the ash remaining from the burn often is laden with toxic metals that must be carefully isolated from the biosphere as it is disposed.

Resource depletion quotas have been recommended by Daly (1977, pp. 61-68) to deal with the problem that resources may be cheap to extract but finite in quantity. Oil in the Middle East, for example, can be extracted very cheaply, but the supply will run out after a few decades. The cost of extraction does not reflect its ultimate scarcity, and its low price encourages wasteful consumption. Daly would have government declare a monopoly on certain resources and ration the amount available so as to stretch supply over a long time. The government would auction off depletion quotas—only a specified amount of the resource would be available each year—thus ensuring that prices would reflect scarcity. The excess of sale price over extraction cost would be captured by the government for public income. Higher prices would compel more efficient and frugal use by both producers and consumers. The quotas not only would limit depletion and encourage conservation but they also would limit pollution by encouraging recycling and by limiting consumption.

The modern communications net provides many structural opportunities to serve NEP values. Alvin Toffler (1980) predicts that in the not too distant future many people will work in an "electronic cottage": it will be wired into such a complete communications net that the persons who live there will not need to leave home in order to go to work or to participate in many other aspects of community life. Persons who work at home while still being plugged into instant communication with their coworkers are called *telecommuters*. The personal computer is the key technology that makes this mode of working possible. It can send, receive, and process massive amounts of data as well as a wide variety of other messages. It would be unrealistic to think that all, or even most, workers will become telecommuters; but if several million did so, this mode of working would conserve resources, because less energy would be consumed commuting to and from work, and fewer materials and energy would be required to construct and maintain workplaces. Dispersed electronic work sites are more adaptable to new tasks, new work teams, and new institutional structures; hence, there would be fewer and smaller obsolete structures to dismantle and dispose of when the useful life of a structure has passed. As usual with any structure of action, there also is a downside. Some workers may be thrust into a piecework mode that would provide none of the benefits of normal employment. Some workers may feel isolated from their workmates and miss the nourishing interaction that characterizes the best work settings.

There is reason to believe that modern communication technologies will have a democratizing effect. Technologies that enable people to work at home in dispersed sites should help to reverse the "massifying" effect of technologies that require people to work on assembly lines in large factories. Telecommuters are likely to feel more autonomous and freer to move on to new jobs or to start a new business. Personal computers can link, via a phone line, to a network computer that maintains a "bulletin board"; messages left on the board can be read at one's convenience and one can leave a message to be read at the receiver's convenience. Persons so linked are less dependent on the mass media for their information and can carry on a dialogue with people in far places; this effect also should be democratizing. Some networks have been developed for environmentalists. *Econet* is a worldwide network primarily to link ecologically-oriented organizations: it is designed to handle electronic mail, bulletin boards, and teleconferencing. *Environet* is oriented to the local grassroots level and deals with environmental, technological, and ecological topics–especially current political issues relevant to these topics.

The mass media in the United States, and in most other countries, are controlled, for the most part, by "the establishment." Those in control are unlikely to encourage media messages that would have the effect of undermining the establishment. How could ordinary people claim some access to their own mass media? Citizens could insist that some portion of *prime* air time (television or radio) be given to airing public issues. Recent action by the Federal Communications Commission (FCC) in the United States to diminish, rather than enlarge, public service air time is a step in the wrong direction. More public service air time will be granted only if the public demands it. Cable television networks provide more channels, thereby more air time for public service, but they also increase the cacophony of messages– thus diminishing the chance that any given message will find many receivers. In order to encourage receptivity, public service broadcasts need to be more relevant and interesting. People who participate in the programs as well as the receivers need to feel that their involvement could make a difference for a better community.

The development of *Electronic Town Meetings* could give people a sense of real participation. An organization called Choosing Our Future in Menlo Park, California, is developing an electronic system to revive and revitalize the traditional town meeting. Even though television is extremely popular and is the main source of news, it provides communication in only one direction. Viewers can learn what the television producers choose to put on the air but they have little opportunity to find out what other people are thinking. Establishing and maintaining a community discussion is very difficult. Elected officials also should not judge what people are thinking by simply attending to the media. Occasionally, a television station will pose a simple yes or no vote on a question and invite viewers to phone a number to register ther vote (the telephone company usually charges $.50 for the call). Because the people who choose to respond are not representative of the public, the results do not reflect the public will. An additional drawback is that the results of the poll are not known for several hours; many viewers may never know the results.

Duane Elgin and Ann Niehaus, who codirect Choosing Our Future, have designed a system for two-way communication that restores many of the aspects of the traditional town meeting. Instead of asking everyone in a community to phone in, they randomly select a sample of about 700 citizens whom they request to participate in the town meeting. Prior to the meeting, members of the sample are sent a folder of information on the topic and a list of the questions likely to be posed. At the appointed time, the telecast presents the basic issue, and pro and con arguments are presented by advocates. Several questions can be presented, in succession, to the audience (including those in the sample). For each question, only those in the sample would know the specific phone numbers to dial to register their vote (the people in the general audience would not know these numbers, but might be allowed to register their vote in a parallel tally). For example, those responding "agree" would dial a different number than those responding "disagree." A computer would instantly register the dialing of the number and the entire vote of the 700 people in the sample would be known in a minute or two. The result then could quickly be presented on screen. The town meeting could proceed through ten or more votes in one-half hour. If the votes were structured in a tiered fashion (for example, the outcome of vote A would determine the alternatives presented in vote B), the town meeting could "develop" an issue and move it to a new stage of understanding. I am not suggesting that electronic town meetings would carry all of the affective riches of the face-to-face interaction of a town gathering in a hall, but, when such gatherings are impossible, the electronic town meeting may be an acceptable substitute—and certainly better than any other alternatives we now settle for.

The electronic town meeting not only would be useful for current issues but also could be used by a community to think through complex problems about its future (for example, development patterns in the community, limits on growth, provision of public goods, structure of its government, educational emphasis in its schools). If the town meeting engaged the interested participation of many people, elected officials would have to pay close attention to the results and ordinary people could begin to play a more significant role in guiding their communities.

Bioregionalism, discussed in Chapter 10, is a way of structuring a community to make it more sustainable. Because the world is so populous and fragmented, guiding it by a single structure is impossible. We are finding it necessary to decentralize some of our activities to bring them under our control. Bioregionalism's emphasis on producing, selling, and consuming locally would conserve resources, stimulate biological diversity, and encourage local self-reliance. People in a bioregional community are likely to be doing things by themselves and for themselves (for example, using barter to exchange many goods and services) and deemphasizing monetary values, the acquisition of debt, and rampant consumerism.

A health structure that emphasizes wellness would be another characteristic of a sustainable society. Wellness means feeling physically and mentally well; this is similar to the definition of quality of life presented earlier. Many people in contemporary American society already have taken up a wellness program. Several schools of public health now publish "Wellness Letters" that explode popular myths about what makes for good health, as well as provide information to people about pharmaceuticals, practices to avoid, practices to adopt, environmental threats to health, dietary and cooking practices, tips on good food, and exercise.

A sustainable society also would support health maintenance organizations (HMOs) that provide regular physical examinations in order to detect incipient problems in their early stages and encourage subscribers to seek treatment for small ailments before they become major problems. The traditional health industry, in contrast, views health as the absence of illness and expects that patients will consult health professionals only when they feel ill. Its emphasis is on sustaining life, not on wellness. Linus Pauling calls it the *sickness industry.* The health insurance industry that nurtured this posture has, until recently, refused to pay subscribers for utilizing maintenance services; it has reimbursed them only for services treating illnesses.

Robert Blank developed the following contrasts in the trade-offs between these two perspectives:

Trade-Offs in Health Care Allocation

Health Maintenance		Emphasis on Sickness
Preventive Medicine	vs.	Curative Medicine
Improved Quality of Life	vs.	Extension of Life
Young and Old Consumers	vs.	Old Consumers
Incidence Diseases	vs.	Rare Diseases
Marginally Ill	vs.	Severely Ill
Cost Containment	vs.	Individual Choice

Fritjof Capra (1982, pp. 332-34) criticized the biomedical structure of modern society from a wellness perspective and concluded that our present health system has become a fundamental threat to our health. He made the following suggestions to encourage people to adopt healthy ways of living:

1. restrict advertising of unhealthy products;

2. impose "health care taxes" on individuals and corporations who generate health hazards to offset the medical costs which inevitably arise from these hazards;

3. adopt social policies to improve education, employment, civil rights, and economic levels of large numbers of impoverished people; these

social policies are also health policies, which affect not only the individuals concerned but also the health of society as a whole;

4. develop family-planning services, family counseling, day-care centers, and so forth, as part of preventive mental health care;

5. develop incentives for industry to produce more nutritious foods and restrict poor nutrition items offered in vending machines and in schools, hospitals, prisons, cafeterias of government agencies, and so on; and,

6. legislate incentives to develop organic methods of farming.

Medical technology has advanced so swiftly that it threatens to transform the institutions it should serve. We marvel at the success of organ transplants, artificial organs, and artificial life support mechanisms. Regrettably, many of the people receiving these "miracles" cannot be restored to wellness; they can only be kept alive. We all know of humans who are kept "alive" in a coma for many months or years by artificial support systems. Is our goal wellness or merely not dying?

If we consider the full ramifications of that choice, one question surely follows: should we continue to develop technologies that are capable of preserving life but cannot offer wellness? If someone is rich enough to afford an expensive life-preserving technology (an artificial heart, for example), should we allow technologists to provide it? "Of course we should," you say, "it would be a denial of freedom to forbid it." But, then what? Before long we will have a new industry employing technologies to sustain life but not providing wellness. Soon we will have a variety of new ways to keep people alive even though they cannot be well. Persons who are desperately ill, and their loved ones, are hardly the rational shoppers required to make the competitive market function wisely.

Rich people will be able to afford these technologies while poor people will not. Is it fair for rich people to stay alive while poor people die because they cannot afford a life-sustaining technology? Surely someone will sue for equitable access to the technology in question and will ask society to pay for it; (they already have in some cases). The court may decide that the technology must be made available to everyone without regard to ability to pay. Now we are in a terrible dilemma; we recognize the justice of the court's decision, yet we cannot afford to make the procedure free to everyone. The debate will be intense and very divisive. No society that wishes its people to be at peace with each other can try to decide who will live and who will die. Our valuation on justice will drive society to take actions that injure the biosphere.

The problem is posed even more poignantly with the technological capability to keep alive newborn infants who will never be well or fully functional, and who would have died without the new technologies. In the recent "Baby Doe" case, the government intervened to keep alive an infant who would be severely handicapped for the rest of her life, even though the parents had decided not to intervene. Robert Blank looks at the societal consequences:

Until now we have shown more interest in "saving" the life than in the less dramatic but more difficult task of caring for those persons we have rescued. Before we rush into aggressive treatment of newborns, including those with multiple gross malformations and severe prematurity, we must ask whether we have the willingness to expend the essential long-term resources to care for those persons saved. We must ask the difficult question, not only of who should make the decision to rescue and on the basis of what criteria, but also who pays the enormous costs, both tangible and intangible. (Blank, 1986, pp. 25-26)

We may enter a situation in which hundreds of thousands, even millions, of people are being kept alive at great cost to society even though they have little prospect for wellness. The people who must pay taxes, or insurance premiums, to provide this support will resent the burden. Some people will wish to keep on living even though in an unwell condition, but others may resent the societal insistence on sustaining a life that cannot be fulfilling. (A person who will never be well recently had to sue to be allowed to starve herself to death.) These unanswerable questions of who should live and who should die would not have arisen if the life-sustaining technologies had not been developed. Is it not reasonable to ask if society would be better off if we had looked ahead and decided not to develop these technologies? It is ironic that a well-intentioned and presumably benign technology can be tyrannizing in its consequences. We are reminded that we can never do merely one thing; also that we had better learn how to look ahead and anticipate, so far as possible, the consequences of technological development.

The reader will recall from the discussion of nanotechnologies in the previous chapter that they would be capable of making repairs at the cellular level and thus could restore people to wellness—people could expect to live for centuries. Because nanotechnologies could provide wellness, thinking about them may help us decide how to proceed with life-sustaining technologies. Clearly, cell repair that restores wellness could be superior to current technologies that are life-sustaining but cannot provide wellness. Yet, nanotechnologies pose some very difficult dilemmas. We must try hard to foresee the consequences for all forms of life if people (perhaps animals too) could live for many centuries. Would we not have to find some other way than death to limit expansion in numbers? Long life would alter family structures forever. The relationship between people and nature would be altered forever. The listing of consequences could go on and on, but we would still miss some that could shatter our notions of a good society.

I pointed out in the nanotechnology discussion that it will be impossible to prevent their development. Furthermore, I would not like to live in a police state that attempted to do so. However, it certainly would be possible to withhold public support for their development. Witholding that support would apply even more strongly to the *development* of life-sustaining technologies that do not promise wellness. Governments and foundations could divert their resources to more worthy activities. Once the technologies are in place it will be very difficult to turn the clock back. Therefore, the public needs to foresee consequences and think carefully about what it values with regard to developing these technologies.

The difficulty of reforming established social structures is well-illustrated by tobacco smoking and the tobacco industry. It is now apparent that smoking is the greatest contemporary threat to public health. It kills more people worldwide than AIDS, alcohol, wars, famines, automobile accidents, drugs—anything you could name. Yet no one seriously proposes outlawing smoking. Because the growing of tobacco—its processing, its sales, the fact that millions believe they could not live without smoking—is so deeply entrenched, no one knows how to remove it from society. Almost everyone could agree that the practice of smoking tobacco should never have been started and that the cultivation of the plant should never have been allowed; but no one knew then what the consequences would be nor how to prevent the development of such practices. It is obvious that we do not know how to extricate ourselves once such a social practice has become firmly entrenched.

Living Wills

More than one-half of the states in the United States have passed living-will laws. Under these laws, a person who is still competent can authorize medical personnel to withdraw artificial life support should s/he become terminally ill. This provision legally allows a person who has no hope of becoming well to die with dignity and without prolonged suffering. Under the law, this transfer of authority must be done while a person is still competent; if not, physicians and public officials may legally be prohibited from withdrawing life support, even though the patient has no hope of recovery. These wills also could give permission to transplant bodily organs to persons who need them.

I have repeatedly made the point that, *"A society must learn how to think integratively, holistically, if it is to be sustainable."* We need to develop a structure for acquiring and transmitting knowledge that encourages multi-disciplinary, integrative and forward looking modes of thought. Why have we not done this all along? Capra traces the problem to an overemphasis on Cartesian modes of rational analytical thinking.

It is now becoming apparent that overemphasis on the scientific method and on rational, analytical thinking has led to attitudes that are profoundly antiecological. In truth, the understanding of ecosystems is hindered by the very nature of the rational mind. Rational thinking is linear, whereas ecological awareness arises from an intuition of nonlinear systems. One of the most difficult things for people in our culture to understand is the fact that if they do something that is good, then more of the same will not necessarily be better. This, to me, is the essence of ecological thinking. Ecosystems sustain themselves in a dynamic balance based on cycles and fluctuations, which are nonlinear processes. Linear enterprises, such as indefinite economic and technological growth—or, to give a more specific example, the storage of radioactive waste over enormous time spans—will necessarily interfere with the natural balance and, sooner or later, will cause severe damage. Ecological awareness, then, will arise only when we combine our rational knowledge with an intuition for the nonlinear nature of our environment. (Capra, 1982, p. 41)

The call for integrative holistic modes of thinking is not a call to abandon rational thinking nor to abandon the development of specialties. We badly need to think rationally and develop specialties to help solve difficult problems. However, specialists can easily learn how to think integratively and holistically while pursuing their specialty. Working in multidisciplinary teams is a good way to learn a holistic perspective and helps to avoid the tunnel vision that seems to accompany extreme degrees of specialization.

Our modern institutions of higher education disproportionally emphasize specialization and rational analytical thinking. The contemporary job market contributes to that unbalanced emphasis; students who specialize find employment more easily when they graduate than those students who seek a broad multidisciplinary holistic education. When we began our interdisciplinary master's program in environmental studies, we announced that we would emphasize holistic integrative thinking. A professor of engineering wrote a letter to the dean of the graduate school objecting to our program on the ground that graduate work should be devoted to specialization. So strong is the emphasis on rational linear thinking in higher education that students entering my classes experience about one month of struggle before they finally learn to think more integratively in holistic patterns. I repeatedly challenge them with the kinds of questions I have posed to you in this book; especially, "And then what?"

Most of them find the holistic approach mind-expanding and more satisfying than the modes of analysis imprinted on them by the dominant institutions in society.

Quite a number of universities have tried in recent years to place more emphasis on "general education." General education is not the same thing as integrative thinking, but a good general education at least has the potential to teach students how to think integratively. These efforts to develop general education have had a hard uphill struggle against the dominating emphasis on specialization in most universities.

Summary

This chapter reviews the way the competition for power that permeates civilization has focused on science and technology as the keys to power. Technological power is imperialistic in several regards:

1. The power it gives some societies to dominate others has carried technological development and the desire for technology to the farthest reaches of the planet.

2. It has turned most people into worshippers of technology.

3. Our dependence on large technology-based systems is so great that they shape and dominate our lives. The bigger and more powerful the technology, the greater its imperialistic impact.

4. We are structurally thrust into wasteful ways of doing things. Market mechanisms accentuate this wasteful behavior.

The second part of the chapter discussed various characteristics of a society that could be restructured to be more sustainable. So many new ideas are possible that this listing can only be illustrative. How about turning your imagination loose on this problem!

1. Products and processes should be conserving, and emphasize long-run sustainability.

2. Planners and designers should avoid large projects that foreclose other options.

3. "Soft" approaches to producing and consuming energy should be chosen over "hard" approaches.

4. Wasteful resource usage should be curtailed by using depletion quotas.

5. Modern communication technologies should be used to encourage sustainability and promote democracy.

6. Electronic town meetings could restore a sense of direct democracy and citizen control to our communities.

7. Decentralizing our social structures and governments into bioregions would promote sustainability.

8. A health structure that emphasizes wellness would better serve NEP values than our present structure, which emphasizes prolonging life.

9. Our educational systems should encourage people to learn integrative and holistic ways of thinking.

A Governance Structure Designed To Help A Society Learn How To Become Sustainable

Chapter 14

Most contemporary governments were designed for horse-and-buggy days when technology had little capability to change our lives quickly and almost no capability to disrupt biospheric systems. With many fewer people, almost no thought was given to limiting family size. Most people believed the future would not be very different from the world they saw every day. It made sense, then, to fix the design of governments so that people could count on their permanence and stability.

Now, powerful technologies dominate and swiftly change our lives; population growth is a dire threat to ecosystems and social systems; and disrupted biospheric systems are inducing enormous socioeconomic changes; yet, the structures and processes of government have changed very little. We urgently need to think of ways to make governments more responsive to and responsible for the changing ecosphere and changing socioeconomic-technical reality of modern life. This chapter proposes a modest but crucial redesign of government to facilitate its ability to cope with these changes. The redesign would not change the way governments govern (their formal decisional processes) as much as it would change the way governments think and learn.

Defining Governance

Governance is the thing that governments do, but the concept has a somewhat broader meaning that is well-expressed by Donald Michael:

By "governance" I mean those ways by which we agree to be reliable personally, organizationally, and politically. We do this via laws, norms, rituals, shared beliefs, roles, etc. These are incorporated within institutions such as those responsible for education, early socialization, religion, the market, and government. Here we are mostly concerned with government, but clearly the changes discussed would apply to the rest of the societal system as well. (Michael, 1983, p. 267)

We can usefully distinguish governmental functions, structures, processes, and outcome. The *functions* of governments are also their responsibilities. Beliefs about the appropriate functions of government are based on the values people hold as well as on their beliefs about the way things work in our daily lives. We now ask our governments to do many things (outlined in Table 14.1) that previously we had not expected of them. We can expect this list to evolve further as society evolves.

Table 14.1
Evolving Functions of Governments

Functions	Values
1. Traditional (everyone assumes these)	
a. Provide order	
b. Provide security	peace
c. Resolve conflicts	justice
2. Recently established (some people still dispute these)	
a. Preserve human rights	
b. Provide economic justice	freedom
c. Provide services	nourishment
d. Insure welfare	comfort
e. Nourish economic activity	
3. Emerging (people are still learning about these)	
a. Facilitate quality of life	quality of life
b. Preserve the biosphere	ecosystem protection
c. Encourage social learning	survival of *Homo sapiens*
d. Ensure the sustainability of society	

Why have we assigned so many new tasks to government? As world population has grown, our personal space has become more crowded, complicated, and difficult to manage. When we cannot manage our lives as individuals, we turn to a collective societal structure—a government—to bring order and security. We now ask government to assume additional functions because we see that they are not being well cared for by our own efforts or by the efforts of private institutions.

Governmental structures are roles that society agreed to assign to certain people (officials) in order to carry out specified duties. These authorized duties, and their boundaries, usually are spelled out in a constitution. The structure of a government is influenced by: beliefs about how the world works, society's values, societal tradition (which embodies social learning about beliefs and values), and society's choice of functions for government. Parliamentary, presidential, monarchy, and dictatorship are some of the adjectives we apply to characterize governmental structures. Structure shapes process, and vice versa, but the linkage is not tight. For example, both Great Britain and the Soviet Union have parliamentary governments, yet the process for decisionmaking is very different in the two systems.

Governmental processes are the routines that governments follow. For example, political scientists study the legislative process, the judicial process, and the administrative process. These processes derive from both structure and functions of government. There are many similarities in processes across governments but there also are many differences (by level of government, tradition, function, stage of development). Whether a government follows a process of encouraging citizen participation in its deliberations is very important for social learning.

Governmental outcomes are the things governments produce, (presumably to make life in society better, but they could make things worse). Distinguishing *outputs* of governments from *outtakes* is useful. An *output* typically is a governmental decision: a new law, a court decision, bureaucratic enforcement of a regulation, a program (for example, putting a man on the moon), appropriations for institutions (for example, schools). *Outtakes* are societal conditions, affected by government, that citizens take out of the system (consume). We expect to take out, because of governmental actions, a sense of peace, security, justice, freedom, nourishment, comfort – these are outtakes.

People recognize that we need government, but, most are disappointed in the performance of their government. Why? The growing complexity and congestion of the modern world, mentioned above, makes the task of government much more difficult and also has made us realize that we must ask government to take on many new functions. In my discussion of information and knowledge in the chapter on social learning, I cited Elgin's analysis that complexity mounts faster than our ability to learn how to understand and control it; knowledge demand outstrips knowledge supply. Managers cannot make bureaucracies work well because the knowledge in their heads is inadequate to the task.

Another possible reason why we are disappointed in government is suggested by Elgin's analysis of stages of growth in the life cycle of Western industrial civilizations. This analysis is summarized in Figure 14.1. Although Elgin's portrayal is intriguing, it should not be taken as established fact; many scholars dispute this cyclical interpretation. Its main value is heuristic.

Elgin believes that civilizations move through the stages of growth portrayed in Figure 14.1. Industrial civilization has, in many respects, been extremely successful. It has supported a surging growth in human population; furthermore, many of those people now live an affluent lifestyle that was unthinkable for the average person a few centuries ago. Now, however,

Figure 14.1

FOUR STAGES OF GROWTH IN THE LIFE CYCLE
OF WESTERN INDUSTRIAL CIVILIZATIONS*

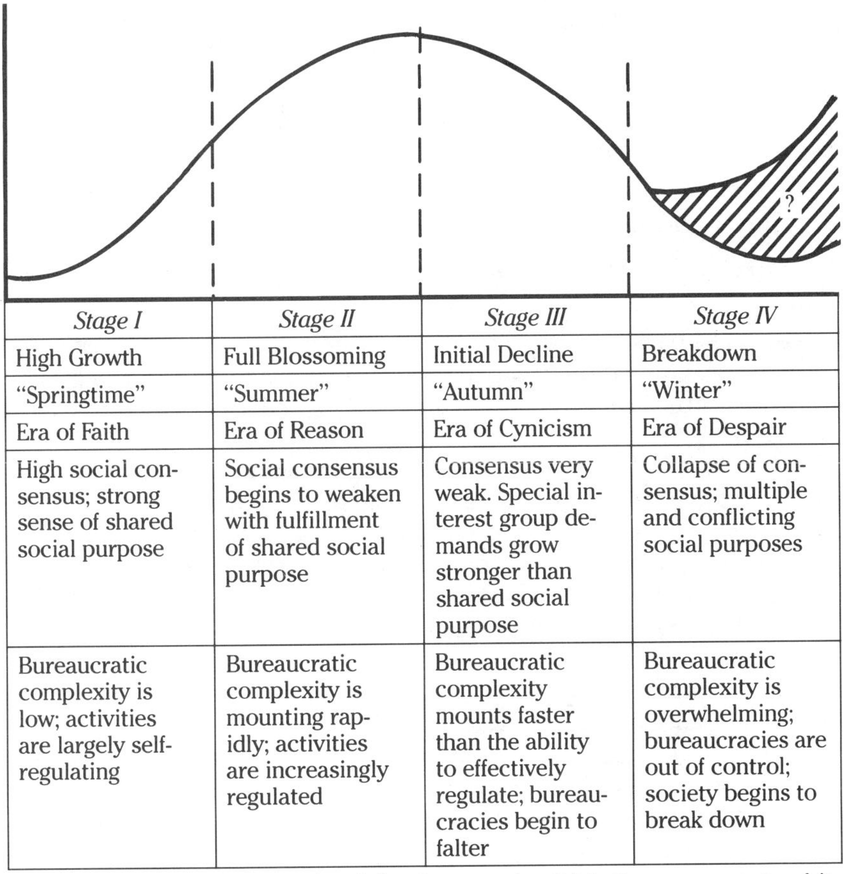

Stage I	Stage II	Stage III	Stage IV
High Growth	Full Blossoming	Initial Decline	Breakdown
"Springtime"	"Summer"	"Autumn"	"Winter"
Era of Faith	Era of Reason	Era of Cynicism	Era of Despair
High social consensus; strong sense of shared social purpose	Social consensus begins to weaken with fulfillment of shared social purpose	Consensus very weak. Special interest group demands grow stronger than shared social purpose	Collapse of consensus; multiple and conflicting social purposes
Bureaucratic complexity is low; activities are largely self-regulating	Bureaucratic complexity is mounting rapidly; activities are increasingly regulated	Bureaucratic complexity mounts faster than the ability to effectively regulate; bureaucracies begin to falter	Bureaucratic complexity is overwhelming; bureaucracies are out of control; society begins to break down

*"A growing civilization may be defined as one in which the components of its culture are in harmony with each other and form an integral whole; on the same principle, a disintegrating civilization can be defined as one in which these same elements have fallen into discord." Arnold Toynbee, *A Study of History.*
From Elgin 1981: 92

we are discovering that our very success is driving both our ecosphere and civilization into failure.

Elgin believes that modern Western societies have moved to Stage III of his posited stages. Many news stories and polls tell us that people are increasingly cynical, especially about government; that special interest group demands are pressed at the expense of our shared social purpose; that our bureaucracies are faltering because they cannot handle the complexity of the world with which they must deal. Our governments are ill-equipped to govern the biophysical and social worlds of our modern day; that is why we find them so disappointing.

If we are to avoid moving into Stage IV, an era of despair and breakdown, we must design governance structures to improve social learning for both people and governments. Our focus will be on structure, as that is the main component of design, but structure will inevitably lead us into some consideration of functions, processes, and outcomes as well.

Design Deficiencies In Extant Governance Structures That Make For Unsustainability

As stated earlier, most governments believe we should quickly and diligently exploit nature to enrich the lives of humans. Any form of government that structurally supports unfettered exploitation of nature cannot sustain itself.

The earlier analysis also showed that an economic system with an unfettered market would be bound to lead to the situation just described. When individuals strive to maximize their own wealth, nature cannot avoid being plundered. Any system of government that upholds an unfettered market is unsustainable. Planned economies have been just as exploitive of nature as have market economies. As stated in Chapter 7, we should think of both planning and markets as useful information systems; we need to find some way to combine planning and market information in managing our complex society and ecosystem.

Some people assert that humans should step back and let ecosystems manage themselves. They fail to recognize that humans are such an obtrusive species, and so densely populate the earth, that all of our actions constitute some kind of management.

This analysis leads to the conclusion that neither capitalism nor socialism, per se, are suitable governance structures for a sustainable society. The Scandinavian countries mix capitalism and socialism, planning and markets, and use a democratic form of decisionmaking to manage their societies. They also have been among the best countries of the world in preserving the integrity of their ecosystems. There is consensus across their political parties that the basic structure of their system is sound, so they maintain it if they gain power. Their societies have been exceptionally stable. Raoul Naroll, a distinguished anthropologist who spent most of his life comparing whole societies, suggested (1983) that we should use Norway as a "guide land" because the Norwegians have achieved the highest quality of life for the most people; they must have worked out, by trial and error, a better system than those used in other lands. Should other countries imitate Norway? How would we get them to do so? Thus far, few countries have chosen to imitate the Scandinavian example.

Would the Scandinavian model work in huge countries, or Third World countries? The Scandinavian societies are more homogeneous than most. Would their model work in highly pluralistic countries like those in North America? It would probably take a lot of experimentation to find out. That could be very important social learning, but also very slow learning. Unless this model could be implemented in many countries, it would not have much impact in preserving the world's biosphere.

What is really required is a world government that would take preservation of the world's biosphere as a central objective. Alas, no one knows how to achieve this. The love of nation is too strong for most people to give up some sovereignty to a world government. Hatred and suspicion of others is too great for them to trust a world government to make decisions that they must obey. It would be unthinkable to obey a world government that limits population growth as strictly as in China, for example. At this time, people simply do not recognize sufficient need for a world government to make the adjustments requisite to its establishment.

Let us look at this problem from another angle. Three prominent writers on environment and politics in the 1970s (Hardin, 1972; Heilbroner, 1974; and Ophuls, 1977) perceived the task of preserving biospheric integrity to be so urgent that people may have to give up their freedom to a strong government which will manage the environment and society for the benefit of all. They have been labelled Neo-Hobbesians* and have been criticized roundly by other scholars for their conclusion that only a Leviathan-like government can save us from our environmental predicament (Orr and Hill, 1978; Leeson, 1979; Holsworth, 1979; Hoffert, 1986; Walker, 1988).

Most of the critics of these Neo-Hobbesians have focused on the premises and internal logic of their arguments. I choose not to enter this argument here; their criticism is far too complicated to be recapped. Ophuls, in any case, has explained that he did not call for a Leviathan; he only pointed to the danger of its possible occurrence. The critics conclude that a Leviathan is not required to save our biosphere and society. Aside from that philosophical conclusion, I believe there also is no prospect that people

---

*Thomas Hobbes, an English philosopher of the seventeenth century, became famous for his philosophy that the selfish greed of people and their desire for power would lead to anarchy. Rather than live in a condition of "war of all against all" the people would accept an absolute ruler. The title of his book, *The Leviathan*, became part of our language; we speak of absolute rulers as "Leviathans."

would give up their freedom and rights to a Leviathan in the hope that it will save their biosphere; I suspect that most people, at least in the West, would rather die first.

If we do not wish to accept a Leviathan, could we decentralize our political structures in a workable fashion? The bioregional vision, discussed in Chapter 10, is based on the idea that community, work patterns, social institutions, and governments should be organized regionally. These units would be small enough so that governance would be manageable. Lifestyles, social relationships, and institutions would deliberately be kept simple. There would be sufficient communication among community members so that social learning would thrive.

The drawback of bioregionalism is that it is too late; the world is already too populated, we are already inextricably linked into national and international networks, our economies are already too interpenetrated. It may be healthy to organize whatever aspects of life we can on a bioregional model, but surely we also will continue to have powerful national governments as well as influential world connections. Social learning must be organized and nurtured at those levels if it is to be truly effective.

So far, we have considered, and eliminated, the following designs: the free market, planning (hence, capitalism and socialism also are eliminated), the Leviathan, world government, bioregionalism. What else is there? *I suggest that we combine the best of what we have and amend it to work better by designing it to facilitate social learning.* As we collectively learn more about governance in the coming decades, we may be able to arrive at a superior design that will more effectively govern a sustainable society. My notion of the best of what we have is to combine democracy with a mixture of planning and market (similar to the design in the Scandinavian countries). This basic design should be amended to maximize its learning capability. If a society is to be truly sustainable, most of its people must learn what that requires of them personally in terms of changes in their values, lifestyles, and institutions.

Donald Michael also sees a transitional period of learning as the appropriate next step for governance:

If anarchic autonomy is not the answer and a global centralized planning state is not the answer either, then we are in an impossible bind—unless we can conceive of a third alternative, one that involves different ways of valuing, thinking, and acting politically. . . . The immediate task is planning for a long period of transition. During such a time, it is predictable that:

1. There will be strong disagreement over the proper balance between the autonomy of individual actors (people, corporations, groups, governments) and the needs of the global society as a whole.

2. There will be a need for transitional strategies and procedures—ways of living, public policies—that are consciously chosen because of their usefulness as bridges between one way of doing things and another. . . . Resolving whole system problems requires a whole system approach with planning and governance that support it. . . . But a whole system solution that denied autonomy would deny sources of innovation, resilience, and meaningfulness—would subvert the goal of a more viable and humane social order. The question—a very big question indeed—is: How can we govern ourselves so as to value autonomy and still retain the integrity of the commons? (Michael, 1983, pp. 254-255)

Overlaying A Learning Structure On A Traditional Governmental Structure

For this discussion, let us assume that we live in a society in which goods and services are exchanged in a market. Some property is owned privately, some by public corporations, and some by governments (the particular mix is not important for our discussion and would, in any case, vary from society to society). The market is somewhat open but is controlled in certain key respects by governmental regulation and action. Further assume that people have some ultimate control of this government (and society): most decisions are made by officials, but support for their retention in office is a decision made by the people.

Surely this kind of system, which can be found in most Western countries, already does some learning, at least by trial and error. Is that not sufficient? Trial and error learning is slow and expensive. Nearly all of the problems discussed earlier in this book have gotten to their near crisis state because our learning has not been sufficiently swift and thorough. Why not design a governance system to maximize learning? The idea is not necessarily to be more rational; trial and error learning is usually quite rational. We need to pursue learning more aggressively in a larger and more holistic context that discloses the location of each specific endeavor within the larger systems that encompass it. Specific actions should not be allowed to interfere with the good functioning of our basic life-support systems.

What about science? Is that not a marvelous system for learning? Of course it is, but, as indicated in previous chapters, scientific learning does not include analysis of values and, due to its emphasis on specialization, typically fails to analyze in an integrative holistic manner. It simply is not sufficient for the kinds of problems we face today. Michael deals with this point as well:

The "separatist" view . . . described a world composed of separate entities: atoms, individuals, school boards, government agencies, corporations, states, nations, "variables" in the behavioral sciences. It was a world wherein effects were separate from causes, facts from values, present from future, humans from nature, private from public.

For some people this description, this social construction of reality—this myth—was extremely efficacious. But it brought us to a world that is now so interconnected that each "entity" unavoidably interferes with the other. Governance premised on this myth is simply not sufficiently responsive, flexible, and systemic—that is, planful—to protect the commons and give real meaning to autonomy . . .

We will have to learn how to govern in ways that allow center(s) of decision and action to shift, fuse, differentiate, enlarge, or diminish according to changing task-priorities and functions. . . . We need to learn procedures that can determine with some accuracy how much time is available to make a decision and what the scope of the decision must be. . . . We must anticipate as well as react. This means that we need to integrate learning and transitional strategies into our fundamental approach to governance. We also need an idea of leadership that values "learning leaders" as well as caricatures of certainty. . . .

This is a difficult piece of political thinking. . . . We are not accustomed to ideologies that look to establishing optimum conditions for learning. Nor is there much in the public domain to encourage whole system thinking: Political dialogue runs strongly toward defining problems and solutions in terms of parts of a situation. (Michael, 1983, pp. 256-57)

How could a learning structure be overlayed on the preexisting structure? What would it do differently from standard practice today? How effective would it be in meeting the governance challenges of our day? These tough questions are addressed one at a time. First, what would such a structure look like? I have deliberately used the word *overlay* because the structure I propose would not displace any of the basic governmental institutions and roles that are well-established. Legislators would still deliberate on and pass laws; judges would still hear and decide cases; executives would still administer their bureaucracies. Public officials would still be held accountable by the people. My proposed amendments are mainly designed to facilitate social learning. They would help public officials to make better informed and more farsighted decisions; some way also must be found to be sure they will use the structure.

Designing learning structures for government is a fairly new idea; being new, we cannot know what the full repertoire of possible structures might be. I can only present my ideas; perhaps others will be stimulated to develop even better ideas. *A learning structure should have at least four basic components: an information system, a systemic and futures thinking capability, an intervention capability, and a motor to make it go.*

Figure 14.2 shows these components and their interrelationships. The "motor" in my proposed structure is a requirement that every major policy initiative be assessed for its impact on the environment, our society, and our values. This assessment would be conducted by the initiator but would

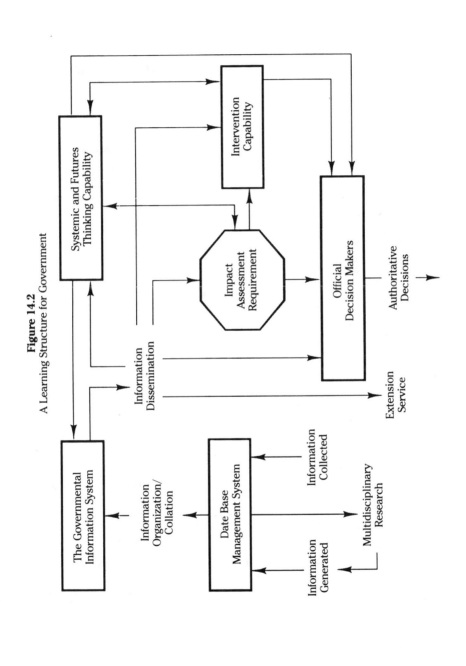

Figure 14.2
A Learning Structure for Government

Systemic and Futures Thinking Capability

Intervention Capability

Impact Assessment Requirement

Official Decision Makers

Authoritative Decisions

Information Dissemination

Extension Service

The Governmental Information System

Information Organization/ Collation

Date Base Management System

Information Collected

Multidisciplinary Research

Information Generated

be reviewed by the systemic and futures thinking unit. Both of these actions would be subject to public review via the intervention capability. All of these actions should be seen as learning opportunities. Before getting into the systemic relationships between the four components of this learning structure, sketching the essence of the components more fully is useful.

An Information System

An information system is the very life blood of a learning structure. Most governments already have elaborate information gathering systems. They have libraries; sponsor research; encourage development of search and retrieval systems, and so forth. This capability should be expanded to put information into machine readable form so that the task of search and retrieval can be made more manageable. Nelson's idea for "hypertext," discussed in chapter 12, would usefully link a given piece of information to related pieces.

Information achieves its full value only when it is utilized for learning. Therefore developing the best possible information dissemination capability is important. Many governments already do this to some extent; the U. S. Agricultural Extension Service, working in collaboration with land grant universities, is the best-known example, but a larger system could be developed for a much wider range of topics. Utilization is much greater if the meaning in research reports can be translated into everyday language and carried into public discourse. It also is enhanced if information is converted into recommended practices for people to use as they go about their daily activities. Television is the best medium for dissemination of information, but it could be used a great deal more creatively. Think how much social learning could be advanced if as much creative effort went into helping people to learn how their world works as goes into advertising.

Information and knowledge are not zero sum; they are not diminished when shared; in fact, their extent and value may actually increase. In another sense, however, information resources are always scarce. The money and talent needed to produce messages that communicate effectively are scarce. Media space (pages in a newspaper or channels on television) is also finite. Intended audiences have limited attention spans. The information dissemination system of this proposed governance structure will need to make thoughtful and farsighted choices about deployment of its informational resources. It might be useful to give highest priority to helping people understand the interconnectedness of their world and the adjustments they must make in order to live in a sustainable relationship with the natural world. According to Michael:

People have learned new norms before; the unique task now is to learn how to go about such learning, to do it intentionally and with greater speed. There are four major immediate learning tasks:

1. Learning not to perceive the work according to the myth of "separateness."
2. Learning an appropriate "systemic" myth.
3. Learning how to translate that myth into new governance norms and procedures, policies and programs.
4. Learning how to cope with the inevitable dialectic between proponents of the old myth and the new.

This is in itself a transitional strategy, and it will not result in more than temporary answers; its primary intent is to discover the right questions: Where do we want to go? How might we get there? Are we getting there? Do we still *want* to get there? ... Governance is the conduct of learning, and vice versa. (Michael, 1983, p. 258).

A Systemic And Futures Thinking Capability

Most governments do not have a designated systemic and futures thinking capability (SFTC). Until recently, most did not recognize that such thinking is required for good government. Although public officials are presumably selected for their good judgment, few people perceive that good judgment includes the ability to think systemically about the future. Officials and the public perceive a world of separate entities (that Michael spoke about); not an interrelated system. I previously have lamented the tendency of people, including officials, to use "linear mechanistic thinking." Additionally, most officials are so busy rushing from one urgent problem to the next that they have little time to think about the future.

If they are too busy to think about the future, why not establish a special unit, staffed by professionals of exceptional learning, to think deeply about the future? (A specific proposal to do this is developed later in this chapter.) Systemic and futures thinking is especially important in government, where policies are made that could well determine whether a society will be sustainable. Because it is so important, one could expect this style of thinking to be used routinely. Ironically, governments make thousands of policy decisions without so much as giving a thought to that question. Similarly, we introduce technologies that will have massive impacts on society without subjecting them to diligent scrutiny.

I am not suggesting that the professionals in such a unit should take over societal leadership. There is little probability that the people would grant them that power. Rather, a way must be found to introduce futures and systemic thinking into public policymaking, while retaining the ultimate power of decision in the hands of the elected representatives of the people. For them to do their job well, they should also stay in close touch with citizens. Provision must be made to encourage citizen participation, not only at election time, but also in many arenas of policymaking: in the bureaucracies and the courts as well as the legislatures. Their participation encourages social learning. I am not suggesting we should try to turn citizens into policymakers. Citizens are unlikely to know the intricacies of complicated policy questions, but they do know their values. Accurate information about what the people truly need and value is most needed and often lacking as policy is being made. Polls can tell us something about people's values but decisionmakers must take other factors into account, as well. Final authoritative policymaking should continue to be the responsibility of elected officials.

An Intervention Capability

Public officials cannot have all the good ideas, foresee all the problems, or keep track of all the transgressors. Explicit right should be granted to citizens and nongovernmental organizations (NGOs) to intervene in the learning structure to obtain a review for a problem they have identified. NGOs could identify failures of responsible officials to conduct an impact assessment. They could bring new knowledge to the attention of public officials or to the professionals in the SFTC. Sometimes governmental agencies are officially assigned a responsibility but given so little money that they cannot carry out the responsibility effectively; therefore, NGOs should have the right to intrude in the governmental budgetary process.

Naturally, NGOs would assume a watchdog role and communicate with the public about what they observe in government. They could teach citizens to target their communications to public officials more meaningfully. Most NGOs also would try to educate the public to a better understanding of how the world works. NGOs probably would form consortia so as to coordinate and share the watchdog and educator roles more effectively. In sum, the main task of the intervention capability is to facilitate learning.

Require Impact Assessments

Environmental impact assessments are now commonplace in many countries. I have urged that the concept of impact assessment be expanded to include social impacts and value impacts. (I will have some suggestions for getting started on this in the next section.) An impact assessment methodology and lore will develop over time; actually a great deal already is known from the experience with environmental impact assessments. These three kinds of assessments (environmental, social, and value) should be thought of as a total package that would be required for all major policy initiatives.

Impact assessments act as a "motor" for the governmental learning structure by initiating study and review of proposed actions. This key ingredient in the learning process, should have many possible points of initiation; this would help avoid knowledge control by a centralized power clique. For example, an NGO might get the motor started by requesting an impact statement. Naturally, the SFTC should have the power to request an impact assessment or to conduct its own impact analysis. Legislative bodies would eventually begin to expect impact statement requests for most of their new policy initiatives.

While this book was being written, both houses of the U. S. Congress had produced and were debating several versions of a tax reform bill. Several organizations (private and public) analyzed what impact each of these proposed bills would have on various kinds of taxpayers. Missing was an analysis of the long-range impact of the proposed reform on the biosphere, society, the economy, governmental debt, and our values.

The Systemic Interconnections In The Learning Structure

The reader can see from Figure 14.2 how the information system would collect data from existing sources as well as conduct research of its own. The SFTC would probably commission much of the research of special concern for the learning structure. The information available in the system would routinely be brought to the attention of the other units in the learning structure: governmental branches and departments, the SFTC, and the NGOs in the intervention capability; it also would be available to the public. The media and the extension service would be the main channels for disseminating information to the public.

The SFTC would be a major consumer of the information in the system. It should assess that information and help direct future information development activities. Many of the inputs to the SFTC would be in the form of impact assessments, but the SFTC should be given the power to initiate or conduct impact assessments if they have not been generated elsewhere in the system. The recommendations and findings of the SFTC must be made public—they constitute inputs to other units in the structure; this will be particularly relevant for the intervention capability.

The persons and groups active in the intervention sector will receive inputs from the information system, the SFTC, and from the impact assessments. The NGOSs in their watchdog role will be searching for possible problems that they will bring to the attention of the SFTC, governmental bodies, and those working in the impact assessment process.

The structure just outlined could not guarantee that government and society would learn enough swiftly enough to become sustainable; but at least we would have a better chance. It will only work if people in government listen to it; that will probably happen only if the public demands that they listen–governmental learning cannot really be separated from social learning. The public must demand that the government support the learning structure with sufficient funds for it to work effectively. With strong public support, and an inclination by public officials to listen, this structure could enable a society to obtain a much better understanding of where it should be going and what steps must be taken to get there.

The proposed changes are not very wrenching. Much of the information system already is in place and we are swiftly learning how to make such systems work better. We already have an intervention capability; but we need to guarantee better access. We already know how to conduct environmental impact assessments; we could soon learn to do social and value impact assessments. The major innovation is to propose a special unit for systemic and futures thinking. Let us take a closer look at how that might be structured and established.

A Proposed New Institution: The Council For Long-Range Societal Guidance*

I propose that a Council for Long-Range Societal Guidance be established as part of every nation's government. In the U. S. government, it would be something like a fourth branch charged with the special responsibility to look ahead to the long-range consequences of proposed actions. It would regularly, as a central aspect of its role, ask the ecolacy question, "and then what?"

In the following discussion, I frequently draw on experience in the United States to illustrate concepts; this type of council, however, could usefully be instituted in any modern democratic governmental system. Governments that are not modern or democratic also might usefully introduce such a council into their system, but it would work differently from the way I describe it in the following pages. It is the major important innovation needed to facilitate governmental learning now missing in most countries.

Canada, Sweden, Norway, The Netherlands, and New Zealand (possibly other nations) have established agencies with the primary mission of looking

---

*Professor Rudolf Wildenmann of the University of Mannheim in West Germany coauthored a paper with me in 1975 that dealt with many of the ideas in this chapter. The council proposed here especially reflects that collaboration.

to the future and providing guidance to leaders and citizens. Those efforts flourished for a while but now seem to have faded as new governments came to power. Russell Peterson, former chairman of the president's Council on Environmental Quality, testified in 1981 before hearings on "National Strategic Planning"* and pleaded with Congress to establish a foresight capability in the U. S. government. He documented numerous proposals that had been made over the past fifty years defining and supporting the need for a foresight capability in the government, and lamented that nothing had been done. Nothing happened then and nothing has been done since.

Most proposals have been to establish a long-range planning office in the executive branch of government. The council proposed here is more than a typical planning agency; rather, it is a new kind of governmental entity designed to meet new major social problems. It is different in that: 1) it is specifically designed to facilitate social learning; 2) it is embedded in a learning structure; 3) it is independent of all branches of government, yet, serves all of them and the public; 4) it is given the specific legal role in policy adoption of reviewing long-range impact assessments.

The various procedures of the council cannot be fully elaborated. Rather, the discussion will concentrate on its main structure, its relations to other governmental branches, the selection procedures for choosing councillors, and some ideas for its modes of operation.

As its principal duty, the council will engage in long-range forecasting and will develop possible future scenarios for thinking about policy. Being part of the learning structure, the council will review the long-range impact assessments required for all major policy initiatives. Although these functions are related to planning, they go considerably beyond the expectation most people have about planning. As the profession of planning has developed, planning has become increasingly thought of as a technical task; planners commonly speak of the technical components of a plan. Michael sees planners playing quite a different role in the governance of a learning society:

> The planner is a storyteller and a teller of stories about storytelling. . . . Technical information, the stuff of conventional planning. . . . will almost never be sufficient to determine a course of action. . . . the future is comprised only of fragmented conjectures and is full of surprises. Therefore, possible outcomes cannot be assigned real probabilities, sophisticated methodologies notwithstanding.
>
> Therefore, the task of the planner is to be an educator. The planner must tell the alternative stories in as rich detail as possible. . . . The political task then becomes

---

*Hearings on National Strategic Planning held jointly by the Subcommittee on Oversight and Investigations and the Subcommittee on Energy Conservation and Power of the U. S. House of Representatives Energy and Commerce Committee, December 1, 1981.

one of choosing which risks to take, in full light of what risks are inherent in the options: In a mythology based on a systemic perspective there are no free lunches.

Villains, then, become not those espousing contrary policies, but instead, those who don't share the learning opportunity, acting instead as if they/we did indeed know what to do and how to do it. ... Part of the planner's task as educator is to help others learn about the mythic nature of their social reality and about the ways they deal with it, including dealing with it through planning. By doing so, planners would empower others to learn to plan and to choose discriminatingly among the stories that planners tell. ... My proposal is intended to change both political and planning activities into complementary contributions to a learning mode. (Michael, 1983, pp. 265-66)

This council will not make authoritative final decisions, it will only recommend. Its recommendations will carry the force of thorough research and deep thought, however. The ultimate effectiveness of the council will depend on its persuasiveness as it presents forecasts and optional scenarios to authoritative policymakers and to the public. *The public, through their elected representatives, shall make authoritative decisions—new laws and policies.* Why make such a strong point of this? When previous versions of this proposal were shared with colleagues, it was criticized for establishing a group of elite decisionmakers who would not be responsible to the electorate. That is *not* the intention; laws and policies will still be made by elected officials. Readers need to be alerted *not* to leap to the incorrect inference.

In order to forecast and plan effectively, the council will regularly monitor conditions and changes in society. This monitoring should include not only the currently used economic and statistical indicators but also subjective indicators of quality of life. Effective measuring systems for all these factors already exist and need only be implemented. Regular monitoring of the extent to which people believe their life goals are fulfilled, as well as of the way they think and behave, will provide necessary feedback to the council so that it can gauge societal success and readjust its projections.

Forecasting, scenario writing, and planning require use of the best and most complete information it is possible to obtain. The search for thorough and reliable information may disclose gaps in knowledge that must be remedied so that the council can perform its task well. Therefore, the council should have the power and the resources to undertake or commission research projects, including basic research.

To ensure that projections and plans are as objective as possible and truly oriented to the long term, the council members must have the freedom (even the obligation) to study and recommend without control by politicians in power. Most currently operating governmental policy planning staffs, even those explicitly oriented to study of the future, are under the complete financial and political leadership of the executive branch. Forecasts and policy recommendations that are not congenial with the policy of the top political leadership usually are not revealed to the public. Another distorting element in the policy studies and recommendations of most current planning staffs is that the personnel who are picked to guide them, and the topics they are assigned to study, are oriented to enhancing the probability that the political party holding power will be able to continue holding power. A policy planning staff that serves an elected official with his eye on the next election is effectively foreclosed from giving long-term considerations an unbiased and prominent place in its recommendations. For the above reasons, the currently operating model for policy planning staffs cannot be made the operating model for the council being proposed here.

The Council for Long-Range Societal Guidance is intended to facilitate social learning, to play a key role in long-range policymaking, and be an integral part of the governmental system. As such, the councillors should receive their appointment by the highest elected official (e.g., the president or prime minister). In order to avoid appointments of persons who are nothing more than partisan ideologists, the enabling legislation could stipulate that persons should be sought for appointment who are trained in systemic integrative analysis oriented toward futures projections. A professional association of such persons could nominate eligible professionals who would be certified as meeting minimum levels of competence. Certification by fellow professionals is quite commonly used, for example, in appointing persons to high judicial office. Just as we would never think of appointing a high judge who was not trained in the law, we should not think of appointing futures councillors who are not trained in their profession. One day we may see professional schools, or national academies, designed to specifically prepare persons for this role—as we now prepare military leaders or judges.

One of my critics incorrectly accused me of proposing a council of experts; narrow specialized experts should *not* serve on the council. *The council would need persons who are generalists, who are richly aware of knowledge from many sectors, who have had a wealth of experience, who can think integratively, holistically, and in the long term.* It would be useful if councillors also were expert in something but that would not be their main qualification for serving; rather, they would be selected for their capability for holistic thinking. In this respect, the council would be different from the science court discussed in Chapter 12, which would be made up of experts.

Once appointed, the councillors should be insulated from political domination. To ensure this independence, the council should have a constitutional basis. It could be made a fourth branch of government which would be independent, in the same way that the judiciary is independent, of direct political control. Of course, no part of government is ever *completely* free of political control, nor would we want it to be. If a society did not wish to go as far as to make it a fourth branch of government, a board of regents model could be followed. Many state educational systems in the United States are governed by a board of regents. The members of these boards are usually appointed by the governor with the consent of the legislature and serve for fairly long, overlapping terms.

The relationship of the council to other branches of government can be clarified by role delineation. The legislature, the executive, and the judiciary all make policy, yet none is assigned the specific role of long-range forecasting and guidance. The existing branches are not foreclosed from conducting careful forecasts, but their current role structures make it extremely difficult for them to do so (as detailed earlier in this chapter). The council will have the specific charge to use the best information available to make long-range forecasts and policy recommendations without being concerned about reelection. Furthermore, the existing branches will be charged to consider the forecasts and recommendations as they go about their regular policymaking tasks.

The kind of council proposed here will encounter the twin danger of elitism and capture—trying to solve one problem worsens the other. Impregnable guarantees against these dangers probably cannot be built into the structure of the council, but it will be useful to give some thought to these problems. I have said that councillors should be appointed by the top elected official to provide some ultimate corrective for elitist perspectives that may be completely out of touch with the people. But being appointed makes them somewhat beholden to the official making the appointment, who may try to capture their ideas for his own purposes. (U. S. presidents are often accused of "stacking" the Supreme Court with people who think as they do.) Because of these twin dangers, council appointments should be made with the greatest care and scrutiny. Selecting councillors nominated by their professional society would lead to wiser choices and would limit the possibility of capture. The selection procedure for members of the U. S. Supreme Court who are appointed by the president but with careful scrutiny and confirmatory review by the Senate could serve as a model. Removal by a procedure similar to impeachment could be used when a clearly wrong appointment has been made.

Giving councillors long terms and staggered terms also would limit the possibility of capture. I propose that the council be made up of twenty-one members who would serve twelve-year staggered terms (one-third to be selected every four years), without the opportunity for renewal. Staggered terms means that two-thirds of the Council will be "old timers" who can provide memory of past deliberations and decisions as well as continuity in developing effective modes of deliberation. The recommended size of twenty-one is large enough to be forged into an effective working team, perhaps several teams, yet not so large as to be unwieldy. Fifteen members is another possibility. A smaller size makes it easier to arrive at decisions, but a larger size would have more members to carry the work load of reviewing numerous long-range impact assessments. Having a long term would give councillors sufficient time for learning, and careful study, with diminished concern for what they will do when their term ends. The very nature of council tasks requires that the appointees be well-educated; the council should include broad-gauged social and behavioral scientists, humanists, physical scientists, life scientists, and persons from such professions as law, medicine and education.

The quality of leadership of this council, particularly in its formative years, will be critical to its success. With this Council, as with any group, there is a risk that Councillors may be poorly informed, misled, or unwise in spite of careful selection procedures. As a partial corrective, separate but competitive forecasting teams should be established to work on major problems within the broad framework of the council. The council would have to choose a staff to comprise these teams. The total number of competitive teams would depend on the council's workload. One could imagine two competitive teams working on each long-range impact review or other major policy problem the council undertakes. Criticism also would be used at all stages of deliberations. Care should be taken so that each team operates separately and has complete freedom to study and report. In this sense, the competitive teams operate like the free press. The Council, therefore, combines two valuable features from other institutions: the independence of the judiciary with the competition of a free press.

It is just as important for the members of the staff (and the members of the teams) to be of the highest caliber as it is to have exceptional persons as members of the council. Therefore, a special effort should be made to find persons who not only are well-grounded in their own profession but also are broad-gauged and extremely knowledgeable in a variety of relevant fields. All teams should be multidisciplinary and large enough to ensure a broad range of relevant talents and perspectives.

As the council gets underway, it will have to develop procedures for arriving at major policy recommendations and for the conduct of its competitive teams. Undoubtedly it will require some experimentation to find ever better procedures. Plenary review is a new policy analysis procedure that holds some promise of being useful in an institution like the council. Stephenson, et al. (1986) define it as "a systematic structured process of interaction which seeks to increase both the depth and breadth of issues considered by decisionmakers during policy review." It is oriented to the production of a decision or a set of conceptual policy choices and is structured to maximize participant interaction and deliberation. The normal first step of a plenary review is to gather the relevant existing information. In the second phase, the participants consult with officials who have worked with the policy under deliberation. They also would invite input from those who are not officials but have special knowledge relevant to the policy. In the final phase there is intense participant-driven interaction and analysis that incorporates estimates of the political, economic, social and moral consequences of the policy as intended and implemented.

Independence from partisan politics requires long-term financial support that is institutionalized and guaranteed; this support could be provided in the establishment law for the council. Another budgetary possibility would be to give the council a financial endowment. When the Canadian National Policy Institute was established in the mid-1970s, for example, it received a $10 million endowment from the federal government which was matched by another $10 million raised from the provinces and private contributions. Another idea would establish a national lottery to support the work of the council. The council would require a modest administrative support base, but it should avoid developing a large entrenched bureaucracy that might hamper its ability to break out of established belief and value patterns. The council must be free to comtemplate ideas and make recommendations that currently do not have much public support.

The assignment of tasks to the study groups could be made by the members of the council *qua* council. Normally a given forecasting, planning, or long-range impact assessment review task would be given to two competitive groups simultaneously. The council would automatically receive all long-range impact assessments, but it also could receive recommendations for study from any other branch of government and possibly from the public at large. It also should have the power to initiate its own policy inquiries and reviews. The council could not study anything and everything, however. Therefore, it must play the role of gatekeeper, accepting some problems for study and rejecting others. This gatekeeping role cannot be avoided. It probably will have to be handled in a similar fashion to the way the U. S. Supreme Court decides which cases it will accept for review.

The public interest should be given a greater chance to become defined by careful, intellectual, holistic, long-term analysis, instead of by simplistic, sloganized appeals to short-term interests. To help ensure careful analysis and criticism, the information that is developed by the study teams, and the recommendations that are made by the council, should be required by law to be made public. No top political leader, no top leader of the council, nor the council itself, should have the power to decide whether the information produced by the council will be made public. So potentially momentous are the consequences of these recommendations, and so broad is the constituency that is being served that it is essential for all information, forecasts, and recommendations produced by the council to go into the public domain. Publication also allows study groups outside the council to replicate council studies and challenge their findings. A society might even decide to fund these challenges.

The council should have a reasonable period of time to forecast probable consequences for each proposed new policy. The council's basic contribution is to develop unbiased information to help guide policymaking; it should not be thought of as a veto group. The influence of the council on policy will be derived partly from the quality of the information and analyses that it produces and partly from its independent status. In such a setting there is a reasonable chance that long range considerations could prominently influence policymaking.

Because the major influence the council will have on policymaking rests on the quality of the information that it generates and on the quality of its analyses, it must have the power to request, or contract for, all kinds of information and data. Effectively, that means it should have the power to subpoena the information it needs; it may have to promise confidentiality with respect to information that could violate individual rights or threaten national security. Even though much of the information it uses will come from other sources, the council should have full data analysis capability in house so that it can perform separate analyses when independent verification is needed. University research, while useful, cannot be counted on to be relevant or unbiased, because it is so strongly guided by the choices of people in government or business who control the funds granted for research.

If this idea for a Council for Long-Range Societal Guidance were adopted in several nations, there would be a potential for cooperation across national councils. Especially they could share in the development of new methodologies for forecasting, monitoring, and planning. Basic data on national and world trends also could be shared. It is probable that recommendations coming out of one council would stimulate or challenge the work in process at other councils. Given the development of cordial relationships across councils, there would be considerable potential for growth and sharing across national councils that would be to everyone's benefit.

It is obvious that the council will have a heavy workload. Hence, it must guard against getting so caught up in daily tasks that it loses the capability to think in a wide-ranging fashion. The councillors, that is those persons who jointly have the responsibility to make final reports, must have the leisure to think, debate, and reexamine their premises and conclusions continuously. This leisure would help to overcome one of the serious difficulties for top policymakers in most contemporary societies; they are under constant pressure from daily problems and have little or no opportunity for serious study and reflection on long-term prospects. Implementation of the council's mandate will probably rule out involvement of the council in policy questions that everyone recognizes as being short-term. Current political policymaking processes probably make short-term decisions about as well as they can be made in most cases.

Long Range Impact Assessments

The enabling constitutional provision (or legislation) establishing the council should require that all major policy initiatives–whether they emerge from the legislative, executive, or judicial branch–be submitted to the council for a long-range impact review before they go into effect. In essence, this review adds an important new check in a system of checks and balances. Its ability to check other branches does not lie in its power to veto their actions but rather in its capability to shed light on the long-range consequences of proposed actions and to appeal to officials and the public to make indicated changes. It also is the "motor" that would keep this learning structure active.

This expanded long-range impact assessment might be thought of as an estimate of the impact of the proposed policy on the quality of life of the people over the long term. Presumably this is what all policymakers do when they evaluate policy proposals; but they seldom do it thoroughly because they typically ignore systemic and long-range consequences. They do not have sufficient time, or the appropriate staff support, to do the job well.

As indicated above, the council would have a variety of policy review tasks; asking it to undertake long-range impact assessments could add substantially to its workload. One way to relieve the strain would be to require institutions that initiate a policy to do the initial spadework by preparing a preliminary impact assessment. It may appear that this is akin to asking the fox to guard the chicken coop, but the council would have to take care that its staff conduct its own independent investigation of key aspects of the assessment. Furthermore, having proposing institutions conduct the preliminary assessment could stimulate their own learning, which is one of the major intentions of this new entity. They should take the assessment into account as they adopt the policy.

The council could not veto a proposal submitted to them, but they would require some time to prepare a thorough review. Policymakers would have to learn to adjust their policy deliberation schedules to anticipate these delays. To ensure that the council does not become obstructive, the proposed policy could be allowed to go into effect without the long-range impact review if it is not completed within a reasonable length of time, say six to nine months.

Requiring that only major policy questions be automatically submitted to the council for review would be another filter to relieve the load. But what is a major policy question and who decides where the line is between major and minor? In the United States, the Congress and the major executive departments probably would first decide this question. As a corrective to the possibility that a policy agency would try to evade submitting a policy proposal for a long-range impact review, citizens might be given the power to bring suit requesting the courts to require submission of a policy proposal to the council for review. In order to guard against the possibility that this power would generate many insubstantial suits, the courts could weed out frivolous requests and require a long-range impact review only in those cases where it has been demonstrated that there is reasonable potential for significant long-range detrimental effects. As a second control, the council itself might also be given the power to decide whether the citizen-initiated request for a long-range impact review is within its capability and worthy of the time and attention such a review would require.

Chapter 12 discussed the fact that certain technologies are so powerful, and shape our society so deterministically, that their deployment constitutes something akin to legislation. It became obvious that society must try to find some way to estimate the long-range impact of proposed powerful new technologies. Review by experts appeared not to be sufficient to elicit the needed information or perspective. The structure proposed here for long-range impact assessments and reviews, within the Council for Long-Range Societal Guidance, might be suitable also for reviews of new technologies such as recombinant DNA and nanotechnologies. When technologies are developed to provide military superiority, they are often kept secret. Firms developing new technologies also may claim secrecy for their proprietary information. Review of technologies that may shape every aspect of our future lives is so important, however, that every effort must be made to avoid their being kept beyond the purview of the council. Surely some way can be found to conduct most impact assessments without conveying secrets to adversaries. Society cannot afford to allow claims of secrecy to foreclose its ability to forecast and plan effectively for its own future.

The council should give priority to developing improved theories and methodologies for making long-range impact statements. The accumulated experience with environmental impact statements in the United States and elsewhere (they are used in many countries) would provide some guidance. The impact statements proposed here, however, are designed to be much broader in scope to incorporate social and value impacts as well as environmental impacts. There is a developing literature on social impact assessments (Branch et al. 1984; Finsterbusch and Wolf, 1981; Finsterbusch, 1982; Murdock et al, 1982; Rickson, Western and Burdge, 1986), but the proposal for value impact assessments seems to be original.*

Conducting A Value Analysis

How could a Council for Long-Range Societal Guidance assess value impacts? The first step would be to identify as clearly and comprehensively as possible all the values likely to be affected by the proposed initiative. It is especially important that the inquiry be broadened to include not only first-order but also second-, third-, and fourth-order consequences of the proposed action. It also is important that noneconomic values be accorded just as much consideration as economic values. Both of these recommendations are intended to help overcome the present bias toward considering only first-order economic values in policy analyses.

Identifying values requires thorough knowledge of the way people feel, think, behave, and organize themselves. Studies of the way people experience or fail to experience, quality of life are especially insightful for identifying values. Value analysts should avoid assuming they know what people value; most of us are caught up in a belief system that tells us how people think, value, and behave. The beliefs in that structure are partly mythical—we believe them to be true but they may not be. Therefore, it is important to test such beliefs against the reality of carefully crafted social scientific research focusing on quality of life.

Once the impactable values hae been identified, the analysis should focus on two aspects: first, the extent of the impact of the proposed policy on the various values that have been identified, and, second, on the weighting to be given to the values (for example, how should the loss of the recreational value of fishing be compared with the gained value of the electricity that could be made if the stream was impounded for waterpower). As long as the imperative values of protecting the integrity of the ecosystem and social system are observed, values can be weighted according to the importance assigned to them by the people. Naturally, various individuals assign differing levels of importance to values, but the variation from person to person is not as wide as might be supposed. My own empirical studies of values have shown, for several different populations, that there is substantial public agreement on the weighting of a great number of values (Milbrath, 1984; see also Rokeach, 1973 and 1979).

---

*Shrader-Frechette (1980 and 1982) has urged broadening impact assessments to include considerations that have value components.

Means and ends tend to get confused in most political discourse and policy analysis; therefore, it is vital to distinguish means from ends. Many values proclaimed by politicians as ends will clearly be disclosed as means by careful analysis. For example, maximizing economic growth is often proclaimed as an end goal of public policy (it was so proclaimed by both 1988 U. S. presidential candidates). Yet, economic growth, per se, is not valuable; it cannot be consumed or enjoyed. It is a means to obtain goods and services that we can enjoy or consume. Mindless pursuit of economic growth can destroy or interfere with enjoyment of other values such as climatic stability, a clean environment, or sufficient energy for our children and grandchildren. Careful means-ends value analysis requires analyzing the structural relationships between a whole nest of values before deciding whether economic growth should vigorously be pursued.

Some people may object that value impact assessment is too difficult and controversial; we would be better off continuing with our present forms of policymaking. They fail to recognize that, in fact, value impact assessment is unavoidable. It is an indelible part of every decision to adopt a new policy that presumably will improve things. Routinely making value impact assessments would render this process thorough, explicit, and open in contrast to the present process which usually is partial, implicit, and hidden. In sum, we have no choice but to value; we have no choice but to weight values; we serve our societal values better by making explicit the values at stake and measuring as effectively as possible how people weight these values. "More important, perhaps, we need to learn for ourselves that only when we explicate the values on which we premise our choices will we be able to use such partial information to help select options that further those values. (Michael, 1983, p.264)

The Council for Long-Range Societal Guidance, then, would provide a basic institutional capability for bringing thoughtful consideration of future probabilites into present policymaking. Establishing it and making it work is probably the most effective thing we could do to program our government to learn. Good communication with officials and with the public is essential to its success. The council should employ skilled writers and media communicators who could take the knowledge it acquires and portray it in such a way that the people will listen and understand, thereby accelerating social learning.

Even if some of its forecasts prove to be wrong, or even if some of its recommendations are rejected, it will generate new knowledge and stimulate vigorous public discourse which will accelerate the social learning process. This process is a total societal effort and the council should be seen as a stimulant and catalyst to accelerate the learning. As the council gains more experience, it will be modified and improved; that is also part of our social learning. We need not ask for, and should not expect, everyone to agree that it is a perfectly crafted new institution before we decide to adopt it. Encouraging public discussion, and taking a little time to think things through before we act, can do very little harm. The council will provide considerable potential for avoiding some bad mistakes and for creatively preparing for predictable problems. Because the council would require only modest support, it certainly would be worthwhile to establish it and see if it can be made to work.

The council has been proposed here as a useful addition to a variety of national governments. Even though nation states have been a dominant governmental form for centuries and are likely to continue for much longer, we can anticipate that society will learn to develop governments at the world level and/or bioregional level at some future time. The council proposed here could also be used effectively in the new governments of bioregions, if we manage to move to bioregional socioeconomic-political structures, or in a world government. The council also could be usefully employed in existing state or provincial governments.

Freedom In A Sustainable Society

I have repeatedly argued that short-term demands should give way to long-term imperatives. Because most people think of the short term rather than the long term, the council's recommendations to prepare for the future could be seen by some as denying the current urgent needs of poor people– as elitist. Asking people to make what they perceive as sacrifices now for a better future could be seen as a diminution of freedom. Hence, some consideration must be given to the question of freedom and civil liberties in a sustainable society, and of the role of the proposed council in a democracy.

I emphasize that in this proposed structure *the power of authoritative decisionmaking rests with elected public officials.* Professional futures analysts on the council would develop options as thoroughly as possible and offer them to public officials and to the people: but the final decision in selecting among options lies firmly with the elected officials. In making policies anticipatory of the future (all policies ostensibly do that), it is very risky to select options that are mere illusions, or which are based on unexamined assumptions, or that are not supported by rigorous analysis of values and their structural relationships. Making bad mistakes robs us of future options and diminishes our freedom. We will *lose less freedom* if we take the time and put in the effort to think things through carefully before we proceed.

Freedom should not be confused with the power to have one's way on every issue. In a large society such as the United States, millions of voices want to be heard on every issue; and, in an ultimate sense, the millions of voices are listened to. But one voice in several million seems tiny and, in fact, it is so perceived by most people. Yet, these same people with the feeble individual voices, many of whom perceive they cannot get their way, still feel that they have their political rights and freedom. They can speak their minds, their votes will be counted, their rights will be protected. The council proposed here would in no way interfere with political freedom and rights.

As we think about freedom, we usually focus on our own freedom to do as we wish or on our right to be heard on an issue. We forget that many others will not have an opportunity to be heard but will strongly be affected by the policies we make. Our children and grandchildren will inherit the consequences of the wise and unwise policies we make; yet they have no opportunity to be heard as the policies are debated. People in other lands are excluded from U. S. policy councils, yet U. S. policies have a large impact on them. Other plant and animal species also are profoundly affected by our policies and our day-to-day actions; who speaks for them? The rights of these "excluded others" probably would be better protected by a Council for Long-Range Societal Guidance than they would be under present modes of policymaking.

Diminution of choices should not be confused with deprivation of political freedom. It is important to recognize that choices are diminished more by the kind of society we have developed than by governmental planning and regulation. Why is this so? As more and more people have joined the human race, and as societies have tried to accomodate them, our lives have become more and more intertwined. The actions of one individual nearly always affect other individuals. Growth of population and limits on the carrying capacity of the earth are additional major limitations on freedom of choice. This crowding can only result in diminished freedom of choice. For example, the Chinese population is so large for its resource base that they have had to institute a policy of one child per family. The cost of not accepting that diminution of freedom would be an even greater loss of freedom when people could not be fed, clothed, and housed. There will be little meaning to political freedom, or freedom of choice, if our society fails to prepare for predictable crises and our outmoded socioeconomic-political structure crumbles about us. Well-deliberated long-range policies offer better solutions than hasty patchwork actions.

Conclusion

The learning structure proposed in this chapter has the potential for strengthening government in three ways. First, if government can better anticipate the future, it has a better chance of creatively dealing with such problems as climate change, overpopulation, resource shortages, species extinction, and ecosystem damage–all the problems highlighted in Chapter 2. Second, government could better undertake the new functions which we are beginning to assign to it that were identified early in this chapter: encouraging social learning, facilitating quality of life, and helping the society to become sustainable. Third, the learning structure could improve the processes and outcomes of government by helping to avoid mistakes or to avoid governmental aid to special interests at the expense of the general welfare.

Although one could spin out scenarios showing how this proposed structure could do those things, critics could spin out equally plausible scenarios showing that this reform will not achieve as we hope it will. The only way to find out if it will work is to try it. Would it be very costly to try it? The Council for Long-Range Societal Guidance would not be a costly unit to operate; salaries for 100 to 200 people would be the main cost. Certainly, the cost for five years would be much less than the cost of a new space shuttle.

The studies the council might commission and the impact assessments that would be required would be more expensive; perhaps as much as one-half billion to a billion dollars per year. Compare that to the $5 billion to $10 billion dollars a new nuclear power plant would cost, or to the estimated $300 billion dollars required to clean up hazardous waste dumps, a problem that might have been avoided had an adequate foresight capability been in place forty years ago.

The learning procedures proposed here would create some delays. "Time is money" we are often told. Is that true? Sometime yes, but often no. If a time for careful reflection is built into the planning, it need not be perceived as a delay; it could become a part of the normal way we expect to do things. Hasty actions that later turn out to be failures waste money and create even greater delays. The chances are that it would be more costly *not* to adopt and utilize a learning structure than it would be to do so.

When I have presented these ideas at professional meetings they have been criticized for being impractical in one way or another. One critic declared that the proposal is "soft" because the Council for Long-Range Societal Guidance would only have the power to recommend. But, would he want to give the council decisive power? Would political authorities give away some of their power by establishing such a council? Would the people accept this? If the answer to those questions is no, would he recommend that we continue in our present ways? I remind the reader of the discussion early in this chapter in which I considered, and discarded as inadequate, all of the current forms of government. As Michael says, "we are in a bind" and we will find our way out of it only by learning.

I return to Elgin's point that we are in an era of cynicism; nearly all of us can see evidence of this in our daily lives. Cynicism can become oppressive, especially if many aspects of life are not going well. When such a mood has settled on a people, even good ideas will be thrust aside. People will be so convinced that nothing will work that they will not even make an effort at a solution. Let us hope that we can get the social learning process moving, and some new structures in place to assist it, before we settle into deep cynicism.

Some cynics will emphasize the well-known resistance of humans to new ways of doing things, and declare that great social trauma will be necessary to make people willing to accept the new learning structure. There is little doubt that times of crisis are opportune for sweeping away old dominant social paradigms and may result in the gradual establishment of new and more adequate ways of thought and styles of life. However, there is also a danger that, when traumatized, people may regress to earlier belief forms and tenaciously hang on to old structures and behavior patterns, despite overwhelming evidence that they are failing. At times of crisis, it could easily happen that people will accept demagogic ideas, and the society could plunge further into chaos.

This is all the more reason that constructive ideas for meeting future problems should be carefully advanced and debated now. A complex public discussion of these matters will help provide a basis for evaluating new ideas should the time come when peoples' minds are more open. The realistic—neither optimistic nor pessimistic—approach would be to change governance forms in order to provide a better institutionalized framework for social learning, opinion formation, and decisionmaking. We should do everything we can to facilitate our common social learning in the years ahead.

Defending A Sustainable Society Against Aggressors

Chapter 15

The Basic Dilemma

The dominant social paradigm (DSP) in modern industrial society emphasizes the accumulation of power and wealth. This book is grounded in the understanding that societies emphasizing power and wealth are not sustainable. I have argued that a sustainable society must operate according to a new environmental paradigm (NEP) that emphasizes humans living in harmony with nature rather than trying to dominate it. Such a society must cease growing in human population, cease striving for power (especially military power), renounce dominator/submissive relationships, must preserve and protect the biosphere, deemphasize consumption of material goods, and live lightly on the land.

Even though I am confident in predicting that DSP societies will not be sustainable, they are far from collapsing at present. To most people, they are dynamic, powerful, and the wave of the future. Even when they suffer strains from their unsustainable characteristics, they will not collapse quickly or all at once. On the other hand, NEP societies are likely to arise slowly, in only a few places at first. Thus, there is likely to be a transition period wherein a few NEP societies are struggling to become viable while some very powerful DSP societies threaten their existence. How can an NEP society, that eschews power, defend itself against aggressive DSP societies? If the NEP society offers no defense, it is likely to be conquered and swallowed up by a DSP aggressor. If it offers an adequate defense, it must become (or continue to be) like the power-maximizing DSP society it is deliberately trying not to emulate.

The reader will recall that this is the basic dilemma posed by Schmookler in his *Parable of the Tribes* (see Chapter 3). According to the parable, if a group of peaceful societies are threatened by even one society that seeks power and domination, all must take up the ways of power. The struggle for power creates societies that dominate and destroy nature, waste resources, injure human and other species, encourage competition and domination between people, and threaten the very continuance of civilization. In other words, modern civilization forces nations to become power-seeking even though the power maximizing emphasis cannot be sustained in the long run—we cannot seem to escape from this fate, yet we cannot live with it.

The power-maximizing craze was ruthlessly expressed in the colonial era during which dozens of LDCs were conquered by aggressive industrial European powers. That experience seems to have taught us a lesson; we have now entered an era in which there is nearly universal disapproval of colonialism and we find it hard to understand how colonial powers could have justified their behavior to themselves or to the rest of humankind. Yet, at that time, their behavior was considered normal and good—at least by Europeans. Why have we turned away from slavery and colonialism which were widely accepted two centuries ago? I believe it is the result of social learning.

Part of the learning came from weighing costs and benfits. Colonial domination was undertaken to enhance the wealth and power of the Mother Country, yet, careful analysis and the movement of history disclosed that empire was an economic drag on the imperial power. Nations that stayed at home and "tended to their knitting," such as Sweden after 1860 and Japan after 1946, became better off than those that went out conquering and pillaging, such as England, France, and Spain. After studying this phenomenon, Kenneth Boulding coined the aphorism, "wealth creates power and power destroys wealth" (1986, p. 32). I think we also learned over those two centuries that slavery and colonial domination are morally unacceptable.

But, we are still caught up in the parable of the tribes and believe we cannot give up the struggle for power. Over the past quarter century, commerce in arms has risen dramatically: "As of 1984, world arms imports totaled $35 billion per year, compared with $33 billion worth of grain, putting guns ahead of butter in world commerce." (Brown, et al., 1986, p. 198) Global military expenditures in 1985 totaled $940 billion; this amount exceeded the combined gross national products of the poorest half of humanity (China, India, and the African countries south of the Sahara). More of the world's resources are devoted to weapons research than to the combined research efforts to improve health care, develop new energy technologies, raise agricultural productivity, and control pollution. (Brown, et al., 1986, Ch. 11)

How can societies that wish to follow an NEP emphasis ever break out of the power-maximizing strait jacket that civilization seems to have imposed on all societies? This is one of the most difficult questions I have faced in writing this book. We do not lack research on problems of defense and peace; thousands of researchers have addressed the problem and the literature is voluminous. Most of this research assumes the continuation of DSP societies and the struggle for power that it imposes. I shall not try to recap that literature here; rather, I confine my inquiry to projecting how an NEP society could deal with a potential aggressor. Finding an answer requires us to project a future social learning process with which we have almost no prior experience. We must stimulate subtleties of learning that lie beyond what can be achieved by most rational discourse and even beyond what we can apprehend with our five senses.

Clearing Away Some Underbrush (So We Can See The Problem More Clearly)

Because wars and aggression have been known through most of history, many people conclude that being warlike and aggressive is part of human nature. Even though there is plenty of aggressive behavior observable in the world, there is good evidence to suggest that aggressiveness is learned and not innate. We socialize most females not to be aggressive. Why could not most males also learn not to be aggressive? Archaeological research suggests that many primitive societies and even some "high civilizations" were not aggressive: the men and women in them felt like partners (Eisler, 1987).

The desire to be cooperative is just as powerful and normal as being competitive. Actually, most people would prefer to live in a society that emphasizes cooperation over competition (Milbrath, 1984, p. 38) The next section discusses a game-theoretic study demonstrating that, over the long term, cooperation is a more rewarding and robust way of relating to people than competition. Eventually, people may learn that being aggressive is not necessary in order to be secure or to obtain life's rewards, that, in the long run, being aggressive actually is destructive of achieving those ends.

The ostensible purpose of the present arms race (perhaps all arms races) is to make each side feel secure. We in the West believe we must keep a highly armed military in readiness because, if we do not, the East will conquer us; naturally, those in the East believe the same. Each side has acquired the power to obliterate the other, many times over. Yet we are regularly told that we must develop more arms—the military feels it never has enough. Most ironic of all, neither side feels secure. Rifkin (1985) likens this situation to two individuals standing in gasoline up to their necks and debating how many matches each should have—foolishly thinking that if they have an equal number of matches they will be secure. Seeking security through mutual terror in an armaments race not only does not provide security but also is not sustainable.

Many people argue today that deterrence through mutually assured destruction has kept the peace for forty years since the end of World War II. Boulding has shown that, while deterrence may have worked in the short run, it is an illusion to think that deterrence can be stable in the long run:

If the probability of nuclear weapons going off were zero, they would not deter anybody. There must be a positive probability of breakdown, therefore, for deterrence to be stable even in the short run. If there is a positive probability of anything, however, if we wait long enough it will happen.

One of the propositions about the future of which I feel fairly certain is that there will be a devastating earthquake in California in X years. X could be anything between tomorrow and 100 years from now. There is a similar proposition that if the system of unilateral national defense continues, the Soviet Union and the United States will destroy each other and perhaps the whole human race in X years. National defense, therefore, has become the greatest enemy of national security. It can no longer defend us, it can only ultimately destroy us. (Boulding, 1986, p.32)

We do not lack efforts to disarm or to find peace. There are endless negotiations, countless conferences, thousands of studies; yet, peace seems so remote. Everyone wants peace but no one knows how to find it. Despite the impeccable logic of Boulding's argument, many people continue to be certain that the only way to be safe is to continue to add to our armaments. Persons with that perspective refused to believe that the Russians could be trusted and vigorously opposed the treaty banning intermediate range missiles that President Gorbachev and former President Reagan signed in Washington, D. C. on December 7, 1987. I doubt that peace will be attainable as long as our present DSP, with its emphasis on wealth and power, continues to dominate the way people view their world. We need a new perspective on our social system accompanied by a new view of our relationship to nature and to each other.

A simple way to imagine a new system would be to wipe away much of the present. Ideally, population growth would cease and we would gradually reduce the world's population to about 3 billion; gross inequalities of wealth would be abolished; memories leading to suspicions and hatred would vanish; we would see each other as one world family; nationalism would cease to be a powerful emotion; and we would join in a world government that would provide the security we all so desperately seek. Realistically, none of these things will happen. Nationalism, our love of nation, is still the most powerful idea in the world—many people will gladly give their fortune and their life for their nation. Populations will continue to grow; inequalities in wealth are likely to become greater rather than smaller; the call for world government will not rally public support. How are we to find peace amidst that grim reality? How could a society that is trying to become sustainable, and live according to NEP values, manage to survive in that world?

Developing New Subtleties Of Social Learning

As we have found out so often when we have pursued deep questions, our search will lead us back to rethinking our values. NEP beliefs and values are likely to promote evolution of a social system that not only will coexist in greater harmony with nature but also will incline societies to live peacefully with their neighbors. What are the most relevant values for peace and security and how might we give them greater prominence in our lives?

I believe it is most important for people (and nations) to *learn to cooperate rather than compete.* I indicated earlier that most people prefer a society that emphasizes cooperation; yet, the DSP constantly reiterates that society should be structured to encourage competition. Axelrod (1984) presents convincing evidence that individuals (or nations) in a continuing interaction tend to evolve toward a cooperative rather than a competitive relationship. It will be worthwhile for us to review his basic argument and evidence.

Axelrod utilizes a game called "the prisoner's dilemma." (The language defining the game and used in discussing its dynamics can put people off. I retain the language because the game is so prominent in current research. I urge readers to put their aversions aside to see what they can learn.) In the original version of the game, two accomplices in crime have been arrested and are interrogated separately. If they stick to their original story (that is, cooperate), they can only be held for a minor crime. If either of them breaks down and tells the whole story (one cooperates and the other defects), the defector will be freed and the cooperator will be convicted of a major crime. If they both defect, they both are sentenced to the major crime. Hence the dilemma: defecting can lead to a large payoff (being freed) or it can lead to severe punishment. Cooperation has a smaller payoff but is safer.

This game has been played extensively in numerous research settings, many of them oriented toward trying to understand disarmament and peace negotiations. Usually the game is now structured to pay off in dollars or points and has nothing to do with prisoners. Frequently, live subjects play the game and are rewarded with money. Many of these studies have shown that cooperation arises between subjects even though the individual reward would be larger if the subject "defects." In many recent versions of the game two subjects (or groups) play the game repeatedly, developing strategies that increase the overall reward–this is learning.

Axelrod's used a computer to play hundreds of iterations of the game. Instead of pitting live subjects against each other he pitted strategies against each other in a tournament. The rewards were structured so that mutual cooperation produced three points for each player. If one "defected" and the other cooperated, the "defector" got five points and the cooperator nothing. If they both "defected," they each got one point.

He invited fourteen game theory specialists to submit their favored strategy; each strategy was run through 200 iterations against each of the other strategies. The winner of the tournament would be the strategy that accumulated the most points against all the other strategies. The strategies could roughly be differentiated into "nice" strategies and "mean" strategies. Nice strategies began with cooperation and generally favored cooperation. Mean strategies favored defection and tried to take advantage of the cooperative behavior of opponents. To Axelrod's surprise, nice strategies generally scored better than mean strategies. The clear winner was the simplest strategy, "*tit for tat*," which opened with a cooperative move and thereafter did what its opponent did on the previous move. *Tit for tat* worked better than other nice strategies because it would not allow mean strategies to take advantage of it. "So while it pays to be nice, it also pays to be retaliatory. *Tit for tat* combines these desirable properties. It is nice, forgiving, and retaliatory. It is never the first to defect, it forgives an isolated defection after a single response; but it is always incited by a defection no matter how good the interaction has been so far." (Axelrod, 1984, p. 46)

Axelrod reported the results of the first tournament to the "community" of game theorists and invited them to submit strategies for another round of the tournament; sixty-two persons did so; *tit for tat* won that tournament as well. In addition, he played six hypothetical tournaments which magnified the effects of different types of rules from the second round; *tit for tat* won five and came in second on the sixth. *Tit for tat* also performs well in laboratory experiments with human subjects.

Tit for tat is difficult for an opponent to exploit because there is a high possibility of encountering the strategy, it is easily recognized, and it is soon perceived as nonexploitable–it is nice, retaliatory, forgiving and clear. It is especially interesting that these abstract rules, played out on an abstract computer, seem to "learn" from each other. A rule that "recognizes" *tit for tat*'s strategy learns that it pays to cooperate with it. Axelrod speaks of these rules almost as if they were persons; he concludes with respect to "mean" rules:

If a rule defects to see what it can get away with, it risks retaliation from the rules that are provocable. In the second place, once mutual recriminations set in, it can be difficult to extract oneself. And, finally, the attempt to identify and give up on unresponsive rules. . . . often mistakenly led to giving up on rules which were in fact salvageable by a more patient rule like *tit for tat*. (Axelrod, 1984, p. 54)

Computers do not have emotions such as liking or disliking, trusting or distrusting; hence, these emotions could play no role in the learning. As a matter of fact, trust is not a necessary condition for cooperation. Axelrod cites several studies showing that the Germans and the Allies, opposing each other across trenches in World War I, learned to cooperate even though they were antagonistic. (They avoided shelling the other side, hoping it would reciprocate by not shelling them.) The crucial condition was the duration of their relationship. If units at the front were changed frequently, cooperation was unlikely to develop. Enemies that learn to cooperate need a memory of the past as well as projection of their future relationship. Caldwell (1984) also cites many instances of cooperation in pollution control between political and military enemies (such as the pact to protect the Mediterranean Sea); he calls it *antagonistic cooperation.*

Axelrod noticed that, in the computer game, cooperation seemed to just evolve as a survival-positive strategy; there was no built-in bias in rules. He wondered if cooperation could be found to evolve in nature, even among creatures with no conceptual ability at all, such as bacteria. He teamed up with biologist William Hamilton and they found plenty of instances in nature where cooperation evolved among creatures with no conceptual ability. A crucial consideration was repeated interaction among the creatures, which eventually evolves into cooperation. The "learning" among these creatures is trial and error; those committing errors do not survive, while those practicing cooperation have a greater chance of survival. We often forget that Darwin's principle of "survival of the fittest" was supplemented by a second principle saying that creatures that learn to cooperate have a better chance of survival. "The evolutionary process needs more than differential growth of the successful. In order for it to go very far, it also needs a source of variety—of new things being tried. In the genetics of biology, this variety is provided by mutation and by a reshuffling of genes with each generation." (Axelrod, 1984, p. 170)

As long as the interaction is not iterated, cooperation is very difficult. That is why an important way to promote cooperation is to arrange that the same two individuals will meet each other again, be able to recognize each other from the past, and to recall how the other has behaved until now. This continuing interaction is what makes it possible for cooperation based on reciprocity to be stable. (Axelrod, 1984, p. 125)

Axelrod gives advice (1984, p. 110) to players of the prisoner's dilemma game: "(1) Don't be envious. (2) Don't be the first to defect. (3) Reciprocate both cooperation and defection. (4) Don't be too clever." His summary of cooperation theory is cogent:

The main results of cooperation theory are encouraging. They show that cooperation can get started by even a small cluster of individuals who are prepared to reciprocate cooperation, even in a world where no one else will cooperate. . . . two key requisites. . . . are that cooperation be based on reciprocity, and that the shadow of the future is important enough to make this reciprocity stable. But once cooperation based on reciprocity is established in a population, it can protect itself from invasion by uncooperative strategies.

... But what is most interesting is how little had to be assumed about the individuals or the social setting to establish these results. The individuals do not have to be rational: the evolutionary process allows the successful strategies to thrive, even if the players do not know why or how. Nor do the players have to exchange messages or commitments: they do not need words, because their deeds speak for them. Likewise, there is no need to assume trust between the players:the use of reciprocity can be enough to make defection unproductive. Altruism is not needed: successful strategies can elicit cooperation even from an egoist. Finally, no central authority is needed:cooperation based on reciprocity can be self-policing. (Axelrod, 1984, pp. 173-74)

Axelrod's findings apply specifically to two-person (or two-party) games that persist for many iterations. Additional research is needed to test how far the findings are applicable to games with more players, other rules, and different situations.

The finding from biology is perhaps most interesting for its conclusive showing that cooperation is survival positive for all kinds of creatures, even those with no conceptual ability. This finding that cooperation is just as vigorous a part of nature as competition denies the assertion by some that aggressive competitiveness is rooted in human nature and, therefore, we should design our society to accentuate those characteristics.

Compassion and love clearly are part of our biological makeup; we should more forthrightly live by those values in our day to day interpersonal relationships, in our business relationships, and even in our international relationships. All the world's great religions emphasize love as a core value for humans. But how do people learn to practice love? For most people the urge to do so already is present; the striking popularity of Leo Buscaglia, and his love message, testifies to that. The problem does not lie with our values and inner desires but rather with social structural barriers which interfere with our desire to express love.

For example, it is much easier to practice love in a society that emphasizes cooperation than in one that emphasizes competition. Competition stresses winning, defeating the opponent, coming out on top. Loving someone who has just defeated or injured you is difficult. Competition may bring out some of the best in people, as is claimed, but it also brings out some of the worst. Cooperation more consistently brings out the best in people. In similar vein, it is more difficult to express love in a superior/inferior relationship than in a partnership relationship.

Expressing love and compassion is presumed to carry its own reward; our modern society only modestly rewards acts of love and compassion. It reserves its great rewards—money, fame, and esteem—for winners. The whole social structure, especially in the United States, orients us in that direction. In school, children are taught to work alone and to try to be better than the next child. Schools reward excellence, achievement, and winning, but seldom love and cooperation. Competitive sports, especially professional competition, accentuate values that suppress the expression of love. Our competitively oriented social structure must give way to a more cooperative one if we wish love to play a greater role in our lives.

Love and compassion usually find expression in caring concern for others, especially those less fortunate. Compassion becomes linked to a sense of *justice* because persons feel they should share life's needs more equitably. If compassionate sharing is extended to people in other lands, future generations, and other species, such persons develop a belief in the value of *self-restraint*. They will consume sparingly and so design their lifestyle that it will be sustainable for centuries. Their decisions about the size of their families should show similar restraint, but the connection to compassion is not so obvious—families may grow in size as people express love and compassion.

Persons who highly value compassion, justice, cooperation, and self-restraint also are likely to value *nonviolence*. A society in which most people hold those values would have less crime, terrorism, and inclination to go to war. The Scandinavian countries and Switzerland are contemporary examples of societies that accentuate those values.

Axelrod says that *trust* is not required for people to learn to cooperate, but trust surely would facilitate cooperation and disarmament. Trust is likely to be engendered in settings where love, cooperation, justice, and nonviolence are practiced. Trust cannot be taught by simple admonition. Trust is experientially based; we develop trust when we repeatedly experience reliability from someone and rarely experience exploitation, competition, and defeat.

The reader will have perceived that these values (cooperation, love, compassion, justice, self-restraint, nonviolence, and trust) form a cluster; they are interrelated and are nourished by the presence of each other. They are the central values of the NEP. If they were accentuated by all societies in the world, wars would be rare.

But how could future NEP societies defend themselves against societies holding the values of the DSP? No one has a powerful irrefutable answer to that question. My own tentative answer is to use *tit for tat* to teach aggressive societies to cooperate and to use nonviolent resistance as the method for retaliating to their defections from cooperation.

Tit for tat already is extensively utilized in international relations. While this chapter was being written, the United States imposed a high tariff on cedar wood products, mostly imported from Canada, in order to preserve its own cedar wood products industry. The Canadians protested loudly, claiming the tariff would force them to shut down their cedar industry.

They retaliated by placing a high tariff on selected goods exported from ι. United States to Canada. These and other painful experiences led the two parties to the negotiating table, and an overall free trade agreement, which took about one year to negotiate, has recently been signed. It took another year for both countries to ratify it and put it into effect.

A whole series of *tit for tat* exchanges recently transpired between the Soviet Union and the United States (one wonders if they had read Axelrod's book). Former President Reagan announced in early 1986 that the United States would no longer observe the SALT II treaty when it expired later in 1986. The Soviets retaliated by being obstructive about agreeing to a summit meeting. Reagan then announced he might reconsider if there was good faith from the Soviet Union side. Additional *tit for tat* exchanges followed that eventually led up the summit meeting in Washington, D. C. in December 1987, during which the treaty banning intermediate range missiles was signed; and to the Moscow summit in June 1988 when the ratified treaty was placed in effect.

It is interesting that the *tit for tat* play in the Canada/United States episode was very similar to that in the Soviet Union/United States episode, even though Canada is a good friend of the United States and the Soviet Union is its most prominently perceived enemy. Note also that none of the retaliatory moves in these two sets of interactions was violent.

An important finding from Axelrod's analysis is that it may take many iterations of *tit for tat* interactions to teach actors (nations) to be cooperative. It also is important for learning cooperation that the same leaders be able to deal with each other over an extended period. Frequent changes in leadership interfere with cooperative learning.

Learning how to cooperate might be accelerated, and sustained, if the *tit for tat* process were made official policy and conducted very publicly. Axelrod's findings are so unambiguously clear, and have such general applicability, that it would be safe and wise for nations to *declare their intentions to follow the* tit for tat *strategy.* As each instance of cooperation, of defection, occurs, a nation should announce publicly that its next step will be taken in pursuit of the strategy and with the purpose of fostering continuing cooperation. In other words, a nation would make cooperation its explicit foreign policy goal and declare that the *tit for tat* strategy is its vehicle to achieve the goal. Any other nation dealing with such a nation would know what to expect and, presumably, would act so as to avoid retaliation. *Tit for tat* would not apply in all situations but, even if it only applied in many, it could have a powerful learning impact.

There would be instances when a nation would choose to defect, even though it knows that retaliation will occur. If the *tit for tat* strategy is explicit and public, the nation choosing to defect will recognize that long-range cooperation may be lost because of the defection. The restraining effect of *tit for tat* is weaker if a large powerful nation is contemplating aggression against a small weaker nation. The small nation simply cannot hurt the larger nation as much as the larger can hurt the smaller. The only corrective for this would be the outrage of the family of nations (really the outrage of humanity), most of whom would be incensed at this unjust behavior. The visibility of the aggression, and the resultant outrage, would affect the willingness of other nations to cooperate with the aggressor, thereby raising the stakes that the aggressor must confront.

The outrage of humanity is a significant deterrent to aggression and domination. It is the major force that has driven slavery and colonialism from the world scene. The Soviet Union's occupation of Afghanistan in the late 1970s became so costly in this regard that the Soviet Union has withdrawn its forces. This same outrage will force the downfall of the repressive dominator regime in South Africa.

Supposing a powerful DSP nation chose to attack and occupy an NEP nation that has renounced power as a national goal and has declared its intention to be nonviolent. The NEP nation probably would resist, but would not do so violently. Fortunately, the human family has had some experience with nonviolent resistance and is familiar with its efficacy; it drove Great Britain from occupation of India, for example.

There were many who equated non-violence with either passivity or cowardice. It is neither, Gandhi said. On the question of passivity, Gandhi counseled his critics that non-violence "does not mean meek submission to the will of the evil-doer, but it means the pitting of one's whole soul against the will of the tyrant. Working under this law of our being it is possible for a single individual to defy the whole might of an unjust empire. . . . and lay the foundation for that empire's fall. . . . " On the question of cowardice, Gandhi remarked that, on the contrary, the fully armed man is more at heart a coward. "Possession of arms implies an element of fear, if not cowardice. But true non-violence is an impossibility without the possession of unadulterated fearlessness."

. . . Gandhi's triumph lies in the means he chose to secure human welfare. He called on his countrymen and women to choose a different path to security. He asked them not only to renounce power over each other but also to accept responsibility for each other. And they responded; by the millions they responded. Gandhi liberated the long-imprisoned empathetic consciousness of the race. . . . Gandhi once remarked that "the religion of non-violence is not meant merely for the . . . saints. It is meant for the common people as well. . . . Man often becomes what he believes himself to be. If I keep on saying to myself that I cannot do a certain thing, it is possible that I may end by really becoming incapable of doing it. On the contrary, if I have the belief that I can do it, I shall surely acquire the capacity to do it even if I may not have it at the beginning. (Rifkin, 1985, pp. 104-05)

Cooperation, *tit for tat*, and nonviolent resistance are the three components essential for an NEP nation to defend itself against aggression; let us call it a *power-renouncing defense*. If this defensive posture was adopted by a cluster of societies, and pursued for some time, it might sufficiently advance social learning in all societies to make world government a possibility. We first must learn how to accentuate the cluster of NEP values that undergird the NEP society.

How Might This Learning Come About?

Several proposals to accelerate the learning of these values were presented in previous chapters. Learning also occurs in many subtle ways; some are so subtle that we may not even be cognizant of the experience. Is it possible for people to share images, and learn from their sharing, even though they are not verbally communicating? We sense this possibility, but only dimly, making it very difficult for us to talk or write about it. The following are some threads of evidence to suggest that subtle learning may be taking place.

It has now been quite well-established that we hold many perceptions, beliefs and feelings at an unconscious level (Harmon, 1984, 1988). These unconscious psychological states undergo change and learning. Because these states are not manifest, it is difficult for us to consciously change them; but they do seem to be affected by the psychological fields of the people around us. There is learning taking place that we simply cannot account for at the conscious level. Harman (1984, 1988) further suggests that we probably are connected at the unconscious level to a "perennial wisdom," a stream of knowing that seems to lie at the core of all the world's great religions.

Biologist Rupert Sheldrake (1984) has studied phenomena that he believes can only be accounted for by a learning process in which our five senses play hardly any role. He implies that our unverbalized thoughts somehow connect with the thoughts of others; he calls the phenomenon "morphic resonance." (Sheldrake's ideas are developed more fully in Chapter 18.) If this is true, it is possible that, at times of heightened awareness, a community or nation could rethink its position or make up its mind about something rather swiftly. The total communication would far exceed verbal exchanges.

The strong desire for peace that has become manifest in numerous contexts all over the world in the last few years (demonstrations, polls, referenda) must constitute a strong pressure speeding up this unconscious learning. Erhard (n.d.) raises an interesting question, "What causes an idea's time to come?" His answer is that we help an idea to become preponderant by creating a context. If enough people believe in the idea, it reaches a critical threshold and an entire society realigns. What had formerly been seen as impossible now becomes possible. The new context generates a new and accelerated process.

That which previously did not work, that which was stuck and not moving, suddenly began to move and start working. When you create a context, it's not that you are now doing something very much different from what you were doing before or even that you now know something very much different from what you knew before. It is that there is a shift in the climate, the space–specifically the context–in which you work, that makes things suddenly workable. (Erhard, n.d., p. 23)

It was fairly obvious that there was a sharp decline in negative attitudes and a sharp rise in positive attitudes of the Russians and Americans toward each other that evolved within a few days during the visit of Gorbachev to Washington for the December 1987 Summit. That was social learning. The learning progressed further during former President Reagan's June 1988 visit to Moscow. Those events were viewed by hundreds of millions of people, probably stimulating billions of discussions among them. People seemed to reach out to each other across national boundaries and spanning oceans. Clearly there was a psychological readiness to accept a new perspective that was ignited by the spectacle of events.

Gorbachev was asked during his press conference in Washington, "What happened to bring about the remarkable change?" He thought for a moment and said it was the intellectuals in each country that made the leaders understand that they could not go on in the old way. He also repeatedly said it was the deep desire of the people.

Senator Daniel P. Moynihan (D-NY), who is a member of the Senate's Arms Control Observer Group, reported on a visit his group made to the arms control talks in Geneva in June 1987. He marveled at the remarkable change in the attitudes and behavior of the delegation from the Soviet Union. It seems that the Chernobyl disaster had been a mightily impressive lesson to the Soviets. "You haven't had your Chernobyl" said the wife of one diplomat. A member of the Soviet delegation said, "The time is past when big nations can intimidate small nations. People, not power, decide. You can't get your way in Nicaragua. We can't get our way in Afghanistan." Gorbachev said in a May 1987 interview with *L'Unita*, the Italian Communist party newspaper:"Over the centuries politics has remained the preserve ... of the strong of this world. Even now it is still largely the lot of states. ... But now, by no means only states. The leading social trend–the shift toward democratization–is declaring itself ever more loudly and convincingly in international political activity." All these snippets of evidence suggest that sweeping social learning was/is underway.

Erhard's notion of developing a context is similar to Hofstadter's (1985) notion of phase transitions. A phase transition occurs when there are sufficiently strong and numerous interactions between the people in a system. For example, the presidential campaign of Corazon Aquino in the Phillipines had, at first, seemed utterly hopeless in opposing the entrenched power of the Ferdinand Marcos regime in early 1986. The election was held in early February and Marcos was declared victor. Yet, within the space of two weeks, nearly the entire Phillipine society realigned. That which had seemed hopeless only a few weeks before suddenly became possible; the Marcos regime was forced to leave the country with hardly a shot being fired; nonviolence had triumphed over violence because the context had changed. Hofstadter suggests (1985 p. 56) that persons caught up in a phase transition (context) are haunted by the idea, it will not leave them alone. It has seized some power inside them and will not let go. Strangely, their old personal "government" that has to some extent been usurped is not entirely displeased.

Conclusion

My most general conclusion is that the same social learning that will bring about an NEP society will carry with it the potential for that society to maintain its integrity against possible aggressors. An NEP society that emphasizes love, compassion, cooperation, justice, self-restraint, and nonviolence is not helpless before its neighbors who may still be following the DSP model of seeking power, wealth, and domination of nature. The power of NEP ideas and convictions are not insignificant. An NEP society could announce and pursue a strategy of tit for tat in relating to other nations—seeking to help them learn to cooperate. Even though it renounces violence and war, it can resist aggressive acts (defections) with nonviolent actions. It should join hands with NEP groups in other nations (for example, green parties) in pressing for worldwide nuclear and conventional disarmament.

Another effective next step would be to build transnational institutions to deal with problems that spill across national boundaries. It requires international action, for example, to deal with transnational environmental problems like climate change, deforestation, acid rain, and depletion of the ozone layer. In September 1986, the General Assembly of the International Council of Scientific Unions established the International Geosphere-Biosphere Program: A Study of Global Change. It involves natural, life, and social scientists from all parts of the world. Governments support most of this scientific work. Even political and military antagonists can cooperate on problems of this type that can only be approached cooperatively. Many such actions will slowly have the effect of building a world community that may, someday, lead to a world government.

Throughout this long and difficult transition, we must avoid the mind-poisoning belief that peace cannot be achieved. The possibility of success cannot be judged solely on the basis of what has happened in the past, or on the seeming intransigence of the present. We can rather swiftly learn a new perspective and new attitudes toward each other. As Erhard says, nothing is more powerful than an idea whose time has come. The time when the ideas leading to peace will prevail draws closer as a result of billions of individual thoughts, desires and convictions among people who are committed to making peace possible.

One Biosphere But A Fragmented World

Chapter 16

The Earth is one but the world is not. We all depend on one biosphere for sustaining our lives. Yet, each community, each country, strives for survival and prosperity with little regard for its impact on others. Some consume the Earth's resources at a rate that would leave little for future generations. Others, many more in number, consume far too little and live with the prospect of hunger, squalor, disease, and early death. ... Today the scale of our interventions in nature is increasing and the physical effects of our decisions spill across national frontiers. The growth in economic interaction between nations amplifies the wider consequences of national decisions. Economics and ecology bind us in ever-tightening networks. (World Commission on Environment and Development, 1987, p.27)

The nation-state is becoming too small for the big problems of life, and too big for the small problems of life. (Bell, 1987, pp. 13-14)

Our Global Commons

The biosphere is that thin layer of land, water and air on the Earth's surface that is occupied by humans and all other living things. We humans claim land territory for our own, but of course it belongs to other creatures, too. Air and water easily cross and interpenetrate our landed areas. Whatever happens to the air and the water, even on the far side of the planet, also happens to us. The nuclear power plant explosion at Chernobyl in the Soviet Union in 1986 sent radioactive particles around the globe. Scientists now reliably estimate that a major nuclear exchange in a war would be likely to put enough smoke into the atmosphere to bring on a nuclear winter that would kill most plant and animal life on Earth (Ehrlich, et al., 1984).

Such other global environmental problems as climate change, acid rain, depletion of the ozone layer, spread of toxics, ocean pollution, desertification, and deforestation originate from human activities in more than one nation and have global consequences. Tropical forests, especially in the Amazon, are being cut down for agriculture and other development; if this trend persists, millions of species of plants and animals will become extinct and the carbon dioxide buildup will increase, worsening climate change. The Amazon is a world resource and a world problem.

The high rate of human population growth, combined with poor ecosystem management, is expanding the size of deserts, especially in Africa. It changes the climate, kills off plant and animal life (including humans), and places a famine relief burden on the rest of the world. This same population growth swiftly uses up natural resources such as soil, water, forests, minerals, fossil fuels,and wildlife. Swift growth also will force many people to live in huge megacities.

Humans are globally connected another way through their economies and their communications net. World markets in natural resources, raw materials, manufactured goods, finance, even waste disposal facilities have some impact on nearly all economic choices. The Organization of African States is currently working with the U. N. Environment Programme to draw up a convention to prevent industrial countries from dumping their toxic wastes in Africa. Such seemingly local decisions as selling one's labor or buying land are affected by considerations of the world market. Mutinational corporations, some of which are quite diversified with widespread holdings, accelerate the movement toward a world economy.

These developments constitute a kind of technological inperialism flowing from the developed countries to engulf the world. This enveloping presence is most visible in worldwide communications such as television, radio, movies, and telephones which put people everywhere into instant contact with events worldwide and have begun to create a world culture. Computers can trade data, via satellite, across continents. Conferences among participants from several continents can now take place with none of the participants having to leave home. The relative ease, and comparatively low cost, of international air travel has brought people from all nations into direct personal contact. Ecologically we have always been one world and we are swiftly becoming so economically and culturally—our politics lags far behind. It is not clear that this outcome is desirable but it is a fact of current life that we all must deal with.

Crisis In Our Global Commons

When the Stockholm Conference on the World Environment was held in 1972, many LDCs perceived environmental degradation as mainly a problem of the rich nations experiencing the side effects of industrial wealth. They sought quick industrial development and postponed worry about environmental protection until later. Their learning was swift and painful, however; many of these LDCs now recognize that they are trapped in a downward spiral of economic and ecological decline and are learning that environmental protection and sustainable development are insepar-able issues.

The U. N. General Assembly passed a resolution during its fall 1983 ses-sion creating a World Commission on Environment and Development (WCED) to study and report on this crisis. Dr. Gro Harlem Bruntland, leader of the Norwegian Labor Party (now prime minister) was selected by the U. N. Secretary General as chairperson and Dr. Mansour Khalid, former minis-ter of foreign affairs for Sudan, as vice chairperson; the two then appointed the other members of the commission—a majority from the Third World.

The commission held deliberative meetings/hearings in eight countries, commissioned seventy-five studies, appointed three advisory panels, and received more than 900 written submissions. Their final report, titled *Our Common Future*, was released in April 1987 and soon became known popu-larly as the Bruntland Commission Report. They provided considerable evi-dence that our global commons is in crisis.

The WCED has several main messages to the world: *First*, that poverty is an inseparable part of the environmental problematique:

Poverty is a major cause and effect of global environmental problems. It is therefore futile to attempt to deal with environmental problems without a broader perspec-tive that encompasses the factors underlying world poverty and international equal-ity. . . . Those who are poor and hungry will often destroy their immediate environment in order to survive: they will cut down forests; their livestock will overgraze grasslands; they will overuse marginal land; and in growing numbers they will crowd into congested cities. The cumulative effect of these changes is so far-reaching as to make poverty itself a major global scourge. (WCED, pp. 3, 28)

Second, development and economic growth must accelerate in order to relieve grinding poverty, but the emphasis must be on development that is sustainable. Developmental planning must be integrated with planning for environmental protection and sustainable yields. The commission decries the practice of assigning economic development responsibilities to one ministry and environmental protection to another. The interlocked eco-nomic and ecological systems of the real world will not change, so institu-tions and policies must. Those bodies whose policy actions degrade the environment must be made responsible to prevent that degradation.

Third,

these are not separate crises: an environmental crisis, a development crisis, an energy crisis. They are all one. . . . We are now forced to accustom ourselves to an accelerating ecological interdependence among nations. Ecology and economy are becoming ever more interwoven—locally, regionally, nationally and globally—into a seamless net of causes and effects. (WCED, 1987, pp.4-5)

Deforestation by highland farmers causes flooding on lowland farms. Factory pollution robs fishermen of their catch. Dryland degradation sends millions of refugees across national borders. Internationally traded hazardous chemicals enter food that also is internationally traded.

Two aspects to the change have brought on the crisis:

When the century began, neither human numbers nor technology had the power radically to alter planetary systems. As the century closes, not only do vastly increased human numbers and their activities have that power, but major unintended changes are occurring in the atmosphere, in soils, in waters, among plants and animals, and in the relationships among all of these. The rate of change is outstripping the ability of scientific disciplines and our current capabilities to assess and advise. It is frustrating the attempts of political and economic institutions, which evolved in a different, more fragmented world, to adapt and cope. It deeply worries many people who are seeking ways to place those concerns on the political agendas. (WCED, 1987, p.22)

Fourth, sustainable development must become a priority goal for *all* nations.

The onus for action lies with no one group of nations. Developing countries face the challenge of desertification, deforestation, and pollution, and endure most of the poverty associated with environmental degradation. The entire human family of nations would suffer from the disappearance of rainforests in the tropics, the loss of plant and animal species, and changes in rainfall patterns. Industrial nations face the challenges of toxic chemicals, toxic wastes, and acidification. All nations suffer from the releases by industrialized countries of carbon dioxide and of gases that react with the ozone layer, and from any war fought with the nuclear arsenals controlled by those nations. All nations will also have a role to play in securing peace, in changing trends, and in righting an international economic system that increases rather than decreases inequality, that increases rather than decreases numbers of poor and hungry. (WCED, 1987, pp.308-09)

The report is far-reaching and bold in its recommendations. Yet, the commission had a difficult time facing some important realities, especially trying to come to agreement about the impact of population pressure and the need for population control. The final language is vague on recommendations with respect to population. Apparently, many commission members believe that ecosystems and economies must be made to accommodate whatever additional people parents choose to bring into the world.

If that accommodation is to be achieved, economic growth and development must be accelerated, but this time with more care to preserve the integrity of the biosphere. This development will be achieved with the help of new scientific knowledge and technology, as well as financial assistance, to be proferred from the developed world. Apparently, complete faith in science and technology is widely and deeply held in the Third World. In my judgment, a strong element of wishful thinking is in that aspect of the report. Does *sustainable development* in the sense taken by the WCED lead to a sustainable society? I do not believe that it can.

Unwillingness within the WCED to face the need for limits to growth displays a clear bias in favor of humans. Accelerated development in order to accomodate continued population growth will surely take over more and more of the ecosphere for humans and drive other species from their niches. On the other hand, if grinding poverty is not eased, desperate populations will destroy their ecosystems thereby killing off other life anyway. If caught in a life or death struggle, humans will choose human life over other life. But what a tragic dilemma—humans will lose either way they choose, for we cannot escape this truth: "There is no habitable abode for human life unless there is a habitable abode for other life. We are dependent, totally dependent, upon other life." (Domain, 1987, p. 9) Unless the whole world can somehow learn to limit human population, there is no escape from this tragedy.

The WCED report is a political document oriented toward social learning. In order to be effective, it must work within currently dominant belief structures. Therefore, it could not speak the truth about limits to both population and economic growth. Only when humankind painfully learns that biospheric integrity cannot be maintained while humans refuse to restrict their growth will it be possible for the truth to be heard.

Can We Overcome Our Fragmentation?

It is ironic that, even though we are one world ecologically, and swiftly becoming so economically and culturally, the world is becoming increasingly fragmented politically. The original United Nations formed in 1947 had fifty nations as members, but its membership now approaches 160. *World Resources 1987* reports data tables for 146 countries. The morning news carried a report that the Tamil separatist group in Sri Lanka had blown up several buses in their campaign for separate nationhood. There is a clear trend for more and more and smaller and smaller nations. Nationalism is still the most powerful idea in the world.

When I was a student in college forty years ago, I joined the United World Federalists; we hoped that our efforts to bring about a world government would avoid another war such as World War II that we had just come through. Our dream was much closer to reality then than it is now. The fanatical drive for nationalism has so poisoned our psyches that no one now proposes world government with any expectation that he will be listened to–college students would not even respond.

People despair that nations will ever effectively solve common problems. This attitude leads to efforts to strengthen national economies, national military might, and national technological capability so that each nation can compete more effectively in international competition. It leads to an insidious emphasis on keeping up, on winning, on a desire for power and wealth, that needlessly depletes our resources and injures the biosphere. It could, as we all fear, lead to a nuclear war that might destroy nearly all life.

The Bruntland Commission is asking all of us to move away from our preoccupation with our own nation and recognize the internationalism that our ecological common future thrusts upon us. Caldwell (1984) shows that we have already started down that path and should move quickly on global ecoprotective projects. The WCED calls for the United Nations to prepare a Universal Declaration on Environmental Protection and Sustainable Development that would be followed by a convention (treaty) to commit nations to this global effort. The commission's report seems to be generating considerable attention and social learning. Even after acceptance of those documents we still have a long way to go to overcome our fragmentation and achieve workable action plans.

A Continual Emphasis On DSP Beliefs And Values Exacerbates These Problems

The DSP belief structure continually emphasizes gathering more power and wealth to protect against aggressors, to be competitive, to be in control. To the extent that this belief structure is acted upon, the gap between the rich and the poor will widen. Power and wealth tend to accrue more power and wealth. Even when rich nations try to help poor nations, the gap often widens. A development loan will only be made if the project can be expected to make money. The poor have little money to pay for a product or service so projects usually are twisted to serve the desires of those who will buy the output; typically it is exported to more developed countries (MDCs) (for example, a Central American land owner using his land to grow carnations for sale abroad instead of growing food for local consumption).

People and leaders in LDCs, many of which were former colonies of MDCs, have long argued that it would be only simple justice for the MDCs to divert some of their resources to help them develop. It is clear, that would be the just thing to do. Why should some countries be rich, and get richer, while others cannot escape their poverty no matter how hard they try? Probably most people in MDCs are genuinely sympathetic and would like to help.

Simply making the argument and expecting it to shape policy in MDCs has proven to be naive, however. Governments in MDCs have been preoccupied with their own problems, and helping poor nations has almost always been low in their priorities. Furthermore, DSP adherents believe that concentrating on maximizing one's own wealth helps everybody and that if we just pursue that policy, everybody will be better off. There is little room in that philosophy for a strong obligation to help other countries.

The belief that market forces and economic growth will raise LDCs out of poverty is invalid in two respects. First, think for a moment like an entrepreneur from an MDC who is contemplating investing in a plant in a LDC. The main attraction for doing so is to gain a competitive advantage in the world market by extracting cheap resources, exploiting cheap labor, and escaping the costly pollution controls that MDCs require. Where will most of the goods be sold? Not in LDCs, because few citizens will have the money to purchase them but, instead, in MDCs where people already are buying and will like his inexpensive prices. People in the LDC will not be seen as consumers/buyers but rather as cheap labor. This process exploits the resources of the LDC and increases rather than decreases the gap between the rich and the poor. If the LDC government requires the entrepreneur to clean up his act or to pay higher wages or resources prices, he might pick up his machinery and move elsewhere. Of course, governments could intervene and redirect these forces—but then it would no longer be a market process.

The second invalidating consideration is that there simply are too few resources and there is too little waste-absorbing capacity in the biosphere to support economic growth of the magnitude needed to bring all, or even most, of the LDCs up to the current standard of living of the MDCs. Chapter 2 presented my reasons for this position. Even if we found the resources for a spurt in economic growth, the inevitable increase in carbon dioxide levels would stimulate such swift climate change that all national economies would plunge into catastrophic economic decline. Instead of expecting that LDCs will rise to the standard of living of the MDCs, we should expect that MDCs will soon experience their own decline resulting from the unsustainable characteristics of their socioeconomic structures. As MDCs demonstrate their own unworkability, less and less development aid for the Third World will be forthcoming.

Leaders in LDCs are surely beginning to recognize that the extreme economic disparity between MDCs and LDCs is likely to continue for a very long time, and that they will get little bilateral voluntary help from MDCs. Have they no recourse other than to carry their plea to the United Nations? And what good does that do? Most power, especially military power, is held by MDCs, so LDCs have little chance of using military power to force the MDCs to help them. Some groups in LDCs have turned to terrorism to try to force policy changes on MDCs. This practice is likely to coalesce the MDCs into united action to fight terrorism but is unlikely to produce a policy more favorable to LDCs. The clear injustice of the LDC situation, combined with their continual frustration in obtaining help, will generate anger and hatred that is likely to poison "North-South" relations for many decades. As long as DSP thinking dominates international relations, we can expect terrorism to increase—and the world will become further divided. I state these realistic expectations with sadness for I do not condone the selfish behavior of MDCs.

LDCs will advance their cause more effectively if they recognize these realities and depend mostly on their own talents and resources for addressing their problems. They could count on some modest help with information as well as scientific and technical advice from U. N. agencies; some governmental and private institutions in MDCs also provide such help.

When firms and nations compete in a world economy, they are likely to select a specialty in order to keep up with the game. As specialization progresses, firms and nations will become more and more interdependent. On first glance that may seem desirable, because it will convince many people and nations that we really are one world. A second glance discloses, however, that those who become more specialized also become more vulnerable. Development of a new technology in nation X that gives them a competitive advantage may sweep the people in nation Y into a deep economic depression. A political revolution, or a war, in a nation that supplies an essential factor of production to many other nations could wipe out a large sector of economic activity in many nations. If the United States were to go into a deep economic depression it would carry much of the rest of the world down with it.

It might be wiser to counteract the tendency toward increasing specialization. Social and economic resiliency in households, firms, and nations would be enhanced if those units continually were to emphasize simpler lifestyles, greater self-sufficiency, reduced competition, and living lightly on the land. Such units would be less likely to come into conflict and could much more easily cooperate.

All of the consequences of emphasizing DSP beliefs, that I just reviewed, will swiftly degrade the biosphere. The competition for power and wealth leads to greater consumption, depletes resources, inflicts a devastating waste burden on the biosphere, and will lead to catastrophic socioeconomic decline. The continual growth in population that the struggle for power among DSP nations encourages, also will increase consumption and waste. Wars, terrorism, and depression often result in direct physical damage to the biosphere. A competitive struggle for wealth and power in a politically fragmented world cannot avoid great biospheric damage; if it degenerates to nuclear war and nuclear winter, or to swift climate change, it will be catastrophic.

Changing How We Think About These Problems

More than anything else, the WCED report is a plea for people of the world to change their way of thinking about environment and development; it is also the central plea of this book. Finding new modes of thinking, valuing, and acting will require considerable transformation in beliefs, values, and lifestyles in both MDCs and LDCs. The adjustment will be greater in MDCs than in LDCs because people in MDCs have more to unlearn. People in MDCs are accustomed to affluent lifestyles and will be reluctant to reduce their consumption. (They probably believe it will reduce their quality of life–conceivably it could increase.) Specialists are unlikely to know how to do many things for themselves and conceivably will have to learn how to diminish their use of technology, to become more self-sufficient, to learn the ways of nature, and adapt their lifestyles to be in greater harmony with it.

The people in LDC's already live simpler lifestyles with less specialization; they will not have to go through the psychologically difficult task of changing their whole way of life as will the people in MDCs. Their learning will be considerable, however. They will learn to limit the size of their families. They will learn they cannot adopt the highly consumptive and wasteful lifestyles of people in MDCs (who also will have to give up those lifestyles). They will have to learn how ecosystems work, so that they can better harmonize their lifestyles with their ecosystem. They will have to learn new ways of doing things that will be less ecologically damaging than their traditional ways. For example, they will learn to build and use biomass digesters to supply themselves with biogas for cooking and heating instead of foraging for fuelwood in ecosystems that already have been stripped of most of this resource. They will have to learn how to develop their own appropriate technology.

Human Population

It is imperative for people to learn to limit the size of their families. This would reduce the demand for resources, reduce the need to displace other species from their habitats, reduce the waste load imposed on ecosystems, and reduce crowding, congestion, and competition in human settlements. A dense human population cannot live gently on the earth. Most people in MDCs have already adopted lifestyles and beliefs that keep human popula-

tion in check (not their consumption, unfortunately). Some newly industrialized countries (NICs), such as Taiwan, South Korea, and Singapore, have
enacted an official policy of limiting population, although compliance at
the family level is still very difficult. China is pursuing the most drastic
population limitation policy in the world (it limits each family to one child);
but China did not come to this policy without first teetering on the brink of
ecological and economic disaster. Even with the new policy, China projected
that its population would continue to grow to at least 1,200 million before it
could level off near the year 2025. The most current information from China,
as this book goes to press, is that the drastic sanctions against having more
than one child are not working well because increasing income allows families to pay the fines and go on having children. Current projections are that
it will grow to at least 1,400 million. If growth continues at present rates,
there will be nearly 2 billion people in another forty years.

The people in many LDCs, however, especially in Africa and Latin America,
do not recognize a need to limit population. Their traditional beliefs continue to encourage large families. Parents may feel that they must have
many children so they can be assured of being cared for in their old age.
These nations are growing so swiftly in population that some will soon face
breakdown of their ecosystem (such as that experienced in Ethiopia
recently). Will they learn at the last minute, as with the Chinese, or will
they allow their population to go into overshoot and succumb to dieback?
Helping such countries to limit their population is an extremely complicated and subtle problem that differs from society to society; it would take
us too far afield to discuss it thoroughly here.

As societies learn to limit their population *and consumption* it will have
several advantages, not only for themselves but for the world community.
People who are not so crowded, and for whom resources are not so scarce,
will find it easier to feel and act with compassion toward other people and
other species. If people aspire less strongly for material wealth, it will be
easier for them to share. The disparity between rich and poor would diminish and could, perhaps, reduce North-South tensions.

Appropriate Technology

As people seek self-sufficiency and harmony with nature they will utilize
technology that is more attuned to their needs and capabilities, and to the
ways of nature–appropriate technology. It is difficult to define appropriate
technology concretely because it is tuned to local conditions that vary
from place to place and country to country. Studying and advocating appropriate technology has become something of a mini-movement. One lesson
has become clear from that inquiry:imposition of the "best" western tech-

nology by outside consultants or lending agencies is unlikely to be successful. Indigenous people need to study their own problems and select those technologies that fit with their own locality and culture. Of course, good ideas, some of which will come from MDCs, are always useful.

For this latter task, they can receive useful advice from the Ecodevelopment Program of the United Nations (as well as from a variety of scientific and action organizations). The U. N. Environmental Program defines ecodevelopment as "development at regional and local levels . . . consistent with the potentials of the area involved, with attention given to the adequate and rational use of the natural resources, and to application of technological styles . . . and organizational forms that respect the natural ecosystems and local sociocultural patterns." (as quoted in Bartelmus, 1986, p. 45)

The main features of ecodevelopment are:

1. resource development for the satisfaction of basic needs;

2. development of a satisfactory social ecosystem;

3. rational (nondegrading and nonwasteful) use of natural resources in solidarity with future generations;

4. use of alternative environmentally sound production procedures;

5. use of alternative energy sources, in particular of the regional capacity for photosynthesis;

6. development and use of ecotechniques;

7. establishment of a horizontal authority ensuring participation of the population concerned and preventing any plundering of the results of ecodevelopment; and,

8. preparatory education to create social awareness of ecological values in development.

(Sachs, 1980, as quoted in Bartelmus, 1986, p. 46)

Agencies that lend money for development have begun to insist that environmental impact statements (EIS) be prepared before they will approve a loan. The WCED believed that it was a mistake to bring in consultants from an MDC to prepare an EIS. Instead, development agencies and governmental leaders should insist on building up an "in-country" EIS-making capacity.

The World Bank announced a major reorganization in the spring of 1987 that increased the staff capability in its environmental department severalfold. The bank adopted a new policy, giving environmental considerations equal weight with economic considerations in making developmental loans. It is clear that the World Bank has been listening to the WCED.

The desire for a loan is strong enough that planners and officials in LDCs will go along with the demand that they prepare an EIS. But the lure of economic growth continues to dominate developmental thinking; thus, the EIS may only display the rhetoric of ecodevelopment. "By and large only lip-service has been paid to the 'integrated approach to environment and development' in most development plans. Environmental objectives are simply added to the list of plan objectives without follow-up during plan implementation." (Bartelmus, 1986, p. 65)

Insistence from the top that a procedure must be followed can be expected to produce mainly lip service if the people at the working level still have not developed a deep environmental awareness and thus do not appreciate the need for the procedure. Following the procedure, even in a perfunctory way, may have some learning value, but the development is not likely to be ecologically sensitive without a deeper awareness and concern by project leaders. This experience illustrates that transformation to the NEP mode of thinking must take place at a deep personal level to be truly effective. I know of no shortcut to this deep ecological learning.

A Painful Dilemma

The WCED gave considerable attention to the problem of mass migration of people seeking a better life. Migration from rural areas to cities has been common for a century or more. There is an especially heavy migration to big cities in LDCs; Mexico City, for example, must somehow accomodate about 1,000 new residents each day. If the city forbids their entry or refuses to give them basic services, such as water and sewerage, they come anyway. Famine or war also produce mass migrations of refugees. A recent television interview showed the Interior Minister of Sudan defending his country's policy of not accepting refugees from the Ethiopian famine. He explained that if they allowed many thousands to cross the border, their own ecosystem would soon be destroyed and their own people would soon be destitute.

Finding a policy to deal with migration is extremely difficult. Consider the following scenario: The gap between rich and poor nations will persist for some time, hence it is likely that affluent MDCs will have poor LDC neighbors (for example, Mexico next to the United States). Typically, most parents in the MDC will have limited their family size, but few parents in the LDC will have done so. Job opportunities and other economic prospects will be very depressed in the LDC, but much better in the MDC. Naturally, the people living in the LDC will gaze with envy at those living in the MDC and will seek to migrate, illegally if necessary, to what they perceive to be a better life.

What should the MDC do about this immigration pressure? It could close its borders tightly by fortification; but that seems to be a cold-hearted rebuff to the plight of its neighbors. If it allows the illegal immigration, it will soon find its own population growing swiftly. Its resources will be depleted quickly; crowding, congestion and pollution will rise sharply; other species will be crowded from their niches; open space will disappear – it will lose its own natural treasure.

What is a nation to do that has hordes of poor people on its borders clamoring to be admitted? Can it let in everybody who wants to come? That policy would destroy everything that the citizens, as well as the immigrants, thought was desirable about the society. Should it let in only a quota? The quota would have to be quite restrictive to be effective. There probably is no just way to decide who will be admitted and who will be excluded. If a quota is used, the border would still have to be sealed in order to be sure that nonquota persons do not enter illegally and negate the policy.

The impact on the country from which people are fleeing is another consideration. In the short run it may get some relief, as there would be fewer poor for whom to care; but basic structural problems will persist. What happens to the helpless poor who do not know how to flee? How will the country get along without the talented and enterprising persons who are likely to be the first to flee? Would social learning come about faster and more deeply if people stayed home and tackled their basic social structural problems? I believe it probably would.

Taking all these considerations into account, I cannot escape the conclusion that a nation with hordes of people seeking admission should seal its borders. I realize that this is nationalism rather than the internationalism I believe we should be moving toward but I also recognize that *open migration in a fragmented world of sovereign states will bring ecological ruin to all*. Protecting the integrity of our biosphere will require effective collaboration under binding agreements or a common government. Only when human excesses have been effectively controlled will it be possible to allow free migration. Meanwhile, I expect people will still show compassion for desperate persons who turn up at their door, but their actions will only be a temporary alleviation of a symptom. That posture can never constitute a successful policy; carrying it into full practice could only lead to even more horrendous problems in the future.

This painful dilemma illustrates the tragedy of political fragmentation resulting from the misplaced emphasis on national sovereignty on our planet. The Americans, for example, cannot do anything to limit population growth in Mexico; they have limited capability to help them economically, as well. Yet, Mexico's problems profoundly impact on the American society; and the reverse is true as well – American exploitation of Central American resources has certainly had some devastating consequences. If the DSP stimulated competition between nations is allowed to be played out, the disparity between rich and poor will become greater and poor people will continue to try to migrate to their richer neighbors. This is *not* the way to build a sustainable society.

How Do We Get The Needed Changes Underway?

An autocratic and determined national government can initiate changes to bring about far-reaching socio-ecological consequences (such as the one child per family policy in China). That option is not available at the international level where there is no government and no recognized authority. It makes no sense to go to war to bring about the changes that are needed for sustainable life on the planet. Changes at the global level can only come about through social learning. We should expect a long and painful transition. Learning the NEP way of life will probably have to be forced by ecological reality that is so grim it no longer can be ignored. It will occur in both MDCs and LDCs. Facing a grim reality can create a mental readiness to learn that rational argument cannot create. People can tune out, or dismiss, a rational argument that they do not like; grim realities intrude even when we try to ignore them.

Population growth, climate change, and resource demands have the most far-reaching consequences, so helping people to understand how they affect their lives should be the first task of the social learning effort. All over the world people must learn to desire small families. An anthropological colleague pointed out with dismay that China's one child per family policy will mean that soon children will have no cousins, aunts, or uncles. I replied that people are creative and will find substitute affectional relationships to enlarge their families. My contacts in China tell me that it is common to adopt friends as cousins, aunts and uncles.

Relearning economic matters also will be important in both MDCs and LDCs, although the change will be greater in the MDCs. People and governments must learn to deemphasize the desire to acquire wealth, as I mentioned repeatedly in Chapters 8 through 13. Both MDCs and LDCs will learn to develop and use appropriate technologies. People in MDCs will learn to give up their dependence on high technologies that tend to become enslaving (see Chapters 12 and 13). People in both LDCs and MDCs will learn more ecological and more efficient ways of meeting their basic needs (solar energy, biogas producing digesters, aquaculture, more productive species).

It would be wise for people in some localities to investigate the possibility of organizing themselves into bioregional units (see Chapter 10). People in a bioregion raise as much of their own food as they can, make their own consumer goods, provide their own services, and exchange them in a regional market. The bioregion would become their main structure of government. The key point of bioregional organization is to make the people of a region self-reliant; they look to themselves to make their own quality of life. (I noted in Chapter 10 that in dense urban settlements, or countries that developed an economy that is intimately tied to the world economy, it may be too late to organize into bioregions).

We all hope that eventually the people in the world will learn enough, and will build a sufficient base of cooperation, that there can be real disarmament. Developing NEP societies and making them work will make it easier to disarm. This will be a long and delicate relearning task. Our only choice, however, is to carefully nurture that activity.

Summary Of Recommendations Of The WCED

Keep in mind while reviewing these recommendations that the WCED report, *Our Common Future* is a political document oriented toward social learning.

Policy Recommendations

1. Limit *extreme rates* of population growth. (My emphasis.)

2. Support farmers in LDCs more adequately to increase production.

3. Reform forest revenue systems and curtail deforestation—preserve endangered species.

4. Conserve energy and use it more efficiently. This emphasis should be the cutting edge of national energy policies (rather than increasing production). Focus on environmentally sound energy sources.

5. Emphasize antipollution technology as cost effective in LDCs as well as MDCs. Transnational corporations should take the lead in helping LDCs to move to cleaner technology.

6. Tighten controls over trade in hazardous materials.

7. Develop new human settlement strategies to take the pressure off megacities and build up smaller cities and towns.

8. Expand the notions of international security to include the necessity of relieving environmental stress. Divert some of the funds now devoted to armaments to environmental protection.

WCED Proposed Institutional And Legal Changes

1. Sustainable development objectives should be incorporated into the mandates of national, regional, and global institutions. Economic and ecological responsibilities should be integrated within the same agencies and be given equal weight—all projects should be both ecologically and economically sustainable.

2. National governments should make a "foreign policy for the environment." Nations without environmental agencies should give high priority to establishing one. Most existing environmental agencies need strengthening and appropriate support. Nations should assign higher priority to international environmental cooperation and should send high-level delegations, led by ministers, to international meetings.

3. Responsible national and international agencies should conduct annual audits and make reports on environmental quality and resource stocks. Nation states and transnational agencies should develop comparable economic and environmental statistics.

4. States and transnational institutions should develop monitoring systems for natural resources and processes—they should be oriented to provide early warnings of impending problems. State and transnational institutions should cooperate in drafting contingency plans and develop the capacity to respond quickly to critical situations. These institutions should cooperate in developing a *global risk assessment programme*.

5. Special attention should be given to ecological zones, systems, and resources that lie along national boundaries (there are 200 major international river basins); they need international cooperative arrangements for their management. Elevate ecosystemic sensitivity above national sensitivity. Joint action programs also are required for such common problems as desertification and acidification.

6. All nations should ratify the Law of the Sea Treaty. New conventions are needed for climate control, Antarctica, hazardous chemicals, preserving biological diversity, and outer space.

7. The United Nations must play a key leadership and coordinating role. All agencies in the UN system should have their mandates, programs, and budgets revised to include sustainable development as an important criterion. The UNEP should be the lead agency to provide this leadership and coordination; its mandates, budget, and staff should appropriately be expanded to fulfill this new responsibility. The U. N.

secretary general should constitute under his chairmanship a special *U.N. Board for Sustainable Development.*

8. Community groups and NGOs are rapidly emerging as important and cost-effective partners in protecting and improving the environment—therefore, they should be given financial support (public and private); this support is especially crucial in LDCs. Appropriate NGOs should be given more meaningful participation in international environmental meetings. Scientific NGOs can play an especially useful independent role in a global risk assessment programme.

9. Do not believe that environmental protection is too costly—*if it is postponed, it will be even more costly.* Nations, especially LDCs, that move now to reorient major economic and sectoral policies can avoid much higher future spending—the *unavoidable* costs of environmental and resource management would thus be paid only once. "The need to support sustainable development will become so imperative that political realism will come to require it." (WCED, 1987, p. 341)

Conclusion

The final sentence in the WCED report reads: *"We are unanimous in our conviction that the security, well-being, and very survival of the planet depends on such changes now."* (WCED, 1987, p. 343)

A world divided into 160 sovereign nations is poorly structured for protecting our global commons. Nations can cooperate on ecosystem protection, even if they are antagonistic on other matters; however, that protection is weak—in a real crunch it would give way. Ecosystem protection cannot be ordered by a world government and it cannot be achieved by going to war. The WCED can only urge us to learn; its recommendations for sustainable development would be very helpful if implemented—alas, many states will resist taking the recommended actions—which will lead to more painful learning.

Nature has a power of its own that speaks loudly to humans when they abuse it; nature will be our most powerful teacher. As NEP societies become a more prominent part of the international community, the need to maintain strict national sovereignty and national borders will diminish—we will have a chance to build a real world community. Some of the problems that now seem hopeless may become solvable. Either we learn fast and well, or nature will find some other way to deal with our exuberant growth.

Part III

Transition From Modern Society To A Sustainable Society:

Some Possible Scenarios

So far we have been imagining what a sustainable society might be like: how it would be structured, how it would work, what kinds of behavior it would encourage or discourage, how it might cope with problems posed by the ecological crisis of modern civilization. Parts I and II put forth ideas that would have a realistic chance of working; Part III analyzes the problems likely to be encountered in the transition from our present society to a future society.

The way that the transition is approached, and the way we move through it, will be strongly influenced by the way we collectively define the nested set of problems that we face, as well as by the physical reality with which society must come to terms. If we define environmental problems simply as failures of technology, most people will believe that they can be fixed by more and better technology. Those who believe in technological salvation also are likely to assume that the basic structure of modern society is sound and that we can continue on our present trajectory.

Many others believe that our entire social system is in jeopardy, that we cannot continue on our present trajectory. If that is the case, environmental problems cannot be solved solely by striving for better technology. They can only be solved by developing a new society. Are we dealing with a technical puzzle or with a systemic failure? The definition of the problem shapes the proposed solutions.

The reader who has followed my argument and evidence so far surely knows that I believe we are facing systemic failure, not simply technical failure. We must redesign society in a more sustainable mode. We start by envisioning what that society would be like and trying to understand how we can get from here to there. Even though I feel sure of my diagnosis, I recognize that many people, probably most people, will not agree with me (at least not at first). They believe that modern Western society is working fine and see no reason to make significant changes. That will make the transition all the more difficult. Open-minded recognition of the deep systemic nature of our problem would allow a planned gradual transition, with minimal dislocation and pain. Stubborn resistance to recognizing the need for deep social change means that we will avoid the needed deep changes until our world collapses around us; and then it may be too late for effective action.

A Mini-Theory Of Social Change

The nature of society—its social structure—is rooted in our ways of thinking. Our ways of thinking are rooted in our beliefs about how the world works. Both our ways of thinking and our beliefs about the way the world works are subject to our own control. (This unique ability sets humans apart from all other creatures.) Therefore, they are, in principle, very malleable. Social change could occur very quickly.

Why, then, does social change come about so slowly? Our beliefs about the way the world works are slowly absorbed from earliest childhood— mostly at the behest of our parents and other authorities. They are constantly reinforced by the dominant patterns of socioeconomic-political activity that surround us every day, and by educational, cultural, religous, and political leaders. We are not inclined to rethink our basic beliefs as long as they are working reasonably well. It requires an extraordinary level of awareness and selfconfidence to question one's most fundamental beliefs about how the world works.

Technological innovations, especially if they are widely adopted, do change the way our everyday world works— and they do change the way we think. Before long, they also change our social structure; both the automobile and television testify to that kind of impact.

Growing awareness of our own human potential also changes our ways of believing and thinking. Our abandonment of slavery and colonialism and our growing acceptance of the equality of races and sexes have profoundly changed our social structures and behavior patterns. Our yearning for democracy is having a similar impact.

A deepening understanding of the consequences of war and of our abuse of the biosphere also is changing our beliefs about the way the world works. New beliefs and ways of thinking will replace the old dominator posture of humans toward each other and toward nature; the only question is whether the new ways of thinking and believing will come in time to save our planet from horrible desecration.

Fast social change can occur when the currently dominant ways of believing and acting start to fall apart. When fast population growth chokes our cities and destroys the quality of life, we will urgently rethink our beliefs about family size and birth control. Biospherical changes will awaken us when the ecosystem can no longer accept our wastes without changing the way it works. When shortages, high prices, and lack of opportunities deprive us of our vision of a better future, we are more likely to cast about to find better ways. The more aware and open minded we are, the more likely we can come through a period of socioeconomic-political breakdown by developing a more viable set of beliefs about the way the world works and new ways of thinking that will successfully guide our society to a better future.

Chapter 17 discusses what is likely to happen if we continue on our present expansionist path and treat our environmental problems mainly as technical problems. We are likely to stumble into a crisis that will threaten our civilization. If we did achieve the goals espoused by our current leaders, I doubt that we would like the world we had created. Will we have to experience a crisis before we learn to make basic changes? Crisis learning is chaotic and unpredictable.

The final chapter discusses the prospects for deliberately accelerating our social learning. I am as hardheadedly analytical as I can be in evaluating the readiness for social learning. My conclusion is that our transition will not be easy. If we are unlikely to learn until we move into a crisis, our best strategy for now is to prepare ourselves so that we can wisely direct our learning once our minds are opened up to new possibilities.

Avoiding Change Will Make Us Victims Of Change

Chapter 17

For me, being a prophet does not mean to be a crazy man with a dirty beard; it means to be strongly in the present, so that you can imagine a possible future of transformation—the possibility of going beyond tomorrow without being naively idealistic. "This is utopianism as a dialectical relationship between denouncing the present and announcing the future." *(Future Survey vol. 9, no. 9, Sept.1987, p.6, quoting Paulo Friere)*

Most developed countries, especially the United States, are known to welcome change; progress is an honorific concept. Scientists and advertisers alike tell us that we cannot stand in the way of progress. But progress means different things to different people. Progress means new technology, growth, and wealth among believers in the DSP. Conversely, it means protecting nature, slowing down, and spending more time in loving relationships among believers in the NEP. That which seems progressive to one group will seem regressive to the other.

Actually, even in the United States, most people do not really welcome change; they are much more responsive to exhortations to reaffirm traditional perspectives and values. One of the reasons many people were attracted to Reagan's presidency was his vigorous reaffirmation of our traditional path and traditional values. Former President Reagan constantly told people what they were eager to hear: that the United States is on the right course; that we must press for more and better technology; that there are no limits to growth in population, in economic activity, in power; that the market will solve our problems; that we can all be well off economically.

Believers in the DSP recognize the existence of environmental problems, but they say they can be fixed by our clever scientists and technologists. The basic premise of contemporary leaders is that our current socioeconomic-political-technological system is sound. Our policy should be one of patching up and making adjustments, but not of making any fundamental changes.

But how farsighted and clever are we? Herbert Simon (1945, 1957) and Charles Lindblom (1959, 1965, 1979) both discussed the difficulty humans have in grasping, reformulating, and solving complex problems; we are especially prone to see only part of the picture. This makes it unlikely that our problem solving will be objectively rational. Most people will settle for a solution that is satisfactory rather than maximally the best. Simon called it *satisficing*. Lindblom characterized policymaking in modern societies as mainly a process of "muddling through." Most public officials, and their constituents, are unlikely to question the direction society is taking and to comprehensively evaluate all the possible policy options; rather, they prefer to continue in familiar patterns, making only minor adjustments as circumstances require. Because of this basic difficulty in public thinking, we should expect that most modern societies will react to the ecological crisis by trying to muddle through. This will be seen as being practical, which is a good word to most people. But will it really be practical?

A Scenario For Continuation Of The DSP

What will happen if we continue on our current path? We will mainly rely on technology to solve our problems. Education and research will greatly emphasize technology. We will try to convert many policy problems to technical problems. We will feel good that we are progressing. We will only belatedly and painfully learn that some problems, many problems, are not amenable to a technological fix.

If we continue on our technological binge, we will discover that technological fixes will become increasingly expensive (and uneconomic). This can be illustrated by some contemporary examples: Nuclear accidents at Three Mile Island and Chernobyl have taught managers to delay further deployment of nuclear power plants until they can be redesigned. Making them safe makes them uneconomical. When nuclear plants come online, it is typical for electrical rates to increase significantly. If we were not already so deeply embedded in nuclear technology, it would be wise to conclude that we should never begin. However, even if we were to call a halt now to all nuclear deployments, the necessity to make existing installations and waste disposal sites safe *would still cost hundreds of billions of dollars*. If we continue with nuclear power, the cost can only rise; these intractable problems will burden our descendants for centuries.

Similarly, our practice of burying toxic wastes in landfills will cost at least $100 billion, perhaps $300 billion, to remedy safely. Toxics also are seeping into drinking water aquifers; a recent study estimated that it will cost $30 billion to make our water safe to drink, if it can be done. Despite this evidence of danger, we continue to chemicalize our world.

These are current problems (and only a small sampling of them at that); we can expect that many other unwanted outcomes from our emphasis on technology will tumble into our laps. Our efforts at fixing will often cause additional unanticipated problems; fixes will be applied to fixes that were applied to fixes. A continuing emphasis on and development of technology cannot avoid adding to this spiral of effort, risk, and waste. Economists will calculate all of these repair and remedial efforts as part of our GNP, whereas they really should be seen as payments for past mistakes; they should be deducted from, rather that added to, the GNP.

The most urgent pressure for costly environmental cleanups will occur at the very time that our economies are under great stress from climate change and other ecosystemic effects, as well as lack of resources. Thus, environmental cleanup efforts will address only the most visible offenses. There will be a show of action directed to the symptoms, but seldom to the root causes. Governments and firms will spend the bare minimum needed to placate angry citizens; they will seldom take the long view and change their basic way of doing things. Governmentally directed fixes will drag on for years, with all of the involved parties (firms, governments) acting cautiously to make sure that they are not held responsible for failure. The leadership in the establishment will assure citizens that all is well. They will offer mainly symbolic assurances designed to cover up weaknesses in their actions and in the basic structure of society.

These same leaders will continue to extol a policy of economic growth (that they believe can solve all the major problems of society). People will be encouraged to consume more and more in order to keep the economy growing. This same idea will be pushed in the Third World where economic growth is even more deeply desired. LDC's following the model of the West, will make vigorous but faltering efforts toward economic development. This will make all countries search far and wide for additional resources to fuel their economic growth. Resources will be extracted from more and more remote areas and transported long distances—and the environment will be increasingly devastated.

How Is This Scenario Likely To Be Played Out?

Trying to solve our nested set of ecological/economic problems only with technological fixes is like treating an organic failure with a bandage. The key difficulties, which will be ignored by that strategy, are that biospheric systems will change their patterns and there will be an increasing squeeze on resources. As global human population continues to grow, and these new people demand economic growth to fulfill their needs, there will be unbearable pressure for resources. Soils will be depleted. Farmland will be gobbled up into urban settlements. Water will become scarce, more pol-luted, and very high priced. Forests will be depleted faster than they can regenerate. Wilderness will nearly disappear. The most easily extracted mineral deposits will be exhausted. We will search the far corners of the globe, at very high economic and environmental cost, for more minerals and possible substitutes for those that are being depleted. Fossil fuels, especially petroleum, will constantly diminish in supply and rise in price. Worst of all, biospheric systems will react to our interference by no longer working the way we have counted on.

International competition for scarce mineral and fuel resources could become intense and bloody. The highly developed nations are likely to try using their money and/or military power to garner the bulk of the resources for their own use. (It is difficult to imagine that a big power would allow its supply of critical fuels or minerals to be cut off without putting up a fight.) At best, those actions will only postpone the inevitable adjustment. The poorest nations (usually those with the densest populations) will be unable to maintain even subsistence levels—they are likely to suffer widespread famine and disease.

All of this frantic activity will have devastating impacts on the ecosphere. Climate change will debilitate every ecosystem and economy. Ultraviolet radiation will increase, as will acid rain and toxic poisoning of our air, soil, and water. In addition, we can expect more and more soil depletion, loss of crop land, mismanagement of water resources, oil spills, devastating accidents (Bhopal, Chernobyl), deforestation, spreading deserts, extinction of species, loss of wildlife, and air and water pollution.

With disrupted biospheric systems and severe resource shortages, I cannot imagine that it will be possible to sustain growth in material throughput. We may be able to grow in nonmaterial ways (increasing knowledge, artistic output, games, and so forth), but material growth cannot continue. Our endeavor not to change will have failed to forestall change; instead, we will become victims of change.

The next few decades are crucial. The time has come to break out of past patterns. Attempts to maintain social and ecological stability through old approaches to development and environmental protection will increase instability. Security must be sought through change. The Commission has noted a number of actions that must be taken to reduce risks to survival and to put future developments on paths that are sustainable. Yet we are aware that such a reorientation on a continuing basis is simply beyond the reach of present decisionmaking structures and institutional arrangements, both national and international. . . . Without such reorientation of attitudes and emphasis, little can be achieved. We have no illusions about "quickfix" solutions. We have tried to point out some pathways to the future. But there is no substitute for the journey itself, and there is no alternative to the process by which we retain a capacity to respond to the experience it provides. (WCED, 1987, pp. 22, 309)

And Then What?

What happens when a people are frustrated in attempting to achieve the economic growth that their leaders have told them they can and should have? What happens when their way of life begins to disintegrate? The sociopolitical dynamics will be turbulent and powerful. No one can anticipate all of the consequences, and their complex interactions. Yet, it will be helpful to use relevant insights from contemporary experiences to foresee some possible scenarios.

I have participated in many discussions concerning the tendency of people to ignore information they do not wish to hear and to resist change until they absolutely must. Someone always invokes this folk wisdom, "We will have to wait until we get into a crisis before people will wake up and make the needed changes." Is this really true? Do people finally wake up and make appropriate changes when they are in a crisis?

Let us return to Elgin's conceptualization of four stages of growth in the life cycle of Western industrial civilization (Figure 14.1). He estimates that most modern societies are in Stage III, initial decline, and will soon be moving to Stage IV where they will face a "crisis of transition." They could enter an era of societal breakdown and stagnation, but they also could turn their crisis into an era of learning and revitalization. Let us examine first the possibility that people will persist in avoiding learning and will stagnate even further.

Psychiatrists have known for a long time that many people are their own worst enemy in times of personal crisis. Instead of making appropriate changes that would allow them to deal more effectively with their problems, they become immobilized by regressing to an earlier stage of personality development. Could this also be true of a nation? Harold F. Searles (1979), a psychiatrist, thinks so. He suggests that contemporary humans have turned to technology to soften some of the harshness and unpredictability of Mother Nature. We have grown to be very dependent on technology which has become like another mother, but now we are told that this new mother is poisoning us. Our technology-dominated, overpopulated world has so reduced our capacity to cope with the losses of life, however, that we continue to despoil our world—eventually we have nothing to lose in dying.

In similar vein, could the disintegration of our way of life lead many people to reaffirm the DSP rather than abandon it? Festinger (1957) studied sects that predicted the end of the world; when their prediction did not come true, they reaffirmed their belief, even more strongly, rather than abandon it. He developed his understanding of this phenomenon into a theory he labelled "cognitive dissonance." According to the theory, people cannot tolerate dissonance in their beliefs (if people believe strongly in the DSP, but there is strong evidence that it is failing, they are experiencing dissonance). One way to relieve the dissonance (perhaps a favored way) is to deny the evidence and reaffirm the belief all the more strongly.

There is historical evidence for this phenomenon. During the fifteenth century in Europe it became clear that knights no longer were effective for winning wars;

... as their usefulness diminished, so their pretensions grew. Their armour became impossibly ornate, their tournaments more costly, their social status more jealously hedged around by a heraldic lore which concentrated the more on questions of status as it had less to do with military function. New knightly orders were founded in conscious imitation of the great orders of the twelfth century: The Knights of the Garter, The Knights of Bath. (Howard, 1976, pp. 12-13)

Elgin (1981, pp. 108-10) paints a scenario as to why and how people would resist change and thus contribute to stagnation. Some will deny that it is a time of fundamental civilizational transition, insisting, instead, that the current distress is merely a short-term aberration. Others will feel helpless and assume that someone else must be in control, so why bother. Still others will look for scapegoats and blame some group "out there" for the problems. Some will look for a way for themselves and their friends to escape. Others will fatalistically accept what seems to be an unstoppable process of disintegration. "We cannot spread our minds into a distant future; like alcoholics, we take one day at a time, plodding on, hoping the worst can be averted until we ourselves are dead. We choose the status quo, not in freedom but in despair; we fear action, since any act may entail something worse than the present situation. (French, 1985, p. 495)

The mass media, particularly television, will be used to divert public attention from the pressing realities of the world. Critical issues will continue to be pushed aside in favor of more traditional and more profitable passive entertainment programming. In-depth exploration of alternative ways of living and working seldom appear. People will turn inwardly to their own family and/or indulge in escapism fantasies.

Citizens' organizations, that might have inserted some creative innovations, will become discouraged and thus be too few and too weakly connected to make any substantial impact on the larger society. People will rely on traditional bureaucracies mainly concerned with not being blamed rather than being innovative. Nations will turn away from responsibilities to the global community and focus on their own survival. The social order will be buffeted by competing interest groups battling for dominance and privilege.

Gurr (1985) predicts that as ecological constraints restrict economic growth, the dominant elements in society will use their superior political and economic power to protect their privileged position (pp. 58-59). This will force the poor into greater poverty and most of the middle class will be squeezed into economic decline. Economic inequalities within a nation will widen.

Gurr also predicts (pp. 60-62) that greater conflict will exist among groups. The new political issue will be how to distribute privation. Organizations with the power to shut down the system may do just that in an effort to increase their share of a shrinking pie. Opposition coalitions may form in order to fight more effectively for their share. These conflicts could become violent and even revolutionary. Walsh (1984, ch. 14) points out that most expressions of hatred and attack stem from an underlying fear and sense of deficiency. We could expect many such expressions at a time of societal crisis. Harried governments may respond to this internal turmoil with repression. Elgin (1981, p. 110) suggests that massive though deadening bureaucracies will be used to manage every facet of life. The social order will expend all of its creative energy in just surviving. Totalitarianism may rise and freeze a dysfunctional pattern in place for many years into the future.

Gurr and Elgin were speaking of increased inequality and conflict within a nation, but one could expect increasing inequality and increasing conflict among nations as well. Many nations will be thrust into a hopeless situation, with a burgeoning population and virtually no chance for economic development. Their appeals for developmental aid will be ignored because the MDCs will frantically be trying to salvage their own disintegrating economies. Some few people (perhaps even nations) will react to this hopelessness by using wanton destruction and terrorism to strike back at what they perceive to be their oppressors.

If the transition to scarcity is rapid, the tendencies just mentioned will be more pronounced. If the transition is gradual, then hardship may become an expected feature of life for most groups. That would allow powerful interests and government to solidify their economic and political control. Overall, the tendency would be toward social structures that would be more inegalitarian and more rigidly stratified and segmented—repression would be used to contain or deflect conflict (Gurr, p.63).

There is a strong possibility, then, that we face a future with shortages, economic privation, famine, environmental degradation, increased inequality, greater conflict, greater totalitarian repression. **Is that the way it has to turn out? It all depends on what we do in our own minds.** For example, Gurr's argument presumes that material loss will be interpreted as loss of quality of life. While that presumption probably is valid for the contemporary American way of looking at things, it may not hold true in the upsetting conditions postulated above. People could learn to decouple their material standard of living from the way they view the quality of their lives. That would allow them to deal more creatively with conditions of scarcity and possibly avoid some of the injustices and conflict predicted by Gurr. Decoupling of the two concepts will be unlikely to happen if a society's established leaders continue to believe in the DSP and continue to press for economic wealth and growth. Toppling such leaders from power may be essential for developing a creative societal response to such a crisis.

The next chapter explores the way a society could prepare to meet the crisis that looms on the horizon for modern society. If we can be wise and clever enough, we can avoid some of the worst consequences of system collapse as we move through the transition from the old society to a new one. We could enter an era of swift social learning, leading to a revitalization of our civilization so that it can provide a decent quality of life in a sustainable mode.

Suppose I Am Wrong?

Before moving to the next chapter we should dispose of a few additional questions: What if I am wrong? I have foreseen the consequences of continuing on our present course as clearly as I can but my analysis may be faulty.

Julian Simon (1981, 1986), an economist writing frequently in defense of the DSP, claims that there will be no resource shortages because human cleverness will always find new resources. He asserts (1986) that statistical economic studies show no relationship between population density and economic growth. He invites us to reflect upon contemporary Hong Kong and Singapore, where there is a high rate of economic growth, yet they are two of the most densely populated places in the world and have almost no natural resources. He does not point out that they both have strict population control regimes. Also, their economies thrive because they can buy all their resources, especially energy. Hong Kong is saturated with artificial glitter and excitement (citizens have hardly any contact with the natural environment) and it is highly technologized. It also is one of the most energy-intensive settlements I have ever visited; it must have cheap energy to be able to thrive. The crucial assumption in Simon's analysis is that human inventiveness will find a plentiful, cheap, and safe form of energy to substitute for fossil fuels when they run out. If that kind of energy were found, my own prediction would have to be altered.

Let us suppose that former President Reagan, and those around him, are wise futurists and that their predictions about the validity of the DSP are true. According to their beliefs, there really are no limits to growth. Our economic output, and that of other nations, can grow steadily for several centuries. Science and technology will develop vigorously and are the key to this growth. Scientific discoveries will reveal new resources so there will be no critical long-range shortages. If nanotechnologies were to work out as envisaged by Drexler (1986), ordinary rocks and solar energy would provide all the necessary resources. Population growth is no cause for worry; economic growth will expand to meet human needs. Human life should be valued above all other life, so abortions must be abolished. New medical technologies will keep people living longer and longer. Bioengineering, or nanotechnologies, will create super plants and animals, so there will be no shortage of food. Pollution will not be serious; better technology can take care of those few problems that arise. Climate change and loss of the ozone layer will come on slowly enough that society can make the necessary adjustments.

Now, let us suppose that all the world has pursued this path until 2100. (This span of time would be not quite one second in the year-long movie portrayed in the Introduction.) What kind of world will we have then? My projections are based on the assumption that the DSP will work the way its advocates believe it will work. Of course, the resulting development would have to fit into the physical reality of the way the world works.

Because food and consumer goods would be plentiful, human population would probably have doubled twice, to more than 20 billion by 2100. All of these people must live somewhere, presumably in good housing; that means that either they will live in highly congested conditions in giant cities (such as Hong Kong) or their settlements will sprawl across vast expanses of what used to be farmland—probably both. Most of the old-style family farms will likely have disappeared. Most food will be grown with advanced technology (not yet invented) in huge factory-like complexes, using specially engineered plants and animals.

The natural environment very probably will have been severely exploited and rearranged in order to provide sufficient resources for all these humans and to dispose of their wastes. Most people will live in specially-designed megacomplexes that provide a controlled artificial environment. Air and water from nature probably will be too polluted for humans to use so their megacomplexes supply artificially cleansed and controlled air and water. Wilderness and wildlife will have greatly declined due to destruction of their habitat. Examples of animals in their natural habitat can be seen in a few special reserves.

People's everyday lives will be in an artificial environment; technological artifacts and controls will surround them. The energy required to power these megacomplexes and make all the consumer goods has put such a heat load on the biosphere that it changed the climate. Most of the time it will be too hot for people to be outside their megacomplex. Every once in a while there will be a major technological failure in a megacomplex, resulting in great discomfort and /or death to its inhabitants. Some cities will have been abandoned due to flooding from rising ocean levels and others due to encroaching deserts. Many of the native plants and animals will have died; some animals will have migrated to remote areas.

Is this where we really want to go? Is this the kind of world we want to live in? The defenders of the DSP will object to this scenario. "We will make changes as we go along," they say; "we will not let it get that bad. We will limit population and consumption, we will protect wilderness and wildlife." But that is precisely the point of the environmentalists. If we do not want to get to where the DSP is leading us, why not change course *now* so that we can plan fully to gradually make the transition to a new and more sustainable society based on NEP principles?

Consider another question: Suppose the defenders of the DSP are wrong about their belief that the DSP society is sustainable; what will we lose if we follow their path? Many of us will lose our lives. Our civilization will be devastated. Many species will become extinct. Those who live will struggle with disease and privation. The biosphere will probably evolve new ecosystems—and the remaining humans will play a very different role in them.

Now consider a parallel question: Suppose we had followed the path proposed by advocates of the NEP, but later found sufficient resources and new technologies (nanotechnologies, for example) that would have enabled us to have followed the path of the DSP. What will we have lost by following the NEP path? We would have "lost" the birth of many more people. We would have "lost" some riches, excitement, and thrills. Our lifestyles would probably have been like those portrayed in Part II of this book. The NEP society would probably have required more hard physical work from its members, but would that have made for less quality of living?

As we weigh these two suppositions, following the NEP path clearly poses much less danger to society and the biosphere, and it probably would provide quality in living that is just as personally fulfilling as following the DSP path.

This leads to one final question: Why do we fear the crisis that could result from failure of the DSP? Humans have experienced many previous crises in their history (wars, famines, earthquakes, and so on); surmounting them often made a society more resilient. What is different about this crisis that makes it so much more dreadful? First, this is a crisis in the relationship between humans and their natural world. That fundamental relationship evolved over billions of years and is the basic underpinning of all human societies. No previous crisis threatened that relationship.

Second, humans now have a vastly more powerful capability to destroy their biosphere, their civilization, and themselves. A nuclear war followed by nuclear winter could destroy much of the life of the planet. The destruction from climate change would be immense, second only to nuclear war, and it may already be on our doorstep. Our machines and our chemicals can wreak environmental devastation on an unprecedented scale. Continued swift human population growth could carry us into overshoot, followed by dieback. If that happened, our frantic efforts to stay alive would devastate other plant and animal species; we would turn large areas of the planet into a wasteland. Ecosystems would probably require centuries to recover from the devastation we would cause.

Third, the DSP society is very vulnerable. We are developing a technological dependence that robs us of our autonomy. Also, we are becoming more specialized and interdependent; what do we do if some critical element fails? (Imagine Hong Kong if its major power plants were destroyed.) When millions of people, who know only their small little role, are caught up in a collapsing system, they can only flail about helplessly as their life support fails. The larger and more interdependent the system, the greater the collapse when it happens.

If we follow the DSP path, we run the risk of a calamitous crisis that could destroy all chances for a decent life. Why risk everything for the dubious gains of spreading riches and thrills to more and more people? Why not seek quality of life in a more sustainable way?

Learning Our Way To A New Society

Chapter 18

Survival of the world as we know it is not possible. The world will have to be transformed and evolve for continued survival. This is the necessity and the imperative of our time and will continue to be so long into the future until this transformation has been achieved, or until there is no longer any hope that it may be possible. (Jonas Salk, 1983)

There is nothing more difficult to carry out nor more doubtful of success, nor more dangerous to handle, than to initiate a new order of things. For the reformer has enemies in all who profit by the old order, and only lukewarm defenders in all those who would profit by the new order. The lukewarmness arises partly from the fear of their adversaries who have law in their favor; and partly from the incredulity of mankind, who do not truly believe in anything new until they have had actual experience of it. (Machiavelli, *The Prince, 1513*)

The day is short and the work is great. . . . It is not your duty to complete the work but neither are you free to desist from it. (Rabbi Tarphon, quoted in From Poetry to Politics)

As the old saying goes, "Life will either grind you down or polish you up," and which it does is now our choice. Necessity may be not only the mother of invention but also of evolution. (Walsh, 1984, pp. 75, 82)

Anyone who calls for a massive transformation of society is bound to be labelled an impractical dreamer. I am calling for people to transform the most basic of all relationships, their relationship to nature. The changes I have proposed will surely be difficult for society to adopt. Yet, modern society's only choice is to change. As I analyze how we get from here to there, I shall try to be hard-headedly realistic while simultaneously being aware of the consequences of not changing.

Jay Forester, a leading systems analyst, characterizes societies as complex, high-order, multiple loop, nonlinear, feedback structures (1972, p. 153). They act in ways that are counterintuitive to our normal linear/casual ways of thinking. They are insensitive to change in many respects—they have internal dynamics that help them maintain stability. Thus, they are resistant to most policy changes. Yet, societies are highly sensitive to a few types of changes; they seem to have "soft spots" where pressure for change can ramify through a whole structure, but these spots are not self-evident.

The environment in which a society is embedded does change—as I have illustrated many times in this book. Therefore, a dynamic society that learns and changes fairly easily is likely to be more stable than a society that resists change. It is paradoxical that the stability of a society may rest on the ease with which it is able to change.

Even though social change is difficult to bring about, and usually is agonizingly slow, it does happen, sometimes quite dramatically and sweepingly. For example, the revolution led by Mao Tse Tung changed almost every aspect of Chinese social, economic, and political life within a few decades. Even though many of the changes were reversed in the new Deng regime, the magnitude and speed of change wrought by Mao's revolution is very impressive. The Iranian revolution, that displaced the Shah and installed a fundamentalist Islamic regime, reversed the direction in which that society was moving within a few weeks. (Some would say it was returned to the original paradigm from which it had strayed.) Such sweeping reversals—paradigm shifts—are rare and do not necessarily produce desirable change, but they remind us that swift change is possible.

Myriad forces both change and conserve the structure of a society. Some of these forces were discussed earlier, especially in Chapter 4, but it will be helpful to review the various ways that social learning and social change take place.

Physical-Technical Change

Changes in the physical environment as well as new technologies are especially powerful stimuli inducing social change. The steam engine, the automobile, nuclear power, airplanes, birth control pills, computers, and bioengineering are obvious recent examples. For their own wellbeing, humans should try to anticipate the social consequences of technological development, but in practice they do not. Changes in the physical environment are equally powerful. As climates have changed, civilizations have died; others have sprung up in places that formerly could not support human life. Technologically actuated biophysical changes, such as the production of greenhouse gases, are bringing climate changes that will result in profound social changes. A "nuclear winter" would devastate most life on the planet.

Evolutionary Transformation Through Adaptation

For the most part, sociocultural evolution is adaptive. Later in the chapter I distinguish it from a new perspective on evolution called creative evolution. Over human history, many societies have made fundamental mistakes and faded away. Other societies that were more adaptive, or made wiser decisions, managed to survive; typically, they also expanded. If DSP societies fail, they might, over the long run, be replaced by more viable societies; but that would be small comfort for those currently living in them. Even though societal evolution seems agonizingly slow, it is still much faster than the evolution of species in nature.

Strategies For Intentional Efforts To Induce Social Change

Intentional strategies can be analyzed into several categories: (some of the following is drawn from Leff, 1978, pp. 430-32)

1. *Empirical-rational approaches* are very common. In our normal discourse we present facts and arguments and hope that the rationality and self-interest of listeners will encourage them to change. Most scientific research falls in this category, as do governmental study commissions, television documentaries, utopian novels, and educational lectures. The rational approach is seldom persuasive in and of itself—many people, societies too, do things that they know are not wise—but logical persuasion is an important first step. Unless something "makes sense," change is unlikely to follow.

2. *Power-coercive approaches* use physical, economic, political, or even "moral" force to coerce the desired changes. Examples include conquest, violent revolution, new laws, lawsuits, strikes, boycotts, civil disobedience, terrorism, sabotage, spanking, buyouts, school grades, and promotions or demotions. If one thinks about it, these strategies are very widely used and commonly believed to be the most effective (but not the most humane). The dominance of a socialist form of society in Eastern Europe, for instance, was established and is maintained by coercive force. The dominance of the Lutheran religion in Scandinavia and Catholicism in France has its roots in conquest. The Chinese one-child-per-family policy depends on coercion for producing social change.

3. *Elite change* can lead to social change. If the elite of a society can be persuaded to a new point of view, they can bring about social change rather quickly. For example, in most parts of the world the thrust toward environmental protection has arisen from the grass roots; elites tended to respond only when the clamor of the citizenry could no longer be ignored. In China, however, most of the thrust for environmental protection has come from the elites. The populace still is largely

unaware of environmental danger. Elite change may not be very thorough, however, because a true social change must affect the habits and everyday behavior of people. China's environment will not be saved until deep change comes in the daily life of the people.

4. *Social inventions* are deliberately intended to produce social change. The founding fathers who wrote the U. S. Constitution were quite conscious that they were inventing a new way of structuring government and society. Social inventions are not the sole province of government, many are developed in private institutions. The management pattern for enterprises (especially factories) invented in Japan is being adopted in the United States because it has demonstrated its greater effectiveness. The "war on poverty" or "war on drugs" are other recent examples of social inventions intended to bring about social change.

5. When a sizeable group of people decide to work together to try to bring about social change we say they have formed a *social movement*. Most people are familiar with the Civil Rights Movement, the Labor Movement, the Women's Movement, the Environmental Movement, and the Nuclear Freeze Movement. Sometimes a movement finds expression in a political party, giving it more direct influence on governmental decisions—the Green Party in Germany is an important recent example, as are labor parties in numerous countries. Social movements will be more thoroughly discussed in the next section.

6. *Normative reeducative approaches* are directed more to individuals than to societies; they bring the people to be helped into the change process and give explicit attention to their attitudes, values, norms, and social relations (as well as provide them with new information). Examples include encounter groups, values clarification, psychological and/or religious counseling. Schein (1972) perceives a three-stage process of social change: unfreezing, changing, and refreezing. *Unfreezing* opens people to change when they encounter disconfirmations of their beliefs, values or behavior patterns. Guilt or anxiety can arise from comparing one's ideals with reality and can lead to unfreezing. A sense of psychological safety to unfreeze is aided by removal of threats or barriers. *Changing* is the process of developing new beliefs, values, and behavior patterns. *Refreezing* is stabilizing and integrating new beliefs, values and behavior patterns into the culture or person. Reconfirmation by "significant others" is an important aid to refreezing. A good change agent would try to help people (a) see the discrepancies between things as they are and things as they could be; (b) overcome unwarranted fear of change; (c) be creative and thor-

ough in searching out innovations that fit their situation and real needs; (d) effectively integrate the chosen innovations into a stable pattern of behavior.

All of these types of experiences result in social learning. Our problem is to accelerate social learning to produce a more harmonious and sustainable relationship between humans and their biosphere before they destroy it. We do not lack good ideas; I found a cornucopia of good ideas that already are available in the public domain. Not only are they there for the asking but they literally are thrust at people; if only people would listen.

One reason why good ideas are lost or hard to find is because they are buried in a cacophony of messages competing for our attention. This book, no matter how good it is, will probably be ignored because it will be lost in a message blizzard.

Our society already possesses most of the knowledge needed to make our lives both good and sustainable. We know, or can know, what needs to be done but we do not know how to get people to do it. Why is it so difficult to get people to listen and act?

Barriers To Social Learning

Barriers To Social Movements

Because social movements have the avowed purpose of trying to foster social learning and change, most of the barriers to social learning can be examined by analyzing why these social movements have so much difficulty effecting change. For example, the Civil Rights Movement attempts to change the way people think about and act toward minorities, especially blacks. The Women's Movement attempts to change the way people think about and act toward women. The Environmental Movement attempts to change the way people think about and act toward the environment, especially the natural environment. All three of these movements have been partially successful; we do think differently about blacks, women, and the environment than we did three decades ago. Public opinion polls disclose that strong majorities believe blacks and women should be treated equally and the environment should be protected. Yet, the movements are far from completing their tasks; blacks, women, and the environment are still discriminated against in our thinking and especially in our behavior. The following discussion refers most directly to the Environmental Movement, but much is applicable to all movements.

The greatest barrier social movements face is that they challenge the status quo way of perceiving, thinking, and behaving. I listened to a black speaker in the late 1960s who demanded that his audience repeat after him, "Blacks are beautiful." Of course I had heard the phrase before, but I thought it was someone else's phrase; no one had ever commanded *me* to say it; the best I could manage was a mumble. The experience made me aware how deeply prejudice was buried in me and my culture. Now, after some learning, I would have no difficulty saying *and believing* it; but my old way of perceiving and thinking had to be shaken vigorously and emotionally before I could come to that realization.

Similarly, when I first heard about the environmental problem, I thought it was an important and urgent problem, but I perceived it as someone else's problem—that of the engineers, for example. As I participated in many hours of discussion with an interdisciplinary study group in 1970, my growing awareness transformed the problem into *my* problem; it moved from my head to my gut and I became personally committed. Many people, most in fact, are environmental sympathizers—but they perceive the environmental problem only with their intellect. They see it as an important problem but they do not see it as *their* problem; it does not gnaw in their gut. In truth, they really do not have a deep understanding of it.

A key insight developed by our study group was that we had gotten into our environmental predicament by developing our kind of society (the DSP society) and that we would find our way out of the predicament only by changing the way people perceived, believed, valued, and acted toward the environment—it would be necessary to change our society. Perceiving the complexity and interconnectedness of the many parts of the environmental problem can easily lead one to feel overwhelmed. At one point I said to my colleagues, "You want us to change our whole society around; we don't know how to do that."

Despite our recognition of the formidable size of the task, our study group discussed what we might do to stimulate social learning and sociopolitical action to make the needed changes. We realized that the average person would not have the training, time and patience to perceive the complexity of the problem. But we reasoned that scientific, technical, business, and governmental elites would be different; they not only would have the requisite understanding of the physical world and of social forces but they also would have the power and responsibility to take effective action. We reasoned, further, that if we could just explain to them what we had discovered, they would see both the urgency for action and the need for appropriate corrective steps. This would be a fast and effective way for society to learn. Alas, we were wrong; elites were not ready to listen then—they still are not ready to listen.

I have now concluded that social learning and social change cannot be brought about simply by presenting research findings, facts, scientific arguments, reasoning, and lectures—education, alone, is insufficient for social learning. Of course, solid evidence is needed as underpinning for any thought or argument that is to survive challenge. But, by itself, evidence has little impact on social learning. Elites turned out to be no different from the masses; they are just as impervious to evidence and rational arguments as anyone else. Let us examine some of the reasons for this imperviousness.

Beliefs are embedded in belief systems that are acquired and reinforced over long developmental periods. Social scientists refer to this developmental process as the socialization process. Once a belief system is set— usually in a person's twenties—potential new beliefs (new information) are screened for compatibility with the already existing belief system; those which do not fit typically are coded as unbelievable and discarded. Belief systems are the essence of a person, the central core of one's personality, and are defended as though one were defending one's own life. When we use such shorthand terms as *liberal, conservative, Christian, Buddhist*, and *environmentalist*, we are actually talking about a belief system. Once such a belief system is set, this key element of the personality is invested with considerable emotion; it may be amended but it is seldom radically changed. This is a main source of stability of personalities, institutions, and whole societies.

Belief systems are defended in a variety of ways. Dissonant information does not fit into one's belief system and is likely to be perceived as incorrect. It also may be seen as being put forward by evil people (communists and secular humanists are current scapegoats in the United States). We take great pains to inculcate our children with our own beliefs, which we presume to be impeccable. This takes place not only in the home but in social institutions such as churches and schools. Not only do we feed our children a steady diet of our cherished beliefs but we guard against their being exposed to competing beliefs. In Tennessee in early 1986, a group of parents brought suit in court seeking control of the materials their children would be exposed to in public school. They especially wanted to be sure they would not be exposed to information about evolution, which they perceived as theatening their literal belief in the story of creation in the Bible.

A less self-conscious defense of belief systems is embedded in the structure of discourse that is common in a community. Many central characteristics of societal belief systems are so basic they are simply assumed. Knowing key assumed concepts is fundamental to grasping the meaning of the discourse, but the assumptions are seldom stated or defended. Environmentalists have begun challenging some of the assumed values and beliefs of the DSP that underlie the discourse of modern industrial society. These challenges typically are reacted to defensively with derision, disbelief, and dismissal. For example, growth is an honorific word in modern industrial society; it is assumed that everyone wants it. When environmentalists say we must limit growth in population and economic throughput, they are reacted to not in terms of the quality of their evidence but rather in terms of their challenge to a cherished belief. Those "believing in" growth are likely to perceive the idea of limits to growth as dissonant with their cherished belief, not in terms of its validity. They will seek a variety of grounds (poor information, evil messenger) for dismissing the evidence and the argument.

As another example, consider the high valuation modern society places on science and technology. It is assumed that their promotion and development will be good. When environmentalists challenge this assumption, when they question the wisdom of developing nuclear power or bioengineering, for example, they are challenging the whole thrust of the society. They are likely to be labeled as naive, soft-headed, and kooks. They will be seen as being on the margin of society; we could say they will be marginalized.* Marginalizing is a societal mechanism for trivializing and dismissing discordant beliefs. Discordant beliefs upset many people; they may actually fear them. Being able to dismiss them as being on the margin is reassuring to the fearful.

Most individuals, and most communities, seem to have a "latitude of acceptance" for unusual beliefs and values. If a belief is offered that falls within the latitude of acceptance there is a chance that it will be considered. If it falls outside the latitude of acceptance it is likely to be rejected. Attitude research suggests that attempts to persuade people to a belief that clearly falls outside the latitude of acceptance may instigate a "boomerang effect" and the two parties may perceive themselves as being farther apart than they were before (Sherif and Hovland, 1961; Brehm, 1966).

The dominant belief system of modern society is reinforced dozens of times a day by modern advertising. We are constantly told that the purchase of some product or service will make us happier. The advertiser may primarily be interested in selling his wares, but indirectly a belief system is being reinforced—and sold: happiness and quality of life, says the advertiser, can be realized by acquiring goods, becoming richer, winning, being competitive. In addition, to its manifest intention, advertising is a political instrument wielded in defense of the DSP.

Similarly, people who work in an organization encounter certain beliefs and values that are constantly reinforced by the discourse of the work community. Alvin Toffler (1984) did an in-depth study of the breakup of the American Telephone and Telegraph Company (AT&T). He reported, "busi-

---

*Gitlin, (1980), reports the way the media and mainstream society marginalized the Vietnam protests of the late 1960s and early 1970s.

ness has a belief system—and it is at least as important as its accounting system or its authority system" (p. 31). As the AT&T leaders struggled with the breakup, they had special difficulty with obsolete corporate beliefs because the most important corporate assumptions were least discussed. I find, for example, that when I interact with people in a business corporation, and hope to be understood, I must use different words and a different frame of reference from those I use in a university or environmental milieu. Business executives would probably also need different words and a different frame of reference in an environmental community. By the way, Toffler also concluded that a new paradigm is emerging to challenge the dominant worldview "that will threaten all our basic institutions, just as the Industrial Revolution threatened and eventually transformed all institutions of the feudal society." (p. 3)

The people who join social movements and try to overcome these belief defenses find that there is little they can do to open the minds of people who are strongly committed to an alternative belief system. Environmentalists have little hope of influencing strong believers in the DSP. They might concentrate their messages on those who are less committed, but most of those people are not very attentive. Typically they are so wrapped up in personal problems that they seek escape, not immersion in a new set of public problems.

Movement people typically follow several tactics in trying to reach the masses. First, they must keep up their own spirits, so they meet regularly to reinforce each other in their own beliefs. They hold conferences and try to involve new people as well as encourage and revitalize current members. Most of the time they are talking to those already converted. They form networks ("networking") with like-minded organizations and individuals. They typically exhort each other to "organize more effectively." They try to recruit and hold new members. Naturally, they try to attract as much media attention as possible but, even on television, they face the fatigue and boredom that burdens most viewers who tune out unpleasant environmental messages.

Environmental groups, especially, face large barriers in trying to reach people. Their message typically is one of some kind of failure or tragedy: climate change, a toxic spill, radon gas in homes, pesticide poisoning of humans and animals, limits to growth, loss of a wetland, parkland or wilderness, loss of the ozone layer, loss or poisoning of water—the list is endless. Large tragedies like the nuclear accidents at Chernobyl and Three Mile Island, the chemical release at Bhopal, and the gas explosion in Mexico City could make us think of even worse tragedies lying ahead, but we do not dwell on the possibility; we tune it out as unthinkable. How do we get people to think about the unthinkable? Should we try?

Furthermore, the actions environmentalists frequently propose are to stop someone from doing something: don't take parkland for a highway; don't take wilderness or a wetland for development; don't have so many children; don't drive gas guzzling automobiles; stop emitting a pollutant; stop making or using certain products. No matter that they are well-intentioned and probably right, environmentalists are likely to be perceived as more negative than positive. And who likes negative people? Environmentalists are trying hard to overcome the negative connotation of many of their messages by proposing constructive alternatives to prevailing practices. This is important for recruiting new supporters and for maintaining the morale of present workers in the movement. Regenerating a viable ecosphere is a huge project requiring dedication and energy from many people who hold a common vision as to where they are going. (This book offers one such vision.)

Despite barriers to acceptance, there is considerable awareness of environmental problems and willingness to support environmental policies. About 70 percent to 80 percent of the American people think environmental problems are serious and deserve urgent attention. A similar percentage are sympathetic to the environmental movement. The percentages in England and Germany are comparable (Milbrath, 1984). Public sentiment is so favorable to environmentalism that former President Reagan felt he must declare himself to be an environmentalist even though he generally has obstructed environmental protection. Somewhere between 15 and 30 million Americans belong to some kind of environmental group; probably 400,000 to 500,000 people are active environmentalists. In Germany, more people belong to environmental organizations than belong to political parties. What is it that makes for social learning about environmentalism despite all the barriers the movement faces? We will return to this question later in the chapter.

Barriers Faced By New Political Parties

Should a social movement become a political party? Will it be more effective in stimulating social learning and social change if it becomes a party as well as a movement? Running candidates in political campaigns provides occasions for speeches, debates, media attention—all of which are opportunities to stimulate social learning. When a person chooses between parties in an election, the programs of the parties are more consciously considered by the voter—that decision process also promotes learning. Winning an office puts a person into a position where votes can be cast on important questions; that office provides opportunities to file legislative bills as well as opportunities to promote policies in public forums—thus creating more stimulation to social learning. A party that could take control of a government could pursue policies that would help to make its belief system dominant; that would be a strong producer of social change.

Parties are a focus of identity and allegiance, and they are quite important for some people. Most movements try to draw supporters from all parties; that way they will still be heard no matter which party is in office. Therefore, making the transition from movement to party is fraught with difficulties. A new party growing out of a movement may not be able to enlist those movement supporters who are strongly attached to a traditional party. The type of party system also is critical; the "winner-take-all" type of party system used in the United States and Great Britain makes it extremely difficult for a third party to get a foothold. Third parties have tried again and again to win elections over the past 200 years in the United States without any significant success. Currently, a group of "greens" are meeting and discussing the possibility of forming a green party in the United States; most of the public have never heard of them.

Another problem is the necessity for a movement's ideology to be watered down as it appeals for votes and/or if it shares with other parties in forming a government. This problem is probably best analyzed and illustrated by taking a closer look at the German Greens (Die Grunen).

Green parties now field candidates in several European countries (Austria, Great Britain, France, Belgium, The Netherlands, Luxembourg, Italy, the Scandinavian countries, Greece), but the most successful, by far, are the West German Greens. These parties are the newest actors on the political scene in Europe and have been closely watched and studied. (The most thorough study written in English is by Capra and Spretnak, 1984.) The German Greens call for a new society; they reject many of the values and beliefs of industrial/materialistic society. They refuse to place themselves on the left-right political dimension; one of their mottos is, "We are neither left nor right; we are in front." Their ideology (their social paradigm) is based on four principles: ecology, social responsibility, grassroots democracy, and nonviolence. (Some recent literature adds equality of the sexes.)

The Green Party is organized, somewhat autonomously, in all the states of West Germany; it has won some parliamentary seats in nearly all of them. In March 1983, it won 5.6 percent of the popular vote and twenty-seven seats in the Bundestag. It was the first new party to win seats there in more than thirty years. Its electoral success stunned political observers and also the leaders of the major parties. It certainly made the country take notice. Since then, it has experienced factional rivalries and it has suffered some reversals in recent state elections; yet, it managed to increase its percentage of the popular vote to 8.3 percent and its seats to forty-two in the January 1987 Bundestag election. These factional rivalries illustrate some of the difficulties encountered when a movement becomes a political party; they stem from trying to maintain ideological purity while simultaneously appealing for votes and power.

The Green Party challenges capitalism more fundamentally than the Marxian parties do, and almost as vigorously; naturally, some believers in Marxian ideology joined the Greens. The Marxian faction readily accepted the ecological perspective and the valuation on social responsibility (social justice) of the Greens. However, the Marxians wanted to dilute or repudiate the Green emphasis on grassroots democracy and nonviolence because Marxism teaches that parties should be led by a small cadre and should seize power by violent revolution. The Marxists did not advocate violent seizure of power in the near future but they did not wish to exclude it as a possible future tactic. This Marxian challenge was eventually defeated within the German Green Party, but only after considerable internal party struggle. Any incipient party, especially one that shows promise of winning elections and believes in grassroots democracy, must worry about takeover from within by persons with beliefs and values that are different from those held by the original cadre that formed the party.

How strongly should a party value ideological purity? Some Greens valued this so highly that they hoped their party would always be in opposition. Others desired to join in a coalition with the Social Democrats in order to participate in forming a government. (In several states the Greens hold swing votes between the two large parties; joining a coalition could help the Social Democrats to form the government.) This would give them the capability to force environmental problems onto the agendas of other parties and the government. The Green emphasis on limits to economic growth is not very attractive to the proletariat in the Social Democratic Party, so coalition is not easy. Those Greens who emphasized ideological purity became known as the Fundis (fundamentalists) and those who emphasized power came to be known as the Realos (realists). These two factions have battled vigorously for party dominance. The May 1987 National Green Convention elected eight Fundis to the eleven-member national leadership committee. As this book goes to press, the Realos had recaptured the party leadership. There is a possibility that the party will split over this issue. The fight also has damaged their image with the German electorate and may lead to losses in the next election.

The people in the German Green Party who emphasized grassroots democracy were fearful that certain leaders would come to dominate the party. They adopted the principle that Green public office holders should give up their office half way through the term so that other party members could rotate into the office. Rotation provided valuable experience to a wider group of leaders and helped avoid one set of leaders dominating the party. This policy, however, left the party without a clear spokesperson, or leadership cadre, for public statements. Each position had to be thoroughly discussed at the grassroots level before a public stand could be taken. Endless hours were devoted to these grassroots discussions. In a legislative body, it takes some time to develop knowledge of the issues; it also takes time for other legislators to develop trust in what the newcomer commits to. Rotation out of office interfered so seriously with this set of needs that the policy was abandoned.

The internal dynamics of the Green Party are so complex they cannot be thoroughly discussed in these few paragraphs. Nor is it possible to make any clear predictions about their future course or prospects for electoral success. The party seems to have considerable difficulty dealing with power. It is not sure it wants to grasp for power. It is not sure how to keep out disruptive factions. It fears becoming institutionalized and developing an oligarchy. It appears that the Greens are effectively using the party to nourish social learning. They do not present a clear image to the public, however, and this may interfere with their ability to lead the public in the direction they wish them to go.

In the United States, the party system is less amenable to starting a new party; therefore, most environmentalists have chosen to work within the present parties and social institutions. In effect, they have chosen to work for gradual reform while biding their time in seeking sweeping social change. They face all the obstacles discussed above under social movements. They have had considerable success in amending certain deleterious business and governmental practices but they have made almost no headway in changing the basic structure and thrust of society.

Some suggest that U. S. environmentalists should focus their energies on trying to get one of the major political parties (the Democratic Party would seem to be most hospitable) to turn more forthrightly in a green direction. After the defeat the Democrats took in the 1984 presidential election, the party was in considerable disarray. Several nationwide study commissions probed for new directions for the party. This seemed like a good chance to introduce some green ideas into those deliberations and a few environmentalists did just that. Their messages fell on deaf ears, however. Even if party leaders or candidates were favorable to an environmental thrust, they did not seem able to communicate it effectively to the public within the frame of reference that dominated public discourse in the United States in the mid-1980s. The 1988 elections have come and gone, and still little attention is being paid by the political parties to the dire predictions of environmentalists about the future of the planet—even though many people perceived the drought of 1988 to be caused by the greenhouse effect.

Summary On Barriers To Social Learning

1. The most pertinent questions are, "Will our present society be able to sustain itself?" "If not, how are the masses and elites likely to respond to their perception of system failure?" "Are they likely to come to the realization that sweeping social change is necessary?" "If so, how long is that paradigm shift likely to take?"

2. Nature itself will mainly determine the answer to those questions; she will be our most powerful teacher.

3. Most societies and their supporting social paradigms are stable structures and they resist challenge by new ideas that would destabilize the structure.

4. Yet, history tells us that societies do swiftly make sweeping change.

5. Adaptive evolutionary transformation in physical systems is slow, unpredictable, and undirectable by humans. Evolution of a social system is likely to be different; humans can participate in and direct their social evolution.

6. Swift change can be induced by new technologies but we should ensure that the long-range effects are good for the biosphere and society.

7. Intentional efforts to direct and accelerate social change are the most relevant for extricating ourselves from our environmental predicament and are the focus of this chapter.

8. Efforts to "educate" people to social learning are unlikely to be successful, at least in the short run. Only as this education is carried out over a very long time is it likely to have much effect. It is not an effective instrument for displacing a dominant social paradigm with another.

9. Social movements will have some success in reaching a small cadre of people who already are tuned to receive their message; these people can be mobilized and energized. Social movements cannot reach the mass of people, however, because most people do not tune in to their messages and do not have a frame of reference for receiving and deriving meaning from them. The movement will be marginalized. The usual strategies and tactics of social movements will have little impact as they struggle to become a dominant social force.

10. Those political parties that urge a new social paradigm will have no greater success than social movements in reaching the masses to

stimulate their social learning. Should the masses begin to accept the new message, a party could be a useful vehicle for propagating it further.

11. In the present climate of political discourse and frame of political reference in the United States, there is almost no chance for one of the major political parties to turn in a green direction.

12. Therefore, as long as the current DSP-dominated society continues to function reasonably well, it will be impervious to efforts to redirect it toward a new social paradigm.

What Conditions Are Likely To Bring About A Societal Readiness For Social Learning?

The evidence and arguments presented in Part I constitute my case that our present societal thrust is unsustainable. This is not my conclusion alone; many thousands of social, natural, and life scientists have arrived at the same assessment. I also recognize that many thousands of other observers believe that our present societal thrust is sustainable. I tried to show in the previous chapter that, even if they are right, we would not like to live in the kind of environment that would be created. I urge skeptical readers to accept, for the moment, my conclusion about unsustainability in order to follow my argument.

In the previous chapter, I also cautioned against waiting for a crisis before amending our beliefs and behavior patterns. If we wait that long, we may be unable to take corrective actions to avoid even deeper troubles. Yet, I have just concluded that the educational efforts of scientists and participants in social movements are unlikely to have any appreciable effect as long as our present society is working reasonably well. How, then, is social learning going to come about before it is too late? We must be aware of the possibility that people could fail to wake up in time. All I can do is *suggest some possible ways* that will open our collective mind so that social relearning will take place.

With the possible exceptions of nuclear war and climate change, our physical environment is unlikely to deteriorate suddenly. Nature can recover from such major accidents as those at Bhopal or Chernobyl without permanent loss of carrying capacity. A likely scenario is a slowly accelerating cascade of unfortunate developments. In one week, for example, the following news stories were aired: depletion of the ozone layer, renewal of concern about a gasoline shortage, radioactive cesium poisoning of reindeer in Lapland, forest death due to acid clouds and ozone, drought in the midwest, cancer threat from radon gas in homes, dangers from pesticide use in

MDCs and LDCs, population growth and soil loss, famine in Ethiopia, an oil spill poisoning drinking water along the Ohio River. As we struggle to deal with these problems, people will slowly come to realize that technology alone is insufficient to handle them and that we will have to make major social changes in order to cope.

As these stories accumulate, people are beginning to realize that the world no longer works well; that their future is in serious jeopardy. Scientists and environmental movement activists are likely to stand ready with plausible explanations for these phenomena. Minds that formerly were closed and unheeding are more likely to be searching for understanding. Not until then can we expect much social learning to take place. But, even then, we should expect the social system to resist change strenuously.

All social structures try to protect their integrity and continue to exist. Nations view their sovereign independence as their highest value and are always willing to go to war to protect it. All bureaucracies try to protect themselves and to grow if possible. Organizations hate to die and will languish for years after their initial purpose has been fulfilled; or they may transform. Cultures, too, have numerous defenses to protect their integrity, and change ever so slowly.

We need to remind ourselves that despite the natural tendency of social structures to protect their current structure and thrust, they can transform, and the transformation can be rather swift. When a society's dominant social paradigm is displaced by a new social paradigm, we call the process *paradigm shift*. I mentioned earlier in this chapter that both China and Iran underwent paradigm shifts in the recent past. Understanding the dynamics of paradigm shift will help us see how social transformation could occur rather swiftly.

We gain some understanding by examining paradigm shifts within scientific disciplines (Kuhn, 1962, articulated the accepted description of this process). The story of paradigm shift that led to the acceptance of the theory of plate tectonics within geology is illustrative. Wegener had proposed the idea of continental drift in 1912. His critics, operating within the dominant theory of physical forces, dismissed his theory on the ground that no physical forces on the planet were large enough to move a continent. Wegener got bogged down in trying to respond to the wrong question put by his critics; the theory that continents were fixed in place continued to dominate the discipline. Wegener died in 1924 while his theory was still disparaged. Better instrumentation, new studies, greatly improved maps of the ocean floor kept producing findings that did not fit well with the theory of continental fixity. The structure of beliefs within the geological discipline became increasingly perturbed. Eventually, in 1968, a seminal paper by Isacks, Oliver and Sykes once more proposed the theory of continental drift but in the context of a theory explaining that the Earth's crust is made up of plates that slowly drift on a molten core. The theory could now be accepted, partially because the data supporting it was more impressive, but, most important, the theory made more sense of a whole host of studies. Wegener was right after all. The greater power and parsimony of the new plate tectonics theory swept the old aside within the space of a couple of years. A whole discipline turned around.

A related belief change phenomenon, not quite of the magnitude of a paradigm shift, is exemplified by several recent instances in which oppressive dictatorships have been displaced by an aroused citizenry; this type of change took place in Spain, Portugal, Argentina, and the Phillipines. The ouster of dictator Ferdinand Marcos from dominance in the Phillipines is especially instructive. He controlled the police, the army, the election process, and most of the media. Nearly every political observer, as late as mid-January 1986, was confident that Marcos's control was so complete that he would retain office; presumably, most of the common people thought so also. Yet, within one month, he had lost such power that he had to flee the country. In a sweeping and swift transition in perception and belief, the country headed in a new direction. In the United States, a seemingly impregnable President Nixon was swept from office in 1974, even without impeachment proceedings being undertaken; the country literally changed its mind about him in the space of a month.

These illustrations lead to two important conclusions: 1) we should never perceive a belief structure or paradigm as the final and immutable truth; 2) we should never perceive that a paradigm is so entrenched that it cannot be displaced.

Creative Evolution

Using linear thinking, we expect that a social system with many dysfunctional components, and experiencing high perturbations, would fly apart or collapse. Ilya Prigogine, a famous Belgian physical chemist and Nobel Laureate, discovered that the expected outcome does not necessarily occur; even nonliving physical systems display "self-organization." He studied "dissipative chemical structures" and discovered that they display the dynamics of self-organization in its simplest form, actually exhibiting most of the phenomena characteristic of life—self-renewal, adaptation, evolution, and even primitive forms of "mental" processes. The only reason they are not considered alive is that they do not reproduce or form cells.*

---

*Most of Prigogine's writing is in French, an English summary statement was published in 1980; see also Prigogine and Stengers (1984).

Drexler has an exceptionally clear description of the way crystalline order can arise from chaos:

Order can emerge from chaos without anyone's giving orders. . . . Imagine a molecule perhaps regular in form, or perhaps lopsided. . . . Now imagine a vast number of such molecules moving randomly in a liquid, tumbling and jostling like drunkards in weightlessness in the dark. Imagine the liquid evaporating and cooling, forcing the molecules closer together and slowing them down. . . . They will usually settle into a crystalline pattern. . . .

Crystals grow by trial and error, by variation and selection. No tiny hands assemble them. A crystal can begin with a chance clumping of molecules: the molecules wander, bump, and clump at random, but clumps stick best when packed in the right crystalline pattern. Other molecules then strike this first, tiny crystal. Some bump in the wrong position or orientation; they stick poorly and shake loose again. Others happen to bump properly; they stick better and often stay. Layer builds on layer, extending the crystalline pattern. Though the molecules bump at random, they do not stick at random. Order grows from chaos through variation and selection. (Drexler, 1986, p. 22)

Prigogine extrapolated a theory from his work with dissipative structures, which has been elaborated by other theorists into an overall systems theory focusing on the dynamics of self-transcendance. Prigogine and another physical chemist, Manfred Eigen, were contributors from chemistry. Biologists Conrad Waddington and Paul Weiss, anthropologist Gregory Bateson, and systems theorists Erich Jantsch and Ervin Laszlo were other significant contributors. Jantsch's *The Self-Organizing Universe* (1980) is the most comprehensive synthesis of the many aspects of the overarching theory. According to it, physical systems, living systems, and social systems are all creatively evolving according to the same process, one that derives order from fluctuation. This theory can help us to understand how social learning of a new social paradigm can come about, despite the barriers detailed above.

The second law of thermodynamics (the law of entropy) states that matter and energy constantly move from a more organized to a less organized state. When a gallon of gasoline is burned, for example, it does not just disappear, but is transformed and dispersed through heat and waste (pollution); it cannot do any more work because it is no longer organized. A closed mechanical system comes to complete rest when all its elements have been entropically scattered. A mechanical clock, for example, will stop if some external form of energy is not applied to wind it up occasionally. If it were simply abandoned out in the weather, it would disintegrate in a few decades and become so scattered (entropic) as to be useless for any human purpose. It has been asserted that this is the ultimate fate of our planet and the universe. Prigogine's discovery shows that the second law does not apply, as previously interpreted, to open nonequilibrium systems that have some way of importing energy or information.

Prigogine, Jantsch, and others perceive that many systems are not closed, do not seek equilibrium, and are evolutionary; these attributes apply to galaxies, stars, planets, cultures, social organizations, living organisms, and spiritual ideas. If a system of any kind is in a sufficiently nonequilibrium state, has many degrees of freedom, and is partially open to the inflow of energy (information) and/or matter, the ensuing instabilities do not lead to random behavior; instead, they tend to drive the system to a new more complex dynamic order. The system acquires new margins to produce entropy, new possibilities for action. Such systems are characterized by a high degree of energy exchange with the environment and are therefore called dissipative structures.

Thus, nonequilibrium thermodynamics leads to a theory of self-organization of physical systems. Its central principle is called "order through fluctuation" and is seen as the basis of a new general system theory valid in all domains: physical, biological, social, and spiritual. (Jantsch, 1975, p. 37) "The common denominator is always an open system far from equilibrium which is driven by fluctuations across one or more instability thresholds and enters a new co-ordinated phase of its evolution." (Jantsch, 1980, p. 73)

Scientists are rethinking their interpretation of evolution as received from Darwin. The classical neo-Darwinian theory sees organisms adapting themselves ever more perfectly to their environment and evolutionary progression toward an equilibrium. The new systems view calls it *creative evolution* which unfolds through an interplay of adaptation and creation and operates far from equilibrium. In this view, the environment is a living system that is also capable of adaptation and the focus shifts to the coevolution of organism and environment. The surviving entity is the organism-in-its-environment; organisms cannot survive at the expense of their environment.

In the systems view, evolution is open and indeterminate. When a system becomes unstable, it will probably evolve into a new structure; the further a system has moved from equilibrium, the more options will be available. As the unstable system moves to a critical point, it "decides" itself which way to go. The choice is truly free; there seems to be no goal or purpose, yet there is a recognizable pattern of development. Evolution is not blind chance but an unfolding of order and complexity that is a creative learning process. (I drew on Capra, 1983, pp. 287-89, for the perspective in the above two paragraphs.)

Communication plays a large role in the dynamics of self-transcendence. The dissipative chemical structures studied by Prigogine seemed to receive and react to information imported from their environment; yet they have none of the sensory organs possessed by mammals. Communication seems to be occuring everywhere in nature, from organism to organism, between organism and environment, across levels and species, across cultures. If we think about communication solely as messages received by one of the five senses, are we not taking too restrictive a view of its dynamics?

Let us return to the phenomenon of whole societies quickly making up their mind about something and turning in quite another direction. In the cases of China and Iran, they went through paradigm shift by displacing one dominant social paradigm with another. This has happened during previous historical eras as well. French society, during its revolution 200 years ago, experienced severe social turmoil that eventually resulted in deposing their king, permanently displacing the aristocracy, and uniting its various provinces into a single French identity. That identity is still one of the strongest in the world.

Jules Michelet spent many years in the French National Archives deeply immersing himself in the records of the Revolution. The following are brief excerpts from his description of the Festival of the Federations that took place on July 14, 1790 (one year after the storming of the Bastille); to Michelet, this was the zenith of French history. He tells a fascinating story of swift societal transformation. The translation into English was made by Charles Cocks in 1847.

France is born and started into life at the sound of the cannon of the Bastille. In one day, without any preparation or previous understanding, the whole of France, both cities and villages, is organized at the same time. . . .

All that had been believed painful, difficult, and insurmountable, became possible and easy. . . .

Is all this a miracle? Yes, and the greatest and most simple of miracles, a return to nature. . . . The rivers, for instance, which under the ancient system, were scarcely better than obstacles . . . became once more what nature had intended them to be, the connecting bond of mankind. . . .

. . . All at once, and without even perceiving it, they have forgotten the things for which they would have sacrificed their lives the day before. . . .

In the villages, especially, there are no longer either rich or poor, nobles or plebians; there is but one general table, and provisions are in common; social dissensions and quarrels have disappeared; enemies become reconciled; and opposite sects, believers and philosophers, Protestants and Catholics, fraternise together. (University of Chicago Press edition, 1967, pp. 433-51.)

Michelet's language is indeed flowery and passionate but there can be no denying that the people of France experienced in a short space of time an extraordinary mass communication that transcended all conventional modes and caught people up in a new belief about their society that swept away the old order. We know that these societal transformations take place occasionally, but we cannot explain the dynamics behind them.

On a more regular basis we can detect mood swings in public opinion. For example, in the early 1980's, former President Reagan exploited a mood favoring a buildup of U. S. national defenses and convinced Congress to support massive infusions of money into the military establishment. By the later part of the 1980s the mood had swung back and people favored reductions in defense spending. Although we can observe these mood swings, we do not have a good theory that allows us to predict them.

People working on "the Hill" in Washington observe mood swings; they become part of the cultural lore; workers on the Hill speak about an "economy mood" or "defense build-up mood." They philosophically ride with the tide and wait for a mood change before advancing their favorite project. Moods in Washington do not always reflect the mood of the people at large. The insularity of the Washington setting makes it difficult for workers there to know the mood of the country. The messages flowing from the media are not a good guide to a mood change either. Sometimes the public mood swings in a different direction than that advocated by most media messages.

The messages passed in daily conversations with family, friends, and associates are just as important as media messages. We have no way of monitoring such messages, but we can imagine that they vary greatly in subject matter and quality of reasoning. Conversational messages are filled with subtle cues that confirm beliefs and values accepted in the community; this is what we mean when we speak of "folk wisdom." These beliefs and values are so basic that they may not even be articulated, other than being brought in as subtle confirmation of "what everyone knows to be true."

Most of the normal verbal messages confirm and support the present social structure and the currently dominant social paradigm. How does it happen, then, that occasionally a society moves contrary to the message and "makes up its mind" to shift its beliefs about how the world works? The theory of creative evolution through fluctuation provides part of an answer. The theory asserts that open communication across organismic or systemic boundaries is essential to the evolutionary process, but it lacks an explanation as to how this communication takes place. Normal verbal communication, while important, is insufficient to account for the development of new understandings among people who have not exchanged verbal messages—as happened in the French Revolution. There must be something more. Several scraps of evidence provide hints for our search.

Nearly all of the world religions and great philosophical traditions imply, (usually they assert), that there is something more. They believe people can communicate with an entity, external to themselves and other people, and that they can acquire special wisdom by doing so. These philosophical and religious practices produce similar kernels of wisdom that we speak of as "the perennial wisdom." Religious and mystical communications have great meaning for some people. Science cannot confirm that such communications exist (are real) but neither can science prove that they do not exist.

Turning to another clue, Watson (1980) describes what has come to be called *the hundredth monkey phenomenon*. On an island near Japan, biologists introduced freshly dug sweet potatoes as a new food to some monkeys. The monkeys were reluctant to eat the dirty potatoes, as all their other food could be eaten without preparation. Eventually, one eighteen-month old female carried some potatoes down to a stream and washed them before feeding. She then taught this behavior to her mother and playmates who taught it to their mothers. The only adults to adopt the behavior were those who were taught by their children. Suddenly, however, a critical threshold was passed (perhaps the hundredth monkey learned it) and the behavior became universal, even among monkeys who had not observed the washing. The phenomenon could be extrapolated to people; when enough of us come to hold something to be true, it becomes true for most everyone. We even have a name for it; we call it, "An idea whose time has come." Ken Keyes Jr. wrote about the process by which we might learn to control nuclear arms and titled his 1982 book, *The Hundredth Monkey.*

As scientists have delved deeper and deeper in their study of nature, they have begun to question some conventional distinctions and constructs. For example, physical objects (for example, steel) that seem so solid and unchanging in everyday use, turn out to be very dynamic structures when explored at the subatomic level. The boundary between living and nonliving matter becomes more blurred with deeper study. Capra explains how living organisms that seem to differentiate so neatly into a multitude of species turn out to have very similar building blocks when their basic structure is studied closely.

As the notion of an independent physical entity has become problematic in subatomic physics, so has the notion of an independent organism in biology. Living organisms, being open systems, keep themselves alive and functioning through intense transactions with their environment, which itself consists partially of organisms. Thus the whole biosphere—our planetary ecosystem—is a dynamic and highly integrated web of living and nonliving forms. Although this web is multileveled, transactions and interdependencies exist among all its levels.

Most organisms are not only embedded in ecosystems but are complex ecosystems themselves, containing a host of smaller organisms that have considerable autonomy and yet integrate themselves harmoniously into the functioning of the whole. The smallest of these living components show an astonishing uniformity, resembling one another quite closely throughout the living world, as vividly described by Lewis Thomas: (Capra, 1983, p. 275)

There they are, moving about in my cytoplasm. . . . They are much less closely related to me than to each other and to the free-living bacteria out under the hill. They feel like strangers, but the thought comes that the same creatures, precisely the same, are out there in the cells of seagulls, and whales, and dune grass, and seaweed, and hermit crabs, and further inland on the leaves of the beech in my backyard, and in the family of skunks beneath the back fence, and even that fly on the window. Through them, I am connected: I have close relatives, once removed, all over the place. (Thomas, 1975, p. 86)

Marvelously complex structures have evolved from these basic building blocks. We saw in Chapter 4 when discussing the work of Margulis and Sagan (1986) that even such simple organisms as bacteria seem to have the capability to learn even though they have no brain and only very primitive sensory capabilities. They apparently have intense communication with their environment which enables them to transform in cooperation with it—they seem to learn.

This next theory is far from proven fact, and the idea has stirred considerable scientific controversy. I have decided to present it for my readers to consider, but I wish them to keep in mind that I am only *tentatively* entertaining it to explain how a society could swiftly learn to accept a new social paradigm. Our learning will be unnecessarily restricted if we consider only proven facts as possible explanations of phenomena.

Biologists have puzzled for a very long time how millions of diverse forms can emerge from the basic building blocks of nature. How does it happen that one part of our body becomes an ear and another a liver, even though each cell carries the same DNA structure? It is like an architect working with a blueprint to make a complex building from simple boards, bricks, and mortar. Some boards form doors and some frame windows. Similarly, life comes into form according to some blueprint, but where is the blueprint? Biologists cannot find it in the cells; so some of them have postulated that development into life forms is guided by a kind of "field." We have good evidence for the existence of gravitational fields or magnetic fields from the effects they create, even though we cannot physically apprehend them with any of our senses. Fields that guide development of life are called morphogenetic (*morphe* for form and *genesis* for coming into being). Morphogenetic fields, like blueprints, do not contribute any energy, yet they guide structure.

Alexander Gurwitsch first introduced the phrase "morphogenetic fields" in Russia in 1922; Paul Weiss independently arrived at the same concept in Vienna in 1925. Von Bertalanffy and Waddington worked with these ideas in developing general systems theory. British biologist Rupert Sheldrake has taken the concept further to suggest that morphogenetic fields can influence form across time and space. These scientists also speculate that fields can evolve, they can affect one another—they can "learn."

Sheldrake cites considerable experimental evidence in support of his "hypothesis of formative causation" in *A New Science of Life* (1981) and *The Presence of the Past* (1988). The 1981 book aroused considerable controversy. The Tarrytown Conference Center in New York has offered a $10,000 prize for a definitive experiment to refute or confirm Sheldrake's theory; a Dutch foundation has added another $5,000 (Wilson, 1984). The most controversial aspect of his hypothesis is his concept of morphic resonance, which holds that similar structures can communicate across time and space. If this is true, it would mean that people can share thoughts without being in direct verbal communication; one's mental activity influences others and also is influenced by others—the minds resonate. We are not really cut off from others; we are all part of humanity: "no man is an island."

Sheldrake suggests that people have to "tune in" for morphic resonance to occur; "we tune in to the memories of innumerable past members of our species . . . there is a kind of pooled or collective memory that we draw upon" (1984, p.8). If millions of people were to tune in at a time of great societal stress and fluctuation, could they think their way through to a new belief or a new belief structure (read, paradigm)? Could they do this within the space of a few weeks?

We are also responsible for our innermost thoughts. If we despair, feeling that there is nothing we can do, that it is all utterly hopeless and inevitable, this attitude itself may spread and influence others. And despair and hopelessness can only help to bring about what we most fear. But if we have hope, and faith in the possibility of a new order of things, not only will our actions be more positive, but our hope and faith themselves may spread. Our responsibility may indeed be even greater than we had supposed. (Sheldrake, 1984, p. 10)

A Plausible Scenario Of Paradigm Shift

Now we have enough pieces for imagining a plausible scenario for paradigm shift, despite well-known social structural barriers to change. I am not asserting that this is *the* way paradigm shift will come about; it probably could come about in several ways. I only suggest one way the barriers to social change could be surmounted, even in a short space of time, given the right circumstances.

Over the next twenty years or so we should expect extremely strong social forces determined to keep us on our present trajectory. Only major failure of physical systems, like swift climate change, could deflect their domination. The DSP will continue to dominate thinking and behavior in the MDCs and in many of the LDCs as well. The challenge to the DSP by NEP advocates will continue. They will point out the deleterious trends, accidents, and mistakes engendered by the DSP society; they will urge new ways of thinking and doing things. A sizeable minority of people will listen, and will come to agree with them, but the majority will marginalize the NEP advocates and tune out their messages. Environmental organizations will continue to educate, advocate, and struggle politically. Green parties will arise but probably will enlist the support of no more than a minority. Science will continue to develop new findings and interpretations, spurred by deepening problems in nature and by inadequacies in the human response to those problems.

Nature itself will be the most frequent and unsettling spur to new thinking. Climate change is likely to be the most insistent and persistent teacher. We also can expect increasing crowding and congestion; shortages of critical natural resources; frequent revelation of new dangers to human and biospheric health (toxics, pollutants, red-tide algae); increasing occurrence of catastrophic accidents; decline of wildlife and wilderness; and breakdown of traditional institutions (for example, family farms). Markets will fail to solve the linked problems of population growth and resource shortages. Most frustrating, but perhaps most stimulating to new thinking, will be the discovery that technological fixes will not be able to cope with the character and scope of the problems.

Societies facing these problems are bound to become more turbulent. Frustration will deepen as people discover that the society and economy no longer work the way they formerly did. Dissatisfaction with governments and other major institutions is likely to grow. Terrorism and other forms of destructive behavior may increase. Some people will lapse into lethargy and hopelessness. For many, however, this turbulence will heighten their sensitivity; they will be searching for new ideas.

This is the critical time for paradigm shift. In order to understand what is likely to happen at such a time, imagine that each mind is similar to a television set that has a few hundred channels (messages) it might tune to but it can tune in only one at a time. During normal times our minds tune to comfortable traditional channels and we only use a dozen or so of the hundreds of channels that are available. (When speaking of channels here I mean any mode of message transmission, not simply television transmission.) During times of turbulence and breakdown, however, the traditional channels are out of order or send messages that no longer make any sense; therefore, many millions of people will search quickly, even frantically, across hundreds of channels. Some of the messages will make more sense than those they have been hearing, and people will begin attending to them more closely. This searching will be somewhat analogous to the way molecules tumble about in a liquid and form themselves into crystalline patterns (as described by Drexler a few pages earlier). As more and more people tune to a new set of channels, the hundredth monkey phenomenon could come into play and soon everyone would be tuning to this new set of beliefs and values—a new paradigm. If events transpired in this fashion, sweeping paradigm changes could crystallize within a few months or years.

Perhaps what Sheldrake calls *morphic resonance* is the process by which the hundredth monkey phenomenon takes place. We do not need to decide if Sheldrake is right to hypothesize that in times of great turbulence there will be extraordinary communication between people. All of this mental activity could lead people to perceive a new pattern of meaning reflecting a new state of complexity—that could evolve into a new biological, social, economic and political order. Old explanations for phenomena will be discarded and people will wonder how they ever believed in them. Ideas that formerly seemed hopeless to get across now seem to make sense and are eagerly accepted. Certainly, we can expect that all this mental activity will lead to heightened sensitivity and to a cornucopia of ideas. Because the old DSP system no longer will be working well, fewer people will have a stake in its preservation—the rearguard will weaken while the vanguard will strengthen.

Both Jantsch and Elgin have sketched their interpretation of this transformation; their sketches are remarkably similar. In each sketch a societal system goes through a period of turbulence, finds a new structure (perhaps a new belief paradigm), and moves to a higher order with greater potential for good-functioning. (see Figures 18.1 and 18.2)

I mentioned above that a society can change its mind about something in a few weeks, such as in the Aquino displacement of Marcos in the Phillipines. We should not expect the NEP to displace the DSP nearly so quickly. In the Phillipines, they merely changed political leadership and adopted a new constitution; they did not change the basic beliefs of the society. The NEP and the DSP, in contrast, are deep beliefs about the proper relationship between humans and nature—with ramifications for economic structures, work patterns, recreational patterns, consumption patterns, engineering projects, social relationships, education, communication, politics, governance, even religion. Once a transformation of worldview is well underway, these patterns will give way to new ones that are more in accord with the new way of thinking. However, we should expect the physical transformation of our infrastructure and patterns of daily activity to take decades, perhaps a century or two. That transformation cannot make significant headway until our thinking (our belief pardigm) changes. I cannot overemphasize how important our thoughtways are; we need to explore and understand the universe out there as well as the universe within—as the following quotations attest:

Figure 18.1

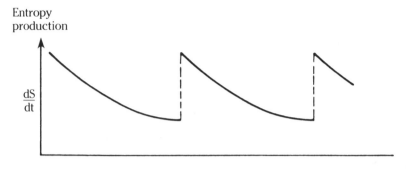

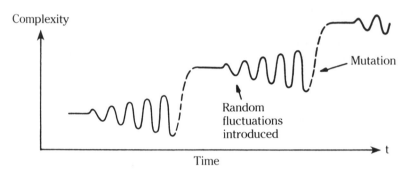

"Order through fluctuation." Sufficiently nonequilibrium systems (dissipative structures) mutate toward new dynamic regimes, which may be at a higher state of complexity, if random fluctuations are introduced. The new regime restores the system's capability for entropy production, which is first high and decreases with rising entropy during each dynamic regime.

Reprinted from Jantsch, 1975, p. 38.

Figure 18.2

SCENARIO II:
A PATH TO CIVILIZATIONAL REVITALIZATION

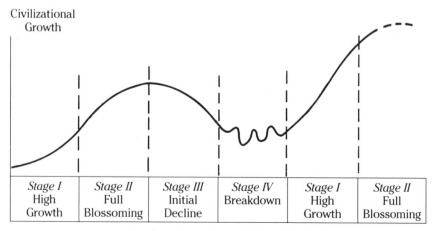

Stage I High Growth	Stage II Full Blossoming	Stage III Initial Decline	Stage IV Breakdown	Stage I High Growth	Stage II Full Blossoming

Reprinted from Elgin, 1981, p. 111.

The thought manifests as the word,
The word manifests as the deed,
The deed develops into habit,
And the habit hardens into character,
So watch the thought and its ways with care,
And let it spring from love,
Born out of concern for all beings. (Anonymous)

In choosing our beliefs we are therefore also choosing the images that will guide, create, and pull us along with our culture, into the future. "The world partly becomes—comes to be—how it is imagined." said Gregory Bateson, thereby echoing the words of the Buddha 2,500 years earlier, who said:

We are what we think.
All that we are arises with our thoughts.
With our thoughts we make the world.

... But there is a more skillful way of using ideals. This is to see them not just as goals that must be reached, but also as guiding images or visions that provide signposts and directions for our lives and decisions. Such images attract us to actualize them and ourselves. ... The world view of materialism says to explore the universe and thereby ourselves; the perennial philosophy says to explore our own minds and consciousness and thereby the universe. "Know thyself" is its central credo. But in practical terms it seems crucial that we do both. Our survival and evolution require no less than that we deepen our understanding of both the universe within and the universe without. (Walsh, 1984, pp. 82, 84)

What Do We Do Now?

The transition from DSP to NEP will not be easy and success is by no means assured. Those who are eager to move to an NEP society as quickly as possible will be disappointed that most people will not listen and that progress will be slow. It may be helpful if they can understand that modern society will first have to experience considerable failure and turbulence so that people's minds can be opened up. People will have to feel deeply the need for new understanding before they will open their minds and search for better ideas. All of this will take time.

But what do we do in the meantime? It is depressing to merely wait for people to open their minds. We do not need to sit idly by. Some people achieve a sense of satisfaction by becoming politically active. Another tactic is to do everything we can to promote social learning. We can try to reorient or redesign our institutions so that they learn more readily. We can study and do research. We can speak up against injury, foolishness, selfishness, injustice, waste, and tyranny. We can try to help our friends and neighbors to think anew about things; we can be a cell of sanity in our community.

We can learn to be more self-reliant; to live more simply, but richly. We can look deeply into our own lives, our own souls, to keep both the material and spiritual aspects of life in balance with each other—allowing each aspect to infuse and inform the other synergistically.

- A progressive refinement of the social and material aspects of life— learning to touch the earth ever more lightly with our material demands; learning to touch others ever more gently and responsively with our social institutions; learning to live our daily lives with ever less complexity and clutter.

- A progressive refinement of the spiritual or consciousness aspects of life—learning the skills of touching life ever more lightly by progressively releasing habitual patterns of thinking and behaving that make our passage through life weighty and cloudy rather than light and spacious; learning how to "touch and go"—to not hold on—but to allow each moment to arise with newness and freshness; learning to be in the world with a quiet mind and an open heart. (Elgin, 1981, p. 234)

Our common journey promises to be challenging and exciting even though difficult. It will be much easier and more likely successful if we face it optimistically with a deep understanding of the pace and character of social transformation. Those given the gift of understanding will become the conscious mind of the biocommunity, a global mind, that will guide and hasten the transformation. Those who understand what is happening to our world are not free to shrink from this responsibility.

References

Abram, David, "The Perceptual Implications of Gaia," *ReVISION*, vol. 9, no. 2 (Winter/Spring 1987) pp. 7-15.

Altieri, Miguel, *Agroecology: The Scientific Basis of Alternative Agriculture*, Boulder, Col., Westview Press, 1987.

Anderberg, Robert K., "Wall Street Sleaze," *The Amicus Journal*, vol. 10, no. 2, (Spring 1988) pp. 8-10.

Anderson, Walter Truett, ed. *Rethinking Liberalism*, New York, Avon Books, 1983.

Anderson, Walter Truett, "The Pitfalls of Bioregionalism," *Utne Reader*, 14 (Feb/March 1986), pp. 35-38.

Andrews, David, *The IRG Solution: Hierarchical Incompetence and How to Overcome It*, London, Souvenir Press, 1984.

Andrews, Frank C., "There is No Entropy Trap," *Environment*, 26, no. 2 (March 1984), pp. 4-6.

Argyris, Chris and Donald Schon, *Organizational Learning: A Theory of Action Perspective*, Reading, Mass., Addison-Wesley, 1978.

Axelrod, Robert, *The Evolution of Cooperation*, New York, Basic Books, 1984.

Bar-Tal, Daniel and Yoram Bar-Tal, "New Perspective of Social Psychology," in Daniel Bar-Tal and Arie W. Kruglanski, eds., *The Social Psychology of Knowledge*, Cambridge, Cambridge University Press, 1988.

Bartelmus, Peter, *Environment and Development*, Boston, Allen & Unwin, 1986.

Bateson, Gregory, *Steps Toward an Ecology of Mind*, New York, Ballantine, 1975.

Bell, Daniel, "The World and the United States in 2013," *Daedalus* vol. 116, no. 3 (Summer 1987), pp. 1-32.

Bellah, Robert N., "The Broken Covenant: American Civil Religion in Time of Trial," Ms.

Berry, Thomas, "The New Story: Comments on the Origin, Identification and Transmission of Values," The American Tielhard Association for the Future of Man, New York (Winter 1978).

————, "Bioregions: The Context for Reinhabiting the Earth," The American Tielhard Association for the Future of Man, New York (November 1984); also in *Hudson Valley Green Times*, vol. 7, no. 4, 1987.

————, "The Viable Human," *ReVISION*, vol. 9, no. 2 (Winter/Spring 1987a), pp. 75-81.

————, "Patriarchy: A New Interpretation of History," The American Tielhard Association for the Future of Man, New York, 1987b.

(Thomas Berry's essays have now been published in a single volume: *The Dream of the Earth*, San Francisco, Sierra Club Books, 1988.)

Berry, Wendell, *A Continuous Harmony: Essays Cultural and Agricultural*, New York, Harcourt Brace, 1970.

———, *The Unsettling of America: Culture and Agriculture*, New York, Avon Books, 1977, 2nd. ed.; San Francisco, Sierra Club Books, 1986.

———, *The Gift of Good Land: Further Essays Cultural and Agricultural*, Berkeley, Calif., North Point Press, 1981.

———, *Home Economics: Fourteen Essays*, Berkeley Calif., North Point Press, 1987.

Blank, Robert, "The Allocation of Scarce Resources: How Much is the Life of a Baby Doe Worth," Paper presented to the annual meeting of the American Political Science Assn., Washington, D.C., 1986.

Bolin, B. and R. Cook, eds., *The Major Biogeochemical Cycles and Their Interactions*, New York, Wiley, 1983.

Bolin, B., B. Doos, J. Jager, and R. Warrick, *Atmospheric Greenhouse Gases and Changing Climate*, New York, Wiley, 1986.

Bookchin, Murray, *The Modern Crisis*, Philadelphia, New Society Publishers, 1986.

Boulding, Kenneth, *The Meaning of the Twentieth Century*, New York, Harpers, 1964.

———, *Ecodynamics: A New Theory of Societal Evolution*. Beverly Hills, Calif., Sage Publishers, 1978.

———, "Confessions of Roots," *International Studies Notes*, vol. 12, no. 2 (Spring 1986), pp. 31-33.

Branch, Kristi, Douglas A. Hooper, James Thompson, and James Creighton, *Guide to Social Impact Assessment: A Framework for Social Change*, Boulder, Colorado, Westview Press, 1984.

Brehm, John W., *A Theory of Psychological Reactance*, New York, Academic Press, 1966.

Brown, Lester R., *Building a Sustainable Society*, New York, Norton, 1981.

Brown, Lester R., et al., *State of the World (1986): A Worldwatch Institute Report on Progress Toward a Sustainable Society*, New York, Norton: annual editions 1984, 1985, 1986, 1987.

Caldwell, Lynton K., *International Environmental Policy: Emergence and Dimensions*, Durham, N.C., Duke University Press, 1984.

Campbell, Donald T., "Reforms as Experiments," *American Psychologist*, vol. 24, no. 4 (April 1969), pp. 409-28.

Capra, Fritjof, *The Turning Point: Science, Society, and the Rising Culture*, New York, Simon & Schuster, 1982.

Capra, Fritjof and Charlene Spretnak, *Green Politics*, New York, Dutton, 1984. Updated paperback edition published by Bear & Co., 1986.

Carson, Rachel, *Silent Spring*, Cambridge, Mass., Riverside Press, 1962.

Cashman, Tyrone, "The Living Earth and the Cybernetics of Self," *ReVision*, vol. 9, no. 2 (Winter/Spring 1987), pp. 25-32.

Catton, William R. Jr., *Overshoot: The Ecological Basis of Revolutionary Change*, Urbana, University of Illinois Press, 1980.

Catton, William R. Jr., and Riley E. Dunlap "A New Ecological Paradigm for Post-Exuberant Sociology," *The American Behavioral Scientist*, 24, 1980, pp. 15-47.

Christie, W. J., M. Becker, J. W. Cowden, and J. R. Vallentyne, "Managing the Great Lakes Basin as a Home," *Journal of Great Lakes Research*, vol. 12, no. 1 (1986), pp. 2-17.

Cheng, Chung-ying, "On the Environmental Ethics of the TAO and the Chi," *Environmental Ethics*, vol. 8, no. 4 (Winter 1986), pp. 351-370.

Cohn, Carol, "Slick'ems, Glick'ems, Christmas Trees, and Cookie Cutters: Nuclear Language and How We Learned to Pat the Bomb," *Bulletin of the Atomic Scientists*, vol. 43, no. 5 (June 1987), pp. 17-24.

Cole, Leonard, *Politics and the Restraint of Science*, Totowa, N. J., Rowman & Allanheld, 1983.

————, "Resolving Science Controversies: From Science Court to Science Hearing Panel," in Malcom Goggin, ed., *Science and Technology in a Democracy: Who Should Govern?*, Knoxville, University of Tennessee Press, 1987.

Cotgrove, Stephen F., *Catastrophe or Cornucopia: The Environment, Politics and the Future*, Chichester/New York, Wiley, 1982.

Cousins, Norman, *The Pathology of Power*, New York, Norton, 1987.

Cronon, William, *Changes in the Land: Indians, Colonists and the Ecology of New England*, New York, Hill & Wang, 1983.

Dahlberg, Kenneth, *Beyond the Green Revolution: The Ecology and Politics of Global Agricultural Development*, New York, Plenum, 1979.

Dahlberg, Kenneth, ed., *New Directions for Agriculture and Agricultural Research: Neglected Dimensions and Emerging Alternatives*, Totawa, N. J., Rowman & Allanheld, 1986.

Daly, Herman E., *Steady State Economics: The Economics of Biophysical Equilibrium and Moral Growth*, San Francisco, Freeman, 1977.

Daly, Herman E., ed., *Economics, Ecology, Ethics: Essays Toward a Steady State Economy*, San Francisco, Freeman, 1980.

Devall, Bill, "The Deep Ecology Movement," *Natural Resources Journal*, vol. 20, 1980, pp. 299-322.

Devall, Bill and George Sessions, *Deep Ecology*, Salt Lake City, Gibbs Smith, 1985.

Domain Biodynamics Research Foundation, "A Global Agenda for Change: Review with Commentary of *Our Common Future*," Breslau, Ontario 1987.

Doyle, Jack, *Altered Harvest: Agriculture, Genetics and the Fate of the World's Food Supply*, New York, Viking, 1985.

Dregne, Harold E., "Soil and Water Conservation: A Global Perspective," *Intersciencia, Journal of Science and Technology of the Americas*, vol. 11, no. 4 (July-August 1986), pp. 166-172.

Drexler, K. Eric, *Engines of Creation*, Garden City, New York, Anchor Press/ Doubleday, 1986.

———, "Technologies of Danger and Wisdom," in Proceedings of: "Directions and Implications of Advanced Computing," Conference in Seattle, Wash., July 12, 1987 called by Computer Professionals for Social Responsibility. Copies available from Foresight Institute, Box 61058, Palo Alto, Calif. 94306.

Dutton, Diana, "The Impact of Public Participation in Biomedical Policy: Evidence From Four Case Studies," in James C. Peterson, ed., *Citizen Participation in Science Policy*, Amherst, University of Massachusetts Press, 1984.

Dunlap, Riley E., "Ecologist vs. Exemptionalist: The Ehrlich-Simon Debate," *Social Science Quarterly*, no. 64 (March 1983), pp. 200-03.

———, "Public Opinion on the Environment in the Reagan Era," *Environment*, vol. 29, no. 6 (July/August 1987), pp. 6-11, 32-37.

Dunlap, Riley and Kent Van Liere, "The New Environmental Paradigm," *The Journal of Environmental Education*, vol. 9, no. 4 (1978), pp. 10-19.

Dyer, Davis, et al., *Changing Alliances: The Harvard Business School Project on the Automobile Industry*, New York, Harper & Row, 1987.

Ehrenfeld, David, *The Arrogance of Humanism*, New York, Oxford University Press, 1978.

Ehrlich, Paul R., Carl Sagan, Donald Kennedy, and Walter Orr Roberts, *The Cold and the Dark: The World After Nuclear War*, New York, Norton, 1984.

Eisler, Riane, *The Chalice and the Blade: Our History, Our Future*, San Francisco, Harper & Row, 1987.

El-Ashry, Mohammed T., "Resource Management and Development in Africa," *Journal 86*, Annual Report of the World Resources Institute, Washington, D.C., 1986.

Elgin, Duane, *Voluntary Simplicity: Toward a Way of Life that is Outwardly Simple, Inwardly Rich*, New York, William Morrow, 1981.

Erhard, Werner, "The End of Starvation: Creating an Idea Whose Time Has Come," Essay distributed by The Hunger Project, 2015 Steiner Street, San Francisco, Calif. 94115.

Festinger, Leon, *A Theory of Cognitive Dissonance*, Stanford, CA, Stanford University Press, 1957.

Finsterbusch, Kurt, "State of the Art in Social Impact Assessment in the United States," Paper delivered at the 10th World Congress of Sociology, Mexico City, 1982.

Finsterbusch, Kurt and C. P. Wolf, eds. *Methodology of Social Impact Assessment*, 2nd ed., Stroudsburg, Penn., Dowden, Hutchinson & Ross, 1981.

Fisher, Jonathan and Norman Myers, "What We Must Do to Save Wildlife," *International Wildlife*, vol. 16, no. 3 (May-June 1986), pp. 12-15.

Fishkin, James S., *Beyond Subjective Morality: Ethical Reasoning and Political Philosophy*, New Haven, Yale University Press, 1984.

Flavin, Christopher, "Energy's Future: Small Looks Beautiful," *The Futurist*, XIX, no. 2 (April 1985), pp. 36-44.

_____ , "How Many Chernobyls?" *World Watch*, vol. 1, no. 1 (January-February 1988), pp. 14-18.

Forrester, Jay, "Planning Under the Dynamic Influences of Complex Social Systems," in G. Kepes, (ed.) *Arts of the Environment*, New York, Braziller, 1972.

Franklin, W. E., M. A. Franklin, and R. G. Hunt, *Waste Paper: The Future of a Resource, 1980-2000*, American Waste Paper Institute, 1982.

French, Marilyn, *Beyond Power: On Women, Men and Morals*, New York, Summit Books, 1985.

Fromm, Erich, *The Sane Society*, New York, Fawcett, 1955.

Gilder, George, *Wealth and Poverty*, New York, Basic Books, 1981.

Gitlin, Todd, *The Whole World is Watching*, Berkeley, Calif., University of California Press, 1980.

Ginsburg, Helen, *Full Employment and Public Policy: The United States and Sweden*, Lexington, Mass., Lexington Books, 1983.

Gordon, Gil, "Telecommuting: What Happens When We Stop Going to Work," *Profiles*, vol. 2, no. 9 (May 1985), pp. 30-34, 76.

Gorz, Andre, *Ecology as Politics*, Trans. by Patsy Vigderman and Jonathan Cloud, Boston, South End Press, 1980.

Gurr, Ted Robert, "On the Political Consequences of Scarcity and Economic Decline," *International Studies Quarterly*, vol. 29, 1985, pp. 51-75.

Handy, Charles, *The Future of Work*, New York, Basil Blackwell, 1984.

Hardin, Garrett, "The Tragedy of the Commons," *Science*, vol. 162 (December 13, 1968), pp. 1243-48.

_____ , *Exploring New Ethics for Survival*, New York, Viking, 1972.

_____ , "An Ecolate View of the Human Predicament," in Claire N. McRostie, ed., *Global Resources: Perspectives and Alternatives*, Baltimore, University Park Press, 1978.

_____ , *Filters Against Folly: How to Survive Despite Economists, Ecologists, and the Merely Eloquent*, New York, Viking, 1985.

Harding, Jim, "Europe Reconsiders Chernobyl," *Alternatives*, vol. 15, no. 1 (December 1987/January 1988), pp. 68-71.

Harman, Willis, "Information Society and 'Meaningful Work'," Institute for Noetic Sciences, Sausalito, Calif., n.d.

_____ , "Peace on Earth: The Impossible Dream Becomes Possible," *Journal of Humanistic Psychology*, vol. 24, no. 3 (Summer 1984), pp. 77-92.

Harman, Willis, "The Learning Society," *Development Forum*, vol. 15, no. 5 (June 1987), p. 11.

_____ , *Global Mind Change: The Promise of the Last Years of the Twentieth Century*, Indianapolis, Knowledge Systems Inc., 1988.

Hawken, Paul, *The Next Economy*, New York, Holt, Rinehart & Winston, 1983.

Hawkes, Jacquetta, *Dawn of the Gods: Minoan and Mycenaean Origins of Greece*, New York, Random House, 1968.

Heilbroner, Robert, *An Inquiry Into the Human Prospect*, New York, Norton, 1974.

_____ , "What Has Posterity Ever Done For Me?", *New York Times Magazine*, January 19, 1975, reprinted in Patridge, 1981.

Heisenberg, Werner, "Remarks on the Origin of the Relations of Uncertainty," in William Price and Seymour Crissick, eds., *The Uncertainty Principle and Foundations of Quantum Mechanics*, New York, Wiley, 1977.

Henderson, Hazel, *The Politics of the Solar Age: Alternatives to Economics*, New York, Anchor/Doubleday, 1981.

Hirsch, Fred, *Social Limits to Growth*, Cambridge, Mass., Harvard University Press, 1976.

Hoffert, Robert W., "The Scarcity of Politics: Ophuls and Western Political Thought," *Environmental Ethics*, vol. 8, no. 1 (Spring 1986), pp. 5-32.

Hofstadter, Douglas, "To Be of One Mind," *Science 85* (May 1985), pp. 53-57.

Holsworth, R. D., "Recycling Hobbes: The Limits to Political Ecology," *The Massachusetts Review*, vol. 20, no. 1 (1979), pp. 19-40.

Howard, Michael, *War in European History*, New York, Oxford University Press, 1976.

Illich, Ivan, *Medical Nemesis: The Expropriation of Health*, New York, Pantheon, 1976.

Iltis, Hugh, "To the Taxonomist and the Ecologist, Whose Fight is the Preservation of Nature," *BioScience*, vol. 17 (1967).

International Labor Organization, "Employment Policy," Report VI, Geneva, International Labor Office, 1982.

Isacks, Bryan, Jack Oliver, and Lynn Sykes, "Seismology of the New Global Tectonics," *Journal of Geophysical Research*, vol. 73, no. 18 (September 15, 1968), pp. 5855-99.

Jacks, G. W. and R. O. Whyte, *The Rape of the Earth*, Faber & Faber, 1939.

Jackson, Wes, *New Roots for Agriculture*, San Francisco, Friends of the Earth, 1980. (A new edition with a foreward by Wendell Berry has been published by the University of Nebraska Press, Lincoln, Neb.)

Jackson, Wes, Wendell Berry, and Bruce Coleman, Eds., *Meeting the Expectations of the Land: Essays in Sustainable Agriculture and Stewardship*,

Berkeley, Calif., North Point Press, 1984.

Jackson, Wes, *Altars of Unhewn Stone: Science and the Earth*, Berkeley, Calif., North Point Press, 1987.

Jantsch, Erich, *Design for Evolution: Self-Organization in the Life of Human Systems*, New York, George Braziller, 1975.

_____, *The Self-Organizing Universe*, New York, Pergammon, 1980.

Johnson, Chalmers, *Revolutionary Change*, Boston, Little Brown, 1966.

Johnson, Warren, *The Future Is Not What It Used To Be: Returning to Traditional Values in an Age of Scarcity*, New York, Dodd Mead, 1985.

Jonas, Hans, "Technology and Responsibility: Reflections on the New Task of Ethics," in his *Philosophical Essays*, Englewood Cliffs, N. J., Prentice Hall, 1974, reprinted in part in Partridge, 1981.

_____, *The Imperatives of Responsibility: In Search of Ethics for the Technological Age*, Chicago, University of Chicago Press, 1984.

Kantrowitz, Arthur, "Controlling Technology Democratically," *American Scientist*, vol. 63 (September-October 1975), pp. 505-09.

_____, "The Science Court Experiment: Criticisms and Responses," *Bulletin of Atomic Scientists* (April 1977), pp. 44-50.

Kelso, Louis & Patricia, *Democracy and Economic Power*, New York, Ballinger, 1987.

Kohlberg, Lawrence, *The Psychology of Moral Development: The Nature and Validity of Moral Stages*, San Francisco, Harper & Row, 1984.

Kohr, Heinz-Ulrich, "Socio-Moral Judgment, Post-Materialism, and Ecologism," Paper presented at the 8th Annual Meeting of the International Society for Political Psychology, Washington, D. C., June 1985.

Kohr, Heinz-Ulrich and Hans-Georg Rader, "Generational Learning and National Security: Paradigms of Military Threat and Socio-Political Orientations Amongst West German Youth," Paper presented at the 7th Annual Meeting of the International Society for Political Psychology, Toronto, June 1984.

Krimsky, Sheldon, "A Citizen Court in the Recombinant DNA Debate," *Bulletin of Atomic Scientists* (October 1978), pp. 37-43.

Kuhn, Thomas S., *The Structure of Scientific Revolutions*, Chicago, University of Chicago Press, 1962, 2nd ed., 1970.

Leff, Herbert, *Experience, Environment and Human Potentials*, New York, Oxford, 1978.

Leeson, S. M., "Philosophical Implications of the Ecological Crisis: The Authoritarian Challenge to Liberalism," *Polity*, 11, 1979, pp. 303-318.

Lewis, C. S., *That Hideous Strength*, New York, Macmillan, 1946.

Lindblom, Charles, "The Science of Muddling Through," *Public Administration Review*, 19 (1959), pp. 79-88.

_____, *The Intelligence of Democracy: Decision Making Through Mutual*

Adjustment, New York, Free Press, 1965.

——— , "Still Muddling, Not Yet Through," *Public Administration Review*, (November-December 1979), pp. 517-25.

Little, Charles E., "The Great American Aquifer," *Wilderness*, vol. 51, no. 178 (Fall 1987), pp. 43-47.

Livingston, John A., "Moral Concern and the Ecosphere," *Alternatives*, vol. 12, no. 2 (Winter 1985), pp. 3-9.

Long, Marion, "The Turncoat of the Computer Revolution," *New Age Journal*, II, no. 5 (December 1985), pp. 46-51, 76-78.

Lovelock, J. E., *Gaia: A New Look At Life On Earth*, London, Oxford University Press, 1979.

Lovins, Amory, *Soft Energy Paths*, New York, Harper & Row, 1977.

Lowrance, Richard, Paul F. Hendrix, and Eugene P. Odum, "A Hierarchical Approach to Sustainable Agriculture," *American Journal of Alternative Agriculture*, vol. 1, no. 4 (Fall 1986), pp. 169-73.

McAllister, D. M., *Evaluation in Environmental Planning; Assessing Environmental, Social, Economic, and Political Tradeoffs*, Cambridge, Mass., MIT Press, 1982.

McLauglin, Corrine and Gordon Davidson, *Builders of the Dawn: Community Lifestyles in a Changing World*, Walpole, N. H., Stillpoint Publishing, 1985.

Madden, Patrick, "Can Sustainable Agriculture be Profitable?" *Environment*, vol. 29, no. 4 (May 1987), pp. 19-20, 28-34.

Margulis, Lynn and Dorion Sagan, *Microcosmos: Four Billion Years of Evolution from Our Microbial Ancestors*, New York, Summit Books, 1986.

Mastekaasa, Arne, "Multiple and Additive Models of Job and Life Satisfaction," *Social Indicators Research*, vol. 14, no. 2 (February 1984), pp. 141-63.

Meadows, Donella, Dennis Meadows, Jorgen Randers, and William W. Behrens III, *The Limits to Growth*, New York, Universe Books, 1972.

Mesarovic, Mihaljo and Eduard Pestel, *Mankind at the Turning Point*, New York, Dutton, 1974.

Michael, Donald N., "Neither Hierarchy nor Anarchy: Notes on Norms for Governance in a Systemic World," in Walter Truitt Anderson, Ed., *Rethinking Liberalism*, New York, Avon Books, 1983.

Milbrath, Lester W., "Quality of Life on the Niagara Frontier," Occasional Paper, Environmental Studies Center, SUNY–Buffalo, 1976.

——— , *Environmentalists: Vanguard for a New Society*, Albany, SUNY Press, 1984a.

——— , "A Proposed Value Structure for a Sustainable Society," *The Environmentalist*, vol. 4, 1984b, pp. 113-24.

Milbrath, Lester and Hsiao-Shih Cheng, "New Beliefs About How the World Works in Modern Industrial Democracies," Paper presented at the 1985 Annual Meeting of the American Political Science Assn., New Orleans,

August 29-September 1, 1985.

Miller, Anne, "The Economic Implications of Basic Income Schemes," Paper delivered at The Other Economic Summit, 1984 (available from 42 Warriner Gardens, London, SW11 4DU, England).

Miller, G. Tyler, *Living in the Environment*, 4th ed., Belmont, Calif., Wadsworth, 1985.

Mishan, Edward J., "The Growth of Affluence and the Decline of Welfare," in Herman E. Daly, Ed., *Economics, Ecology, Ethics*, San Francisco, Freeman, 1980.

Morone, Joseph G. and Edward J. Woodhouse, *Averting Catastrophe: Strategies for Regulating Risky Technologies*, Berkeley, Calif. and London, University of California Press, 1986.

Murdock, Steve H., F. Larry Leistritz, Rita R. Hamm, and Sean-Shong Hwang, "An Assessment of Socioeconomic Assessments: Utility, Accuracy, and Policy Considerations," *Environmental Impact Assessment Review*, vol. 3, no. 4 (1982), pp. 333-50.

Myers, Norman, *The Primary Source: Tropical Forests and Our Future*, New York, Norton, 1986.

Naess, Arne, "The Shallow and the Deep, Long Range Ecology Movements: A Summary," *Inquiry*, 1973, vol. 16, pp. 95-100.

_____ , "Intrinsic Value: Will the Defenders of Nature Please Rise?" Keynote Address, Second International Conference on Conservation Biology, University of Michigan, May 1985.

Naroll, Raoul, *The Moral Order: An Introduction to the Human Situation*, Beverly Hills, Sage Publications, 1983.

Nelson, Theodor, *Literary Machines*, Swarthmore, Penn., Ted Nelson, 1981.

Newell, Norman D. and Leslie Marcus, "Carbon Dioxide and People," *Palaios*, vol. 2 (1987), pp. 101-03.

Odum, Eugene P., *Fundamentals of Ecology*, 3rd. ed., Philadelphia, Saunders, 1971.

Odum, Howard T., *Environment, Power, and Society*, New York, Wiley, 1971.

Office of Technology Assessment, "Automation of America's Offices," Washington, D. C., United States Government Printing Office, 1985.

Office of Technology Assessment, "Technology and Structural Unemployment: Reemploying Displaced Adults," Washington, D. C., U. S. Government Printing Office, 1986.

Ophuls, William, *Ecology and the Politics of Scarcity: Prologue to a Political Theory of the Steady State*, San Francisco, Freeman, 1977.

Orr, David, "In the Tracks of the Dinosaur: Modernization and the Ecological Perspective," *Polity*, vol. 11 (Summer 1979), pp. 562-87.

Orr, David and Stuart Hill, "Leviathan, the Open Society, and the Crisis of Ecology," *Western Political Quarterly*, vol. 31 (1978), pp. 457-69.

Parrish, Michael, "MacCready's Flights of Fancy," *New Age Journal* (June 1985), pp. 36-41.

Partridge, Ernest, ed., *Responsibilities to Future Generations*, Buffalo, New York, Prometheus Books, 1981.

Partridge, Ernest, "Are We Ready for an Ecological Morality," *Environmental Ethics*, vol. 4, no. 2 (Summer 1982), pp. 175-90.

Pedlar, Kit, *The Quest for Gaia*, London, Souvenir Press, 1979.

Perlmutter, Howard V. and Eric Trist, "Paradigms for Societal Transition," *Human Relations*, vol. 39, no. 1 (1986), pp. 1-27.

Pirages, Dennis and Paul Ehrlich, *Ark II: Social Response to Environmental Imperatives*, San Francisco, Freeman, 1974.

Pirsig, Robert M., *Zen and the Art of Motorcycle Maintenance*, New York, Morrow, 1974.

Platon, Nicolas, *Crete*, Geneva, Nagel Publishers, 1966.

Popper, Sir Karl, *Conjectures and Refutations: The Growth of Scientific Knowledge*, New York, Basic Books, 1962.

Postel, Sandra and Lori Heise, *Reforesting the Earth*, Worldwatch Paper, Worldwatch Institute, 1988.

Prigogine, Ilya, *From Being to Becoming*, San Francisco, Freeman, 1980.

Prigogine, Ilya and Isabelle Stengers, *Order Out of Chaos: Man's New Dialogue With Nature*. Boulder, Col. and London, Shambala Publications, 1984.

Putnam, Hilary, *Reason, Truth, and History*, New York, Cambridge Unversity Press, 1981.

Rawls, John, *A Theory of Justice*, Cambridge, Mass., Harvard University Press, 1971.

Rees, Judith, *Natural Resources: Allocation, Economics and Policy*, New York and London, Metheun, 1985.

Rickson, Roy E., J. S. Western, and R. Burdge, *Theoretical Bases of Social Impact Assessment*, School of Australian Environmental Studies, Griffith University, Brisbane, Australia, 1986.

Rifkin, Jeremy, *Declaration of a Heretic*, Boston and London, Routledge & Kegan Paul, 1985.

Roberts, Keith, *Automation, Unemployment and the Distribution of Income*, European Centre for Work and Society, 1982.

Robertson, James, *Future Work: Jobs, Self-employment, and Leisure After the Industrial Age*, New York, Universe Books, 1985.

Rokeach, Milton, *The Nature of Human Values*, New York, Free Press, 1973.

_____ , *Understanding Human Values: Individual and Social*, New York, Free Press, 1979.

Rolston, Holmes III, "Duties to Endangered Species," *Bioscience*, vol. 35, no. 11 (1985), pp. 718-26.

Rorty, Richard, *Philosophy and the Mirror of Nature*, Princeton, N. J., Princeton University Press, 1979.

Roszak, Theodore, *Where the Wasteland Ends*, Garden City, N. Y., Doubleday Press, 1972.

———, *Person/Planet*, New York, Doubleday/Anchor, 1978.

Rubinoff, Lionel, "Beyond the Domination of Nature: Moral Foundations of a Conserver Society," *Alternatives*, vol. 12, no. 2 (Winter 1985), pp. 37-48.

Sachs, Ignacy, *Strategies de l'Ecodevelopment*, Paris, Editions Ourvrieres, 1980.

Sale, Kirkpatrick, *Dwellers in the Land: The Bioregional Vision*, San Francisco, Sierra Club Books, 1985.

Salk, Jonas, *Anatomy of Reality: Merging of Intuition and Reason*, New York, Columbia University Press, 1983.

Schein, Edgar H., *Professional Education: Some New Directions*, New York, McGraw Hill, 1972.

Scherer, Donald and Thomas Attiq, Eds., *Ethics and the Environment*, Englewood Cliffs, N. J., Prentice Hall, 1983.

Schlesinger, Arthur M. Jr., *The Cycles of American History*, Boston, Houghton Mifflin, 1986.

Schleuning, Neala, "The Aesthetics of Work," *The Human Economy Newsletter*, vol. 7, no. 2 (June 1986), pp. 3-8.

Schmookler, Andrew Bard, *The Parable of the Tribes: The Problem of Power in Social Evolution*, Berkeley, Calif., University of California Press, 1984.

Searles, Harold F. MD, *Counter Transference and Related Subjects*, New York, International Universities Press, 1979.

Sheldrake, Rupert, *A New Science of Life: The Hypothesis of Formative Causation*, London, Blond & Briggs, 1981; 2nd. ed., Los Angeles, J. P. Tarcher, 1988.

———, "Form and Origin," *Resurgence*, no. 103 (March/April 1984), pp. 6-10.

———, *The Presence of the Past: Morphic Resonance and the Habits of Nature*, New York, Times Books, 1988.

Sherif, Muzafer and Carl I. Hovland, *Social Judgment*, New Haven, Conn., Yale University Press, 1961.

Shrader-Frechette, Kristin S., "Technology Assessment as Applied Philosophy of Science," *Science and Technology*, no. 33 (Fall 1980), pp. 33-50.

———, *Environmental Ethics*, Pacific Grove, Calif., Boxwood Press, 1981.

———, "Environmental Impact Assessment and the Fallacy of Unfinished Business," *Environmental Ethics*, vol. 4, no. 1 (Spring 1982), pp. 37-47.

Sikora, R. I. and Brian Barry, eds., *Obligations to Future Generations*, Philadelphia, Temple University Press, 1978.

Simon, Herbert A., *Administrative Behavior: A Study of Decision-Making Processes in Administrative Organization*, New York, Macmillan, 1945;

2nd. ed., 1957.

Simon, Julian, *The Ultimate Resource*, Princeton, N. J., Princeton University Press, 1981.

———, "Wealth in Numbers," *Development Forum*, vol. 14, (July-August 1986), pp. 3, 4.

Singer, Peter, *Animal Liberation*, New York, Avon Books, 1977.

Stephens, Sharon, "Chernobyl Fallout: A Hard Rain for the Sami," *Cultural Survival Quarterly*, vol. 11, no. 2 (1987), pp. 66-71.

Stephenson, Max O. Jr., David J. Webber, and David G. Williams, "Plenary Review: A Structured Interaction-Based Policy Decision Process," Presented at the National Conference of the American Society for Public Administration, Anaheim, Calif., April 1986.

Stonier, Tom, *The Wealth of Information: A Profile of the Post-Industrial Economy*, London, Metheun, 1983.

Speiser, Stuart M., *The USOP Handbook*, Council on International and Public Affairs, New York, 1986.

Speth, James Gustave, "Environment, Economy, Security: The Emerging Agenda," in *Protecting Our Environment: A New Agenda*, Washington, D. C., Center for National Policy, 1985.

Swimme, Brian, "The Resurgence of Cosmic Storytellers," *ReVISION*, vol. 9, no. 2 (Winter/Spring 1987), pp. 83-88.

Taylor, Paul, *Respect for Nature: A Theory of Environmental Ethics*, Princeton, N. J., Princeton University Press, 1986.

Thomas, Lewis, *The Lives of a Cell*, New York, Bantam, 1975.

Thurow, Lester, *The Zero Sum Society*, New York, Basic Books, 1980.

Toffler, Alvin,*The Third Wave*, New York, Morrow, 1980.

———, *The Adaptive Corporation*, New York, McGraw Hill, 1984.

Toynbee, Arnold, *A Study of History*, New York, Oxford University Press, 1972.

Trist, Eric, "The Environment and System-Response Capability," *Futures*, vol. 12, no. 2 (April 1980), pp. 113-27.

Tucker, William, *Progress and Privilege: America in the Age of Environmentalism*, Garden City, New York, Doubleday, 1982.

Urbanska, Wanda, *The Singular Generation: Young Americans in the 1980s*, Garden City, N. Y., Doubleday, 1986.

Van de Kragt, Alphons J. C., John Orbell, and Robyn Dawes, "The Minimal Contributing Set as a Solution to Public Goods Problems," *American Political Science Review*, vol. 77, no. 1 (March 1983), pp. 112-22.

Wachtel, Paul L., *The Poverty of Affluence*, New York, The Free Press, 1983.

Walker, Kenneth J., "The Environmental Crisis: A Critique of Neo-Hobbesian Responses," *Polity*, vol. 21, no. 1 (Fall 1988), pp. 67-81.

Walsh, Roger, MD, *Staying Alive: The Psychology of Human Survival*, Boulder, Col., New Science Library, 1984.

Watanuki, Joji, "Politics and Ecology in Japan," Paper prepared for the Tenth World Congress of Sociology, Mexico City, August 1982.

Watson, Lyall, *Lifetide: The Biology of Consciousness*, New York, Simon & Schuster, 1980.

Wattenberg, Ben, *The Birth Dearth*, New York, Pharos, 1987.

Wenz, Peter, *Environmental Justice*, Albany, SUNY Press, 1988.

White, Lynn, Jr., "The Historical Roots of Our Ecological Crisis," *Science*, vol. 155, no. 3767 (March 1967), pp. 1203-07.

Wilson, Robert Anton, "Interview With Dr. Rupert Sheldrake," *New Age* (February 1984), pp. 43-45, 84-87.

Winner, Langdon, *The Whale and the Reactor: A Search for Limits in an Age of High Technology*, Chicago, Chicago University Press, 1986.

Woodhouse, Edward J., "Re-Visioning the Future of the Third World: An Ecological Perspective on Development," *World Politics*, vol. 25 (October 1972), pp. 1-33.

World Commission on Environment and Development, *Our Common Future*, London and New York, Oxford University Press, 1987.

World Resources Institute and the International Institute for Environment and Development, *World Resources 1986*, New York, Basic Books, 1986.

Yankelovich, Daniel, *New Rules: Searching for Self-Fulfillment in a World Turned Upside Down*, New York, Random House, 1981.

Yeager, Rodger and Norman N. Miller, *Wildlife, Wild Death: Land Use and Survival in Eastern Africa*, Albany, SUNY Press, 1986.

Index